CHICAGO PUBLIC LIBRARY
HAROLD WASHINGTON LIBRARY CENTER
R0016495085

REF
QH
212
.E4
R4
1976
cop.1

Recent progress in electron microscopy of cells and tissues

DATE DUE

THE CHICAGO
PUBLIC LIBRARY

FOR REFERENCE USE ONLY
Not to be taken from this building

REF
QH212
.E4
R4
1976
cop. 1

SCIENCE DIVISION

The Chicago Public Library

Received JUN 29 1977

Recent Progress in Electron Microscopy of Cells and Tissues

Edited by

EICHI YAMADA, M.D.
Professor, Department of Anatomy
Faculty of Medicine, University of Tokyo, Tokyo

VINCI MIZUHIRA, M.D.
Professor, Department of Cell Biology
Medical Research Institute
Tokyo Medical and Dental University, Tokyo

KAZUMASA KUROSUMI, M.D.
Professor, Department of Morphology
Institute of Endocrinology
Gunma University, Maebashi

TOSHIO NAGANO, M.D.
Professor, Department of Anatomy
School of Medicine, Chiba University, Chiba

UNIVERSITY PARK PRESS
BALTIMORE   AND   LONDON
IGAKU SHOIN LTD. TOKYO

Library of Congress Cataloging in Publication Data

Main entry under title:
Recent progress in electron microscopy of cells and tissues.

    Includes index.
    1. Electron microscopy. I. Yamada, Eichi, 1922–
[DNLM: 1. Cytology. 2. Microscopy, Electron. 3. Cytological technics. 4. Histological technics. QH581 R295]
QH212.E4R4   1976   574.8'028   76–6132
ISBN 0-8391-0888-5

PUBLISHERS
© American Edition, 1976 by IGAKU SHOIN LTD., 5-24-3 Hongo, Bunkyo-ku, Tokyo.
    Rights in the United States of America, the Dominion of Canada and Latin America granted to
    UNIVERSITY PARK PRESS: Chamber of Commerce Building, Baltimore, Maryland 21202.

All rights reserved. No part of this book may be reproduced in any form, by print, photoprint, microfilm or any other means without written permission from the publisher.

Printed in Japan. Composed and printed by Gakujutsu Tosho Printing Co., Ltd., Tokyo and bound by Kojima Binding Co., Ltd., Tokyo. The blocks for the illustrations were made by Gakujutsu Photoengraving Co., Ltd., Tokyo.

Preface

The use of electron microscopic techniques in the fields of cytology and histology not only greatly increased fundamental knowledge concerning the normal fine structures of cells and tissues of the human as well as the animal bodies, but also opened new research fields in pathology and clinical medicine regarding the interpretation of disease processes. Ultrathin sectioning has been most widely applied to many kinds of cells and tissues among various methods of preparation for electron microscopy of biological specimens, and most intensely stimulated the progress of cell biology, modern histology and histopathology. However, other recently developed techniques in electron microscopy such as scanning electron microscopy, X-ray microanalysis, freeze etching and freeze fracture methods, high voltage electron microscopy of thick sections using a goniometer stage and a modified ultramicrotomy including various histochemical, immunohistological and autoradiographical techniques promoted advancement of the ultrastructural morphology of the cell.

Such newly explored fields of electron microscopy both in technology and in its application to biology are too broad to be surveyed by just one or few authors, thus this book had to be written by many authors. In the past decade, the Ministry of Education of the National Government of Japan provided the financial support to various groups for the investigation of ultrastructural morphology of cells and related fields of physiology and biochemistry. The editors of this book were organizers of such research groups in periods of two or three years each. Results of most of the coordinated research activities by these groups were published separately by each of the constituent members of these groups on various journals and read on many occasions at scientific meetings. However, some of the unique studies which produced very interesting results still remained unpublished. This book titled "Recent Progress in Electron Microscopy of Cells and Tissues" contain not only the results of these studies as they progressed in the above mentioned research groups supported by the National Government, but also contain valuable papers by other authors in Japan and other foreign countries. The authors of this book are all distinguished electron microscopists in the field of medical and biological sciences. Therefore, the theoretical fields of electron optics are excluded, but practical techniques in specimen preparation and observation as well as their medical and biological applications are described.

The connection between the morphology and biochemistry is especially emphasized. Concurrent studies of electron microscopy and biochemistry on certain particulate fractions from the cell as well as the so-called topochemical reactions on thin sections clarified the localization of biologically active substances such as enzymes, hormones and synaptic transmitter substances in the fine structural elements of cells. Recently developed cytochemical methods include autoradiography, immunohistochemical reaction and freeze preservation of soluble substances. The X ray microanalyzer may indicate the localization of biologically significant elements in the tissue under examination. Crystals of protein or other substance may occur within a cell. The molecular arrangement in such crystals is also an attractive field of ultrastructural morphology. Structure and morphogenesis of cytoplasmic and extracellular filaments and fibrils were recently studied. The chemical basis of formation of such biological filaments was also explored.

The second emphasis in modern electron microscopy is the quantitative approach. Morphometry and its computerization have been innovated. Three-dimensional consideration is necessary for a quantitative understanding of the cell structure and its modification associated with the functional dynamics of the cell. Stereoscopic observations on a pair of pictures taken by tilting the specimen at a suitable angle either with the ordinary, high voltage or scanning electron microscopes give surprising three-dimensional images. Freeze-fractured replica methods are useful for the preservation of substances soluble in water or organic solvents, but also give a good stereographic picture. Though it is still impossible to observe the internal fine structure of living cells, recent progress in electron microscopy has made a valuable approach to the understanding of the real structure of the cell.

This book not only gives useful information concerning the recent progress in electron microscopic techniques, but also contains biological data obtained through the utilization of such newly explored techniques. Students as well as researchers in biological, medical, dental and veterinary sciences may obtain effective benefits from this book in learning the most advanced knowledge in morphology and the related parts of physiology and biochemistry of normal and abnormal cells of the human and animal bodies.

December 1975

EICHI YAMADA
VINCI MIZUHIRA
KAZUMASA KUROSUMI
TOSHIO NAGANO

Contributors

Kazuhiro ABE	Department of Anatomy, Hokkaido University School of Medicine, North 15 West 7, Kita-ku, Sapporo 060 (Japan)
Kensuke BABA	Division of Pathology, National Cancer Center Research Institute, 5-1-1 Tsukiji, Chuo-ku, Tokyo 104 (Japan) Present Adress: Department of Pathology, Medical School, Dokkyo University, 880 Kita-kobayashi, Mibu-machi, Tochigi-ken 321-02 (Japan)
Jeffrey P. CHANG	Division of Cell Biology, Department of Human Biological Chemistry and Genetics, The University of Texas Medical Branch, Galveston, Texas (U.S.A.)
Hisao FUJITA	Department of Anatomy, Hiroshima University School of Medicine, 1-2-3 Kasumi, Hiroshima 734 (Japan)
Tsuneo FUJITA	Department of Anatomy, Niigata University School of Medicine, 1 Asahimachi-dori, Niigata 951 (Japan)
Yutaka FUTAESAKU	Department of Cell Biology, Medical Research Institute, Tokyo Medical and Dental University, 1-5-45 Yushima, Bunkyo-ku, Tokyo 113 (Japan)
Kiyoshi HAMA	Division of Ultrastructure, The Institute of Medical Science, University of Tokyo, 4-6-1 Shirogane-dai, Minato-ku, Tokyo 108 (Japan)
Ryohei HONJIN	Department of Anatomy and Neuro-Information Research Institute, School of Medicine, Kanazawa University, 13-1 Takara-machi, Kanazawa 920 (Japan)
Atsushi ICHIKAWA	Department of Anatomy, Yokohama City University School of Medicine, 2-33 Urafune-cho, Minami-ku, Yokohama 232 (Japan)
Misao ICHIKAWA	Department of Anatomy, Yokohama City University School of Medicine, 2-33 Urafune-cho, Minami-ku, Yokohama 232 (Japan)
Harunori ISHIKAWA	Department of Anatomy, Faculty of Medicine, University of Tokyo, 7-3-1 Hongo, Bunkyo-ku, Tokyo 113 (Japan)
Takashi ITO	Department of Anatomy, Hokkaido University School of Medicine, North 15 West 7, Kita-ku, Sapporo 060 (Japan)
Kazumasa KUROSUMI	Department of Morphology, Institute of Endocrinology, Gunma University, 3-39-15 Showa-machi, Maebashi 371 (Japan)
Masahiro MATSUNAKA	Department of Dermatology, Wakayama Medical University, 1 Shichi-bancho, Wakayama 641 (Japan)

CONTRIBUTORS

Shiro MATSUURA	Department of Phisiology, Kansai Medical School, 1 Fumizono-cho, Moriguchi 570 (Japan)
Yutaka MISHIMA	Department of Dermatology, Kobe University School of Medicine, 12-7 Kusunoki-cho, Ikuta-ku, Kobe 650 (Japan)
Vinci MIZUHIRA	Department of Cell Biology, Medical Research Institute, Tokyo Medical and Dental University, 1-5-45 Yushima, Bunkyo-ku, Tokyo 113 (Japan)
Toshio NAGANO	Department of Anatomy, School of Medicine, Chiba University, 1-8-1 Inohana-cho, Chiba 280 (Japan)
Paul K. NAKANE	Department of Pathology, University of Colorado Medical Center, 4200 East Ninth Avenue, Denver, Colorado 80220 (U.S.A.)
Tokio NEI	Division of Medical Science, The Institute of Low Temperature Science, Hokkaido University, North 19 West 8, Kita-ku, Sapporo 060 (Japan)
Yoshiaki NONOMURA	Department of Pharmacology, Faculty of Medicine, University of Tokyo, 7-3-1 Hong, Bunkyo-ku, Tokyo 113 (Japan)
Takuzo ODA	Department of Biochemistry, Cancer Institute, Okayama University Medical School, 2-5-1 Shikata-cho, Okayama 700 (Japan)
Kazuo OGAWA	Department of Anatomy, Kansai Medical School, 1 Fumizono-cho, Moriguchi 570 (Japan)
Teiji OKAYASU	Laboratory of Computation and Electronics, Medical School, Dokkyo University, 880 Kita-kobayashi, Mibu-machi, Tochigi-ken 321-02 (Japan)
John C. RUSS	EDAX International, Inc., Prairie View, Illinois 60069 (U.S.A.)
Takuma SAITO	Department of Morphology, Institute for Developmental Research, 713-1 Kamiya-cho, Kasugai 480-03 (Japan)
Alan O. SANDBORG	EDAX International, Inc., Prairie View, Illinois 60069 (U.S.A.)
Mitsuru SATO	Department of Anatomy 2, Faculty of Medicine, Kagoshima University, 1208-1 Ushuku-cho, Kagoshima 890 (Japan)
Shiro SONODA	Department of Anatomy 2, Faculty of Medicine, Kagoshima University, 1208-1 Ushuku-cho, Kagoshima 890 (Japan)
Yutaka TASHIRO	Department of Physiology, Kansai Medical School, 1 Fumizono-cho, Moriguchi 641 (Japan)
Junichi TOKUNAGA	Department of Microbiology, Kyushu Dental College, 2-6-1 Manazuru-cho, Kokura-ku, Kita-kyushu 803 (Japan)
Koji UCHIZONO	Department of Physiology, Faculty of Medicine, University of Tokyo, 7-3-1 Hongo, Bunkyo-ku, Tokyo 113 (Japan)

Hiroshi WATANABE	Department of Anatomy, Tohoku University School of Medicine, 2–1 Seiryo-cho, Sendai 980 (Japan)
Nakazo WATARI	The First Department of Anatomy, Nagoya City University Medical School, Mizuho-cho, Mizuho-ku, Nagoya 467 (Japan)
Eichi YAMADA	Department of Anatomy, Faculty of Medicine, University of Tokyo, 7–3–1 Hongo, Bunkyo-ku, Tokyo 113 (Japan)
Toshi Yuki YAMAMOTO	Department of Anatomy, Tohoku University School of Medicine, 2–1 Seiryo-cho, Sendai 980 (Japan)
Masayuki YONEMARU	Department of Anatomy 2, Faculty of Medicine, Kagoshima University, 1208–1 Ushuku-cho, Kagoshima 890 (Japan)

Contents

I. MORPHOLOGY OF INTRACELLULAR MACROMOLECULES 1

1. Membranes 3

    Molecular Organization of Cellular Membranes
    T. Oda 3

    Fine Structure of the Synaptic Membrane in the Central Nervous System
    K. Uchizono 24

2. Filaments 40

    Fine Structure of Myofilaments in Chicken Gizzard Smooth Muscle
    Y. Nonomura 40

    Degeneration and Regeneration of Striated Muscle Fibers: Fine Structural Changes of the Rat Skeletal Muscle in Response to a Myotoxic Drug, Bupivacaine Hydrochloride
    A. Ichikawa and M. Ichikawa 49

3. Crystalloids 69

    Crystalloid Structure in Granules of Human Mast Cells
    S. Matsuura and Y. Tashiro 69

    Crystalloid Inclusions in Human Testicular Cells
    T. Nagano 82

    Crystal Structures in Amphibian Yolk Platelets
    R. Honjin 95

    Crystalline Structures Observed in Both the Exocrine and Endocrine Pancreatic Cells
    N. Watari 109

II. ELECTRON MICROSCOPIC HISTOCHEMISTRY 131

1. Enzyme Histochemistry 133

    Ultracytochemistry of Rat Hepatic Parenchymal Cells during the Mitotic Cycle
    T. Saito, J. P. Chang and K. Ogawa 133

2. Autoradiography 147

    Limits of Resolution in Electron Microscope Autoradiography
    V. Mizuhira and Y. Futaesaku 147

Application of Electron Microscopic Autoradiography of Radioactive Iodine for a Comparative Study of the Thyroid Gland
  H. FUJITA 175

3. Immunohistochemistry 189
  Application of Enzyme-labeled Antibody Methods for the Ultrastructural Localization of Hormones: Review
   P. K. NAKANE 189

4. X-ray Microanalysis 201
  Microanalysis of Biological Sections in the TEM and SEM
   A. O. SANDBORG and J. C. RUSS 201

III. FREEZING TECHNIQUE FOR ELECTRON MICROSCOPY OF CELLS 211

Review of the Freezing Techniques and their Theories
 T. NEI 213

Membrane Modifications for Cell Junction in the Mucosal Epithelia of the Rat Small Intestine: A Freeze-etch Study
 T. Y. YAMAMOTO and H. WATANABE 244

IV. QUANTITATIVE METHODS IN ELECTRON MICROSCOPIC CYTOLOGY 251

Cytometric Analysis of Thymic Small Lymphocytes, Studied by a Stereological Method in Electron Microscopy
 T. ITO and K. ABE 253

Quantitative Analysis of Secretory Granules of the STH-cell in the Rat Hypophysis
 M. SATO, M. YONEMARU and S. SONODA 266

Significance of Morphometry in the Cytophysiology of Secretion: Its Application to the Endocrinological Studies of the Rat Hypophysis
 K. KUROSUMI 272

Quantitative and Three-dimensional Analyses of Dendritic Cells and Keratin Cells in the Human Epidermis Revealed by Transmission and Scanning Electron Microscopy
 Y. MISHIMA and M. MATSUNAKA 290

A Study on the Computerization of Quantitative Electron Microscopy: Application of the Distance Function to Ribosome Count
 K. BABA and T. OKAYASU 305

V. THREE-DIMENSIONAL OBSERVATIONS IN ELECTRON MICROSCOPY 317

Scanning Electron Microscopy in Histology and Cytology
 T. FUJITA and J. TOKUNAGA 319

Three-dimensional Observations of the Cellular Fine Structure by Means of High Voltage Electron Microscopy
  K. HAMA 343

High Voltage Electron Microscopy Combined with Molecular Tracers: Observations on the Cardiac Muscle T-system and Renal Epitheliocyte
  E. YAMADA and H. ISHIKAWA 354

INDEX 363

I.

Morphology of Intracellular Macromolecules

1. MEMBRANES

Molecular Organization of Cellular Membranes

TAKUZO ODA

The basic structure of cells is made up of a variety of membrane systems, and the most essential physiological processes are implemented exclusively by these membrane systems. It is important to note that the components which constitute the structure of cellular membranes are the molecules which carry out its physiological functions. Hence, the structure and function of the cellular membranes should not be separated at the molecular level. Investigations to establish the molecular correlation between structure and function will thus afford us an understanding of the basic structural organization of living cells and so help to resolve the mystery of life. I intend to concentrate on this correlation and to illustrate current ideas with the aid of electron micrographs obtained in our own series of studies on the molecular structures and biochemical functions of mitochondrial and plasma membranes. Discussions will also review briefly some representative models of molecular organization and dynamic changes of the basic structure and specific sites of biomembranes.

Basic Functions of Cellular Membranes and Their Specialization

Basic functions and components of biomembranes

One of the most important feature of a membrane is that it is a closed system in a vesicular or tubular form, which localizes specific components within itself by selective permeability and/or energized translocation of ions and molecules. To implement this basic function all biological membranes have a basic structure primarily composed of two major components: lipids and proteins. Lipids of an amphipathic structure can form a membrane by themselves in an aqueous medium by micellar aggregation. These lipid micelles serve as a barrier against ion permeation owing to their hydrophobic properties. Proteins alone, on the other hand, essentially cannot form a membrane. The main constituents which carry out biochemical reactions in the membrane, however, are proteins. There are two classes of membrane proteins; one is hydrophobically associated with and integrated in the membrane and makes up part of the membrane continuum (intrinsic or integral proteins). The other is hydrophilic and attached on the surface of the membrane but does not make up part of the membrane continuum (extrinsic or peripheral proteins).

Specialization of membrane systems

Cells of higher organisms have extensive networks of specialized membrane systems, each with specific proteins or enzymes organized to fulfill their unique functions. For example, the plasma membrane which encloses a cell and delimits the internal from the external environment, contains the most powerful enzyme system for selective permeability and energized ion translocation, and serves in the maintenance of homeostasis within the cell. In addition to this, on the surface of the plasma membrane, there exist specific receptor systems for a variety of external information, which control membrane functions and metabolic processes in the cell. Special kinds of glycoproteins are associated with these receptor systems. Further, special kinds of enzymes or enzyme systems are localized in highly differentiated plasma membranes: the intestinal epithelial microvillus membrane contains a set of enzymes essential for the terminal hydrolytic digestion of carbohydrates and proteins; the plasma membrane of nerve endings contains enzyme systems necessary for the transmission of nerve impulses. Intracellular membrane systems, such as mitochondria, endoplasmic reticula, nuclear membranes, and Golgi membranes, also contain a series of special enzymes or enzyme complexes. The mitochondrion is a membranous organelle which oxidizes pyruvic acid to carbon dioxide and water by way of the citric acid cycle and the electron transfer chain, and couples these oxidations to the synthesis of ATP from ADP and inorganic phosphate. The high energy stored in ATP provides the driving force for carrying out the living processes. Components of the electron transfer and oxidative phosphorylation system are organized in the inner membrane of the mitochondria. The endoplasmic reticulum is composed of two kinds of continuous network of tubular membranes: one is a rough surfaced membrane associated with ribosome particles, which are the loci of protein synthesis. The other is a smooth surfaced membrane with no attachment of ribosomes. Both kinds of membrane systems carry out many important biochemical processes, such as the syntheses of fatty acids, phospholipids, and steroids as well as the hydroxylation of steroids and detoxication. The nuclear membrane encloses the genetic materials, chromatin: its inner membrane is attached to the chromatin and serves in the regulation of gene action, and its outer membrane is similar to the rough surfaced endoplasmic reticulum.

Vesicular membranes, such as those in lysosomes and microbodies, encompass special enzymes in a condensed state interior to the sacs. The Golgi apparatus, which is made up of tubules and expanded sacs, is specialized to concentrate, package, and deliver secretory products to the exterior of the cell.

It is important, therefore, to clarify the basic structure of cellular membranes which is common to all membrane systems and also the structural specialities of individual membranes which are associated with their specialized functions.

Review of the Molecular Models of Biological Membranes

For many years before and after the electron microscope became available for membrane studies, diverse molecular models of biological membranes have been proposed. Main discrepancies among these models lie in the manner of the interaction of lipids and proteins and in the geometrical relationship of proteins to lipids in the organization of biomembranes. The molecular model proposed by DANIELLI and DAVSON (1935), based on bimolecular phospholipid micelles, has been commonly recognized as an approximate representation of the fundamental structure of biological membranes (Fig. 1a). The principal feature of this model is the bimolecular phospholipid micelle layer with globular proteins bound by electrostatic forces at the polar groups on each side of the layer. Later, DANIELLI (1954) modified this model by replacing the globular proteins with extended polypeptide chains and by introducing hydrophilic pores in the lipid layer (Fig. 1b). The unit membrane hypothesis of ROBERTSON (1959), which was based mainly on data from electron micrographs and X-ray diffraction patterns of myelin, is comparable to the Danielli-Davson model with some exceptions. In the unit membrane hypothesis the protein layers are considered to be extended polypeptide chains, and their outer surface is covered with mucopolysaccharides (Fig. 1c). The trilaminal structure of biological membranes (total thickness 60–100 A) can actually be observed in ultrathin sections with the electron microscope. The central layer (thickness about 35 A) is electron transparent and is thought to correspond to the nonpolar hydrocarbon chains of the bimolecular phospholipid micelle layer. The outer dense layers (thickness about 20 A each) are considered to represent the polar groups of the phospholipids and proteins.

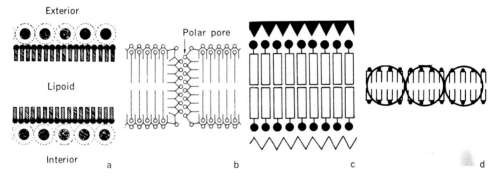

Fig. 1 Molecular models of biological membranes. **a.** The original model of DANIELLI and DAVSON (1935) showing the bimolecular phospholipid layer with globular proteins bound electrostatically on each surface of the layer. **b.** DANIELLI's modification (1954), replacing the globular proteins with extended polypeptide chains and introducing hydrophilic pores in the phospholipid bilayer. **c.** The unit membrane model of ROBERTSON (1959) showing the bimolecular lipid layer with extended polypeptide chains on each surface of the layer and the addition of polysaccharides on the outer surface of the plasma membrane. **d.** The repeating unit model proposed by GREEN and PERDUE (1966) and by BENSON (1966) showing the organization of the membrane structure by hydrophobic association of repeating units of lipoprotein complexes.

The above ideas concerning the basic structure of biological membranes appear to be applicable to artificial phospholipid membranes and to the myelin sheath that has a high lipid to protein ratio (2:1) and a special species of basic protein (EYLAR et al. 1969). Their wider application to biological membranes in general, however, is not completely justified, especially with respect to the association and arrangement of lipids and proteins. In the last decade, evidences have been accumulated from correlative studies in biochemical analysis and electron microscopy, especially with the negative staining technique, of the presence of macromolecular repeating units in plasma membranes (ODA & SATO 1964; ODA & SEKI 1965, 1966b; RAZIN et al. 1965), mitochondrial membranes (GREEN & ODA 1961; ODA & NISHI 1963; FERNÁNDEZ-MORÁN et al. 1964; GREEN 1965; GREEN et al. 1965; ODA 1967), chloroplast membranes (PARK & PON 1961; ODA & HUZISIGE 1965; PARK 1965) and disk membranes of the retinal rod outer segments (BLASIE & DEWEY 1965), etc. Thus, the repeating unit model of a membrane structure has been proposed by GREEN et al. (1966) and by the author and his colleagues (ODA & NISHI 1963; ODA 1968, 1970), the central concept of which is the postulate that biological membranes are composed of repeating units of lipoprotein complexes. Each unit is assumed to be a set of proteins associated hydrophobically with phospholipids within the membranes (GREEN 1965; BENSON 1966; KORN 1966) (Fig. 1d).

Recent progress in electron microscopy using the freeze-etching technique provides evidence that a substantial amount of protein is deeply embedded in most cellular membranes (BRANTON 1966; PINTO DA SILVA & BRANTON 1970; BRANTON & DEAMER 1972). The existence of a bilayer structure of phospholipids in membranes has also been established (STEIM et al. 1969; TOURTELLOTTE et al. 1970; McCONNELL et al. 1972). Reviewing these basic concepts underlying the structure of biological membranes, GREEN (1972) and GREEN et al. (1972) have recently developed a modified repeating unit model of a membrane structure as a logical marriage of the bimolecular phospholipid membrane model of DANIELLI and DAVSON (1935) or the unit membrane model of ROBERTSON (1959) with the original repeating unit model of GREEN and PERDUE (1966). On the other hand, by a variety of physical methods, such as spin-labeling (HUBBELL & McCONNELL 1968; TOURTELLOTTE et al. 1970; McCONNELL et al. 1972), X-ray diffraction (BLASIE & WORTHINGTON 1969; ENGELMAN 1970; CAIN et al. 1972), and differential calorimetry (STEIM et al. 1969; MILCHOIR et al. 1970), it has been shown that the lipids of cellularmembranes are in a fluid rather than a crystalline state. Experimental evidence has been presented by EDIDIN and his associates (EDIDIN & WEISS 1972; FRYE & EDIDIN 1972) that proteins in the plasma membrane are in a fluid state and move freely by diffusion within the membrane. SINGER and his co-workers (NICOLSON et al. 1971a, b, c) have developed electron microscopic techniques using ferritin-conjugated antibodies to visualize the distribution of specific antigens on the surface of membranes, and they have proposed a fluid mosaic model for the structure of cell membranes (SINGER & NICOLSON 1972).

Before giving any further comments on the validity of these molecular models, I would like to present our own series of electron microscopic studies on the molecular correlation of structure and function of biological membranes.

Structure of an Artificial Membrane Composed of Phospholipid Micelles

Lipids account for some 30–40% of the total dry weight of cellular membranes. Large numbers of different kinds of lipids are found in the membranes, but membranes of animal cells in general contain three types of lipids: phospholipid, glycolipid, and sterol, usually cholesterol. All of these are amphipathic molecules with a hydrophilic end and a hydrophobic remainder. Phospholipids are the predominant species among the amphipathic lipids with more than 90% being located in the mitochondrial membrane and 50–70% in the plasma membrane. Cholesterol is particularly rich in the plasma membrane, accounting for about 20–30% of the total lipids in the erythrocyte membrane. The main species among the phospholipids are phosphatidyl choline (lecithin), phosphatidyl ethanolamine, phosphatidyl serine, phosphatidyl inositol and cardiolipin. The amount and distribution of these phospholipids differ among various membrane systems; for example, cardiolipin exists exclusively in mitochondria and partly in heart endoplasmic reticulum, but not in liver endoplasmic reticulum.

All phospholipids with the amphipathic structure form micellar aggregates in water, and by employing this property artificial membranes can be created with them. When purified lecithin is suspended in water with an adequate amount of detergent, such as sodium deoxycholate or tetraglycol, it disperses to form vesicles or particles of approximately 0.05–0.5 μm in diameter. Electron micrographs of the lecithin micelles negatively stained with phosphotungstic acid reveal that each is made up of one or more layers of bimolecular leaflets of lecithin micelles, which appear as vesicular or scroll-like structures of approximately 45 A in width.

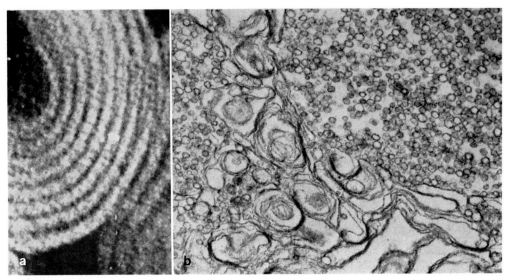

Fig. 2 a. Phospholipid micelles, negatively stained with phosphotungstic acid, showing a lamellar structure made up of multilayers of bimolecular lecithin micelles. ($\times$857,000) **b.** Artificial membranes composed of purified lecithin micelles. The specimen was prepared by suspending lecithin in a 1% tetraglycol solution, subjecting it to ultrasonic treatment, fixing it with glutaraldehyde and osmium tetroxide, dehydrating it with a series of ethanol, and embedding it in Epon 812. The sectioned specimen was stained with uranyl acetate and lead citrate. ($\times$27,400)

The brighter portion of each layer (width 35 A) represents the nonpolar fatty acid residue region of the bimolecular leaflet of the lecithin micelle, while the darker portion (width 10 A) represents the polar region consisting of phosphoric acid and choline residues (Fig. 2a) (ODA 1970, 1972a). Ultrathin sectioned specimens of the artificial membrane composed of lecithin micelles can be prepared by fixing the suspended micelles with glutaraldehyde and then with osmium tetroxide, dehydrating them in a series of ethanol, and embedding them in an epoxy resin. Electron micrographs of the sectioned specimens positively stained with uranyl acetate and lead citrate reveal the formation of vesicular membranes of various sizes (Fig. 2b).

The Molecular Correlation of Structure and Function of Cellular Membranes as Revealed by Resolution and Reconstitution

The energy transducing system of the mitochondrial membrane

The mitochondrion is a prototype of an energy transducing system, and the components of the electron transfer and oxidative phosphorylation system are organized in the inner membrane of the mitochondrion. The electron transfer chain can be divided into four complexes: complex I is NADH-CoQ reductase (f_D); complex II is succinic-CoQ reductase (f_S); complex III is QH_2-cytochrome c reductase (cyt. $b+c_1$ complex); complex IV is cytochrome oxidase (cyt. $a+a_3$) (Fig. 3). The inner mitochondrial membrane has been shown to be composed of a repeating unit of the tripartite form of the elementary particle, which is composed of a head piece, stalk, and base piece (FERNÁNDEZ-MORÁN et al. 1964). The fused base pieces form the membrane continuum of the cristae. In 1963, we reported that the projected head pieces could be situated within the membrane in an actively respiring and phosphorylating state (Fig. 4a) (ODA & NISHI 1963), and we have recently

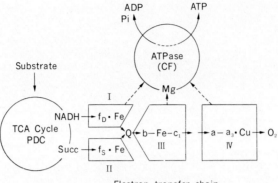

Fig. 3 A diagrammatic representation of the electron transfer and oxidative phosphorylation systems of mitochondria.

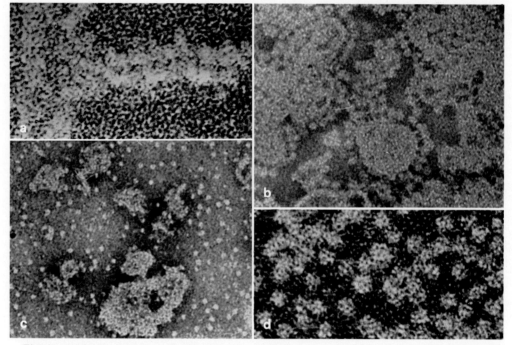

Fig. 4 **a.** Crista membrane of rabbit heart mitochondria negatively stained with phosphotungstate by the spreading method in an actively respiring state in the presence of succinate. Most of the head pieces appear to be situated within the membrane. ($\times 336,000$) **b.** Inner mitochondrial membrane fragments (electron transfer particles) obtained by a mild sonic treatment of beef heart mitochondria, showing the projected head pieces attached by stalks to the membrane. ($\times 224,000$) **c.** Headpiece detachment induced by severe sonication of the isolated inner membrane fragments. ($\times 146,000$) **d.** Purified head pieces with high oligomycin-insensitive ATPase activity. ($\times 537,000$)

demonstrated that the particles undergo conformational changes at the molecular level according to the energy state of the mitochondria (HATASE & ODA 1972).

In order to correlate molecular organization with function, we have carried out biochemical and electron microscopic studies on the molecular organization of the electron transfer and oxidative phosphorylation system in the inner membrane of beef heart mitochondria by a systematic isolation of unit particles and stepwise purification of oligomysin-sensitive and insensitive ATPases and the complexes of the electron transfer chain (ODA & NISHI 1963; ODA & SEKI 1966a; ODA et al. 1966; ODA 1967, 1968, 1970, 1972a; SEKI et al. 1970; SEKI & ODA 1970, 1971). A reconstitution of the membrane structure and function from isolated components and complexes has also been made.

Detachment of the extrinsic protein from the membrane. The inner mitochondrial membrane fragments, electron transfer particles (ETP), were prepared from heavy beef heart mitochondria by mild sonication and repeated washing to remove the soluble components. The isolated inner membrane fragments (ETP) are vesicular in structure, exhibiting the tripartite form of the repeating units. A predominant distribution of the projected head pieces at the periphery of the membranes is considered to be produced by the artificial sliding of the structures on the membranes caused by surface tension in the process of negative staining (Fig. 4b). By severe sonication of the ETP, most of the head pieces are detached from the membrane (Fig. 4c), recovered in the soluble supernatant fraction by centrifugation, and are purified by pursuing the particle with an electron microscope. Oligomycin-insensitive ATPase activity increases with each step of the purification reaching an extremely high activity. The highly purified head pieces are uniform particles, measuring approximately 85–90 A in diameter, which are similar in size and structure to those in the inner mitochondrial membrane (Fig. 4d). Thus the head pieces are identified as ATPase insensitive to oligomycin in their isolated form.

Interaction of the extrinsic protein with the membrane. ATPase activity of the head pieces is originally oligomycin-sensitive in the membrane-bound state. This sensitivity is required for the synthesis of ATP. Therefore, to isolate the head pieces in the oligomycin-sensitive form, the mitochondria or inner membrane fragments (ETP) were treated with a low concentration of deoxycholate (0.1 mg DOC per mg of protein) and potassium chloride (72 mg per ml), diluted with two volumes of water, and fractionated on a discontinuous sucrose gradient. The head pieces connected by stalks to the external layer of the base pieces in a ribbon-like structure are stripped off from the membranes, and recovered in a layer having a density of 1.10 (Fig. 5a). The membranes precipitated in the bottom of the tubes are depleted of the head pieces but contain the whole electron transfer chain with full activities of NADH-oxidase and succinic-oxidase. The isolated head pieces with the associated structure mentioned above contain a high activity of oligomycin-sensitive ATPase but almost no cytochrome components, and is designated as an oligomycin-sensitive ATPase (OSA) particle. If the OSA particles are mixed with an excessive amount of phospholipids, the head pieces become bound by their stalks to the surface of the phospholipid micelle membrane, in which the base portions are embedded or inserted. The OSA particles are distributed almost exclusively at the periphery of the membrane, which may be because of the reason mentioned above and would suggest the fluidity of the membrane (Fig. 5b). If the OSA particles are depleted of phospholipids by ammonium sulfate fractionation, the particles aggregate and lose their ATPase activity. The original OSA structure can be reconstituted and the activity restored by the addition of an adequate amount of phospholipids to the aggregates. If the OSA particles are heated at 60 °C for a few minutes, the head pieces are isolated and easily purified in soluble form as oligomysin-insensitive ATPase (Fig. 5c).

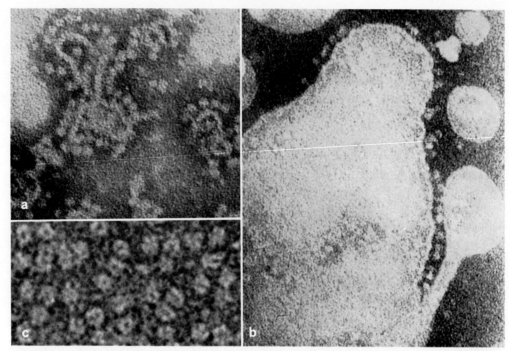

Fig. 5 a. Isolated oligomycin-sensitive ATPase particles (polymeric form), showing head pieces connected by stalks to a thread-like structure. ($\times 282,000$) **b.** Oligomycin-sensitive ATPase bound on the surface of an artificial lecithin micelle membrane. Head pieces are bound by stalks to a thread-like structure which is embedded in some areas of the lipid membrane. Predominant distribution of the oligomycin-sensitive ATPase particles at the periphery of the membrane is considered to be due to an artificial sliding of the particles on the surface of the membrane in the process of negative staining, suggesting the fluidity of membrane. ($\times 276,000$) **c.** Purified oligomycin-insensitive ATPase obtained by a light heat treatment of the oligomycin-sensitive ATPase particles. ($\times 456,000$)

Dissociation and reconstitution of lipoprotein complexes of the electron transfer chain in the inner membrane. When mitochondria or inner membrane fragments (ETP) are treated with a higher concentration of deoxycholate (0.3 mg DOC per mg of protein) and potassium chloride (72 mg per ml), and directly fractionated by ultracentrifugation, a red soluble extract and a green insoluble residue are separated. The red supernatant fraction contains the major portions of complex I, complex III, cytochrome c, and oligomycin-sensitive ATPase. The red extract can be further fractionated with bile salts and salts to isolate complex I and complex III. The purified complex III, that is cyt. $b+c_1$ complex having a high QH_2-cyt. c reductase activity, is a soluble dispersed particle measuring about 100–120 A in diameter in the presence of bile salt. In removing the bile salt, the particles aggregate and form a membrane-like structure.

The green insoluble residue mentioned above contains most of the complex II and complex IV. A cytochrome oxidase-rich green membrane is obatined from the green residue. The surface of the green membrane exhibits dense, regular arrays of small particles of about 60 A with a center-to-center distance of 70 A (Fig. 6a). These particles appear to be densely arranged in a woven paracrystalline structure in the green membrane. They can be extracted with bile salt and ammonium sulfate in a green supernatant fraction leaving an insoluble white residue, and thus purified into a cytochrome oxidase preparation. The purified and highly active cytochrome oxidase appears as dispersed particles measuring approximately 70×95 A in diameter in the presence of detergents, such as bile salt or tween

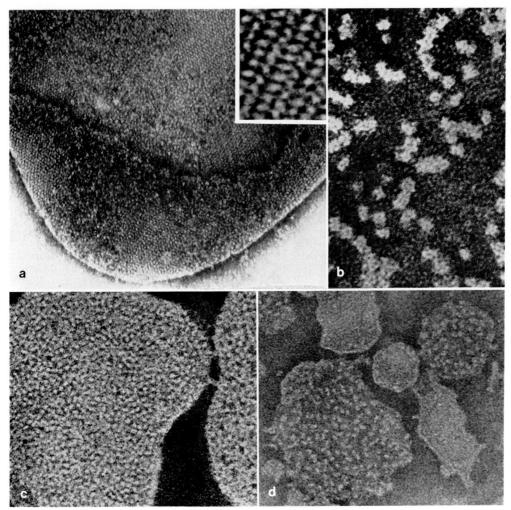

Fig. 6 a. Cytochrome oxidase-rich green membrane. (×122,000) Inset: Magnified picture of a portion of the green membrane. (×448,000) **b.** Purified active cytochrome oxidase, dispersed state. (×375,000) **c.** Vesicular membrane formed by a recombination of purified cytochrome oxidase particles at pH 10. (×218,000) **d.** Cytochrome oxidase particles scattered within an artificial lecithin micelle membrane when mixed with an excess amount of lipid at pH 12. (×260,000)

20 (Fig. 6b). The molecular weight of such active units, containing 2 heme *a* molecules and 2 copper atoms, is about 200,000. This is roughly equivalent to two of the particles observable in the green membrane, each particle corresponding to one heme *a* unit. The dispersed particles form a rod-like or ribbon-like structure and then a membrane structure by dialysis or dilution, presenting a densely packed particular surface structure similar to the original green membrane. The particular structure becomes more distinct by increasing the pH of the medium (Fig. 6c). When the green membrane is treated with a high alkaline pH, the particles with cytochrome oxidase activity are partially extracted from the membrane leaving particle-free smooth areas in the membrane. If this purified cytochrome oxidase is mixed with an excessive amount of phospholipids at an alkaline pH, the cytochrome oxidase particles are scattered within the phospholipid micelle membrane (Fig. 6d).

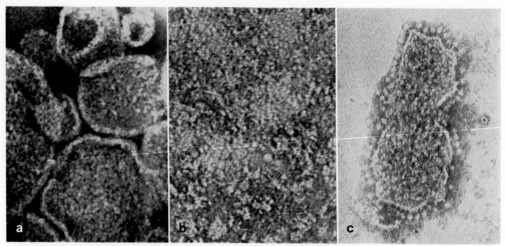

Fig. 7 a. Membranes formed from a mixture of purified and solubilized complex III and complex IV. (×187,500) **b.** Solubilized whole inner mitochondrial membrane components. (×282,000) **c.** Membranes formed by reconstitution from solubilized whole inner mitochondrial membrane components. (×169,000)

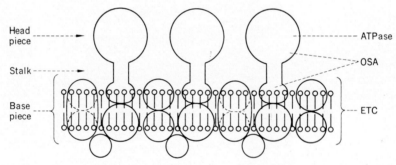

Fig. 8 A molecular model of the inner mitochondrial membrane. OSA: oligomycin-sensitive ATPase, ETC: electron transfer chain.

When purified membranous complex III and membranous complex IV are mixed together, no mixed complex (III+IV) membrane can be obtained. However, if solubilized complex III and complex IV are mixed in the presence of bile salts and then these salts are removed by dialysis or dilution, a mixed complex (III+IV) membrane is formed (Fig. 7a). In the mixed complex (III+IV) membrane, the overall respiratory activity of the QH_2 oxidase is restored by the addition of cytochrome c. This is also true for the solubilized complexes of the whole electron transfer chain. In a similar manner, if the whole inner mitochondrial membrane is solubilized with bile salts and ammonium sulfate (Fig. 7b), and these salts are then decreased through dialysis or dilution, a membrane composed of a combined tripartite form of the repeating units is reconstituted (Fig. 7c), and both overall respiratory functions, such as NADH-oxidase and succinic oxidase activities, and oligomycin-sensitive ATPase activity are restored. Although it has not been possible to restore full oxidative phosphorylation in DOC-treated mitochondria or submitochondrial particles, partial restoration is possible by reconstituting with oligomycin-sensitive ATPase particles and submitochondrial particles which have been isolated by a triton X-100 treatment.

The principle and mode of interaction of lipoprotein complexes and the organization of the basic inner mitochondrial membrane structure are diagrammatically shown in Fig. 8. On the basis of such results, it can be deduced that the inner mitochondrial membrane is an organized assembly of repeating units in which lipoprotein complexes of the components of electron transfer chain are integrated in the phospholipid bilayer membrane and are closely linked with each other. The units themselves basically bind hydrophobically and partly electrostatically, and the head piece ATPase is attached by stalks hydrophobically to the inner surface (matrix side) of the inner membrane. Soluble proteins interact electrostatically on the surface of the membrane; for example, cytochrome c is a highly polar basic protein which interacts electrostatically with the negatively charged polar heads of the phospholipid molecules on the outer surface of the inner membrane. The hydrophobic bonding of repeating units of lipoprotein complexes will have flexibility or fluidity, and the conformational changes in lipoprotein complexes and the phase transition of phospholipid micelles may be closely associated with functional changes in the mitochondrial membrane, such as energy transduction and ion translocation. Stoichiometric analyses in these dynamic processes are required for a final elucidation of the molecular mechanism of oxidative phosphorylation.

Molecular organization and biochemical function of differentiated and specialized plasma membranes

The microvillus membrane of intestinal epithelial cells.
The apical surface of

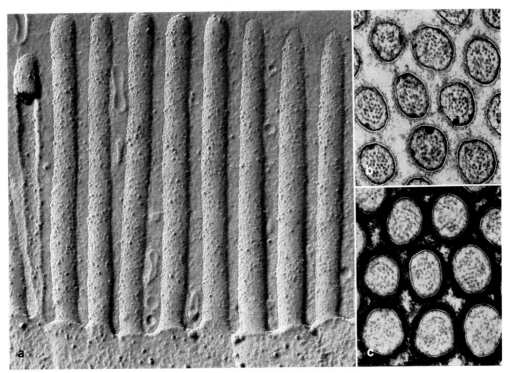

Fig. 9 a. Microvilli of intestinal epithelial cells, prepared by the freeze-etching technique. The surface of microvillus membranes, etched face (Ê), is densely packed with small particles. (×51,000) **b.** A transverse section of microvilli stained with uranyl acetate and lead citrate, showing a distinct trilaminar image. (×81,200) **c.** A transverse section of microvilli stained with ruthenium red, showing that the surface of the membrane is densely covered with acidic polysaccharides. (×70,800)

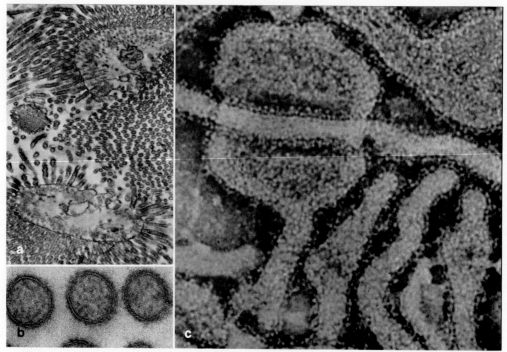

Fig. 10 a. Sectioned specimens of isolated microvillus borders. (×12,000) **b.** High-power transverse section of isolated microvilli. (×106,000) **c.** Isolated microvillus borders negatively stained with phosphotungstic acid. Repeating particles are attached to the surface of the microvillus membrane. (×188,000)

the intestinal epithelial cells is composed of numerous protrusions of the plasma membrane which are called microvilli (Fig. 9a). The microvillus membrane is a prototype of a membrane system highly specialized both in structure and function. The surface of the microvillus membrane is covered with acidic polysaccharides as shown by ruthenium red staining (Fig. 9c). The mucopolysaccharides are associated with fuzzy substances on the surface of the microvilli. The membrane of the microvilli presents a distinct trilaminar image or a so-called unit membrane image composed of inner and outer electron opaque layers and a central electron transparent layer (Fig. 9b). The total width of the membrane measures approximately 90–110 Å. The microvilli carry out two principal functions: one is the terminal hydrolytic digestion of carbohydrates, proteins, phosphate compounds, etc. by the action of disaccharidases, peptidases, phosphatases, and so on; the other is absorption, including the active transport of various ions and degraded molecules, such as monosaccharides, amino acids, etc. We have tried to correlate these functions with the molecular organization of the microvillus membrane using biochemical and electron microscopic procedures (ODA & SEKI 1965, 1966b; ODA et al. 1969; ODA 1970; YAMAMOTO et al. 1971).

The microvillus borders were isolated from the small intestinal epithelial cells of rabbits by scraping the mucosa, disrupting the cells with a hypotonic solution containing 5 mM EDTA, and fractionating by ultracentrifugation. The membranes of the isolated microvilli show morphological intactness in their basic structure (Figs. 10a, b). Enzyme assays in each step of the isolation of microvillus borders indicate that disaccharidases, leucine aminopeptidase, alkaline phosphatase, and ATPase are localized predominantly in the microvillus borders. By negative staining with phosphotungstate, we have discovered

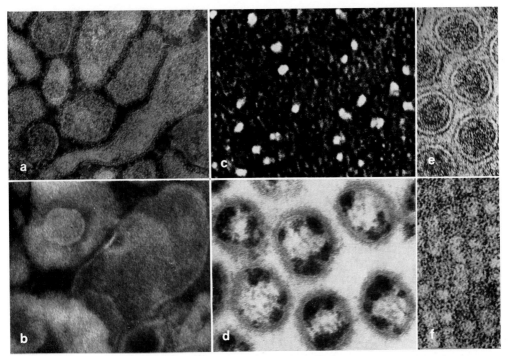

Fig. 11 a. Purified microvillus membranes showing repeating particles on the membrane surface. ($\times 94{,}200$) **b.** Papain-treated microvillus membranes in which the repeating particles have been removed. The membranes exhibit a particle-free smooth surface. ($\times 116{,}000$) **c.** Partially purified repeating particles obtained from isolated microvilli by treatment with papain, negatively stained with uranyl acetate. ($\times 537{,}000$) **d.** Electron microscopic cytochemical demonstration of Mg^{2+}-activated ATPase in microvillus membranes. ($\times 137{,}000$) **e.** Lipid-depleted microvilli whose membrane exhibits a distinct trilaminar image with a wide transparent central layer. ($\times 143{,}000$) **f.** A sectioned specimen of lactate oxygenase crystals, fixed with glutaraldehyde and osmium tetroxide, post-stained with uranyl acetate and lead citrate, showing arrays of electron transparent particles and electron opaque interstices. ($\times 251{,}000$)

repeating particles on the surface of the isolated microvillus borders. These particles measure approximately 60 Å in size with their center-to-center distances approximately 60 to 80 Å (Fig. 10c). They occasionally appear to be arranged as a small ring-like polymer when observed *en face*. To make an enzymatic identification of these particles, the microvillus membranes were purified by disrupting the microvillus borders with sonication followed by fractionating with centrifugation in a glycerol density gradient. The repeating particles are still attached to the surface of the isolated microvillus membranes (Fig. 11a), in which the activities of disaccharidase, leucine aminopeptidase, alkaline phosphatase, and ATPase are concentrated. Treatment of the isolated microvilli or microvillus membranes with papain quantitatively solubilizes all of the disaccharidases and leucine aminopeptidase, which can then be recovered in the supernatant fraction. Neither alkaline phosphatase nor ATPase, however, can be solubilized in any significant amount by the papain treatment. The bulk of these activities remains in the insoluble portion from which the repeating particles are being removed and the membrane exhibits a particle-free smooth surface (Fig. 11b). Sucrase can be purified from the supernatant fraction by DEAE cellulose and Sephadex chromatographies while pursuing the particles with an electron microscope. The purified sucrase fraction, which was partially contaminated with leucine aminopeptidase, contained enzyme particles of a size and structure similar to those of the repeating particles on the microvillus membranes (Fig. 11c). The evidence indicates that each of these repeating particles coincides with or includes an enzyme molecule such as disaccharidase or peptidase, which carries out the terminal hydrolytic digestion of carbohydrates and proteins on the surface of the microvillus membrane (membrane surface digestion). Magnesium ion-activated adenosine triphosphatase was shown by a cytochemical electron microscopic demonstration and by biochemical studies to be organized within the membrane. The active site of this enzyme appeared to be on the inner surface of the membrane (Fig. 11d).

Treatment of the isolated microvillus membrane with deoxycholate results in the destruction of the membrane structure and a marked decrease in the ATPase activity. The membrane structure and the decreased ATPase activity are partially restored by washing and decreasing the amount of bile salts.

When the microvillus membranes are examined with an electron microscope after extraction of the lipids with an aqueous acetone treatment, the characteristic trilaminar feature or so-called unit membrane image is still retained, and the central electron transparent layer of the lipid-depleted membranes appears more transparent and much broader than that of the original membranes (Fig. 11e). Thus, the evidence suggests that the central transparent layer corresponds to the hydrophobic region of the lipoproteins.

To verify the propriety of this interpretation lipid-free enzyme crystals (lactate oxygenase) were observed under the electron microscope using negative staining with phosphotungstate in unsectioned specimens and positive staining with uranyl acetate and lead citrate in ultrathin-sectioned specimens. Since positive staining takes place on the hydrophilic outer surface of the protein molecules but not in the hydrophobic inner core of the molecules, the protein molecules appeared as lighter particles separated by darker interstices in the sectioned specimens of enzyme crystals (Fig. 11f).

The internal ultrastructures of the microvillus membranes were further analyzed by electron microscopy using the freeze-fracture and freeze-etching techniques (Fig. 12a). Both the convex fracture face ($\hat{F}$) and the concave fracture face ($\underline{F}$) of the microvillus membrane are densely packed with globular particles. Fracture faces are interpreted to result from a cleavage of the membrane along its interior hydrophobic face (BRANTON & DEAMER 1972). Fine fibrillar structures interconnecting between both fracture faces can be seen (Fig. 12b). Densely arranged globular particles are also observed on the surface of the membrane, etch face ($\hat{E}$) (Fig. 9a); however, there is a great diversity in the density of particles among various membrane systems in the cell; in some membranes virtually none of the area is occupied by particles whereas in others the greater portion of the area is occupied by particles (Fig. 12a). This varied distribution depends upon the differentiation and specialization of individual membrane systems.

Liver cell membrane. Plasma membranes, isolated from rat liver cells and negatively stained with phosphotungstate, exhibit a fine granular structure with no distinct projected particles on the surface of the membranes. When treated at 37°C before or during the negative staining, a hexagonal pattern appears in some areas of the membranes, presumably in gap junctions (Figs. 13a, b). This pattern is not affected by treatment with trypsin but disappears after the extraction of lipids by treating with aqueous acetone (ODA 1970; SEKI et al. 1971). Thus the temperature-dependent hexagonal pattern seems to be produced by a phase transition of phospholipid micelles of lipoprotein complexes in the membrane.

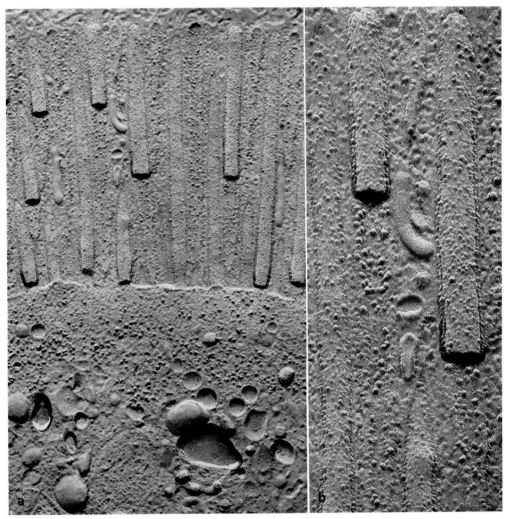

Fig. 12 a. Microvilli and a portion of cytoplasm of an intestinal epithelial cell prepared by the freeze-etching technique. (×33,600) **b.** High-power picture of microvilli prepared by the freeze-etching technique. (×75,200)

A probable model of the molecular organization of the basic structure of cellular membranes

On the basis of the experimental data presented above, the basic structure of cellular membranes may be considered to be represented by a composite of biomolecular phospholipid micelles and integrated globular proteins, which are bound hydrophobically to each other (protein-integrated lipid bilayer model) (Fig. 14) (ODA 1972b). The integral proteins with an amphipathic structure are inserted in the phospholipid layer by their hydrophobic areas and protrude from the phospholipid layer into the aqueous phase by their hydrophilic groups. The integral proteins exist either separately or in a set of organized complexes as repeating units. Some of the integral proteins may exist as "through" membrane bimodal lipoprotein complexes, which span the phospholipid bilayer in their hydrophobic areas and protrude into the aqueous phase on both sides by their hydrophilic groups. For example, in the mitochondrial inner membranes, some of the integral proteins

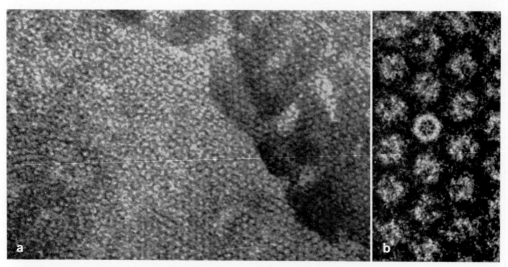

Fig. 13 a. Hexagonal pattern in the plasma membrane of liver cells, negatively stained with phosphotungstate at 37°C. (×300,000) **b.** High-power picture printed by the rotation technique. (×950,000)

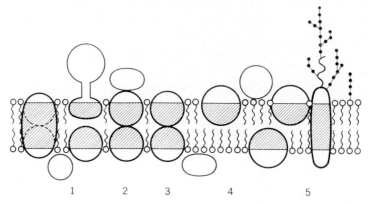

Fig. 14 A molecular model of a cellular membrane (protein-integrated lipid bilayer model), showing various densities of integral proteins with or without peripheral proteins depending upon differentiation and specialization of the membrane.

are semi-organized in sets of lipoprotein complexes of the electron transfer chain in the phospholipid layer. The plasma membrane of the intestinal microvilli is densely packed with integral proteins, which may be responsible for a marked thickening of the microvillus membrane (90–110 Å). The extrinsic proteins or peripheral proteins are bound either by stalks or without stalks directly to the membrane as previously mentioned. These particles can be detached from the membrane in a soluble form by either mechanical, ionic, or enzymatic treatment. The species and density of integral and peripheral proteins in the membrane are diverse among a variety of cellular membrane systems depending upon their differentiation and specialization.

Cell Surface and Membrane Fluidity

A great contribution to the understanding of the distribution and topography of specialized protein molecules on the surface of membranes has been made by SINGER and his coworkers (NICOLSON & SINGER 1971). They developed and applied electron microscopic techniques using ferritin-conjugated antibodies to visualize the distribution of specific antigens over large areas of their membrane surfaces. The Rh_0 (D) antigen on human erythrocyte membranes (NICOLSON, MASOUREDIES et al. 1971) and the antigenic sites of the H-2 histocompatibility antigen (glycoprotein) on mouse erythrocytes (NICOLSON, HYMAN et al. 1971) have been shown to be randomly distributed in a two-dimensional array or in patches on their respective membrane surfaces. Our previous demonstration of the predominant distribution of oligomycin-sensitive ATPase particles at the periphery of the inner mitochondrial membrane and the artificial phospholipid membrane in negatively stained specimens would suggest a sliding of the particles in a fluid state of the membrane (ODA 1968, 1970). Evidence that proteins are in a fluid state in intact membranes and move freely within the membrane has been presented by FRYE and EDIDIN (1970), who have shown by the immunofluorescent antibody technique that membrane components are intermixed in heterokaryons within a short period of time after the induction of fusion of human and mouse cells in a culture. They suggested that the intermixing of membrane components is due to the lateral diffusion of these components within the membrane. Membrane fluidity was further demonstrated by a patch or cap formation on lymphocytes by the fluorescent antibody technique (EDIDIN & WEISS 1972).

Asymmetrical arrangement of membrane components in the outer and inner surfaces has been shown in electron micrographs obtained by negative staining, freeze-etching and shadow-cast replica techniques. One of the most distinct differences between the two surfaces lies in the distribution of polysaccharides and oligosaccharides associated with glycoproteins. Direct evidence of an asymmetry was given by ferritin-conjugated lectin (plant agglutinin such as concanavalin A), which binds specifically to the outer surface of the membrane but not to the inner surface (NICOLSON, MASOUREDIES et al. 1971). Therefore, free mobility of the membrane components could be applied to particular components on the outer surface of the membrane. On the basis of these findings and their thermodynamic considerations SINGER and NICOLSON (1972) have recently proposed a fluid mosaic model of a membrane structure, in which globular proteins are scatteringly inserted in the phospholipid bilayer. The mechanisms of membrane functions, especially changes of surface properties in malignant transformation and pertubation of cells, such as the loss of contact inhibition and changes in agglutinability of cells by plant lectin and in the topographical distributions of lectin-binding sites, have been discussed from the standpoint of the fluid mosaic model (NICOLSON 1971, 1972; ROSENBLITH et al. 1973). MARCHESI et al. (1972) characterized the major glycoprotein of the erythrocyte membrane which carries many blood group antigens and lectin receptors. They suggested that the glycoprotein molecule is oriented at the cell surface with its oligosaccharide-rich N-terminal end exposed to the exterior, while its C-terminal segment is located internal to the lipid barrier of the membrane to form intramembranous particles, and that a part of it may extend into the cytoplasm of the cell.

The modified repeating unit model proposed by GREEN and his colleagues (VANDERKOOI & GREEN 1970; GREEN 1972; GREEN et al. 1972; VANDERKOOI 1972) and by the author (ODA et al. 1969; ODA 1972a) and the fluid mosaic model proposed by SINGER and NICOLSON (1972) are similar to each other with respect to their basic molecular features such as the hydrophobic bonding of integral protein and phospholipids, the bilayer character of phospholipid micelles, the bimodal character of integral proteins, and the mode of interaction of peripheral proteins to the membrane continuum, but differ with respect to the organization of integral proteins. In a modified repeating unit model such as the bimodal protein crystal model (VANDERKOOI & GREEN 1970) or the structure-function unitization model proposed by GREEN et al. (1972), the organization of the functional sets of intrinsic proteins and their repetition are emphasized, whereas in the fluid mosaic model (SINGER & NICOLSON 1972), a random distribution and fluidity of integral proteins are emphasized. These models, however, seem to represent both extremes of the molecular organization of integral proteins in biological membranes and the diversity of the molecular organization could be dependent upon the membrane systems with which they dealt. Most cellular membrane systems seem to fall into categories between these extremes. Analytical data indicate that the ratio of phospholipids to intrinsic proteins in cellular membranes is roughly 1 to 2 in the mitochondrial inner membrane and 1 to 1 in general plasma membranes. On the basis of such data and the density of particles in cellular membranes observed in electron micrographs by the freeze-etch technique, rough estimations of the molecular ratio of lipid to protein and the density of integral proteins in the phospholipid bilayer would be possible. The mitochondrial inner membrane and the intestinal microvillus membrane have been proved to be densely packed with globular particles.

What must further be taken into consideration for the molecular organization of cellular membranes is, which are the principal components between proteins and lipids, in the process of *de novo* formation of membranes and in the determination of differentiation and specialization of respective membrane systems. The asymmetrical distribution of hydrophilic and hydrophobic groups in the tertial structure of membrane proteins and their chemical structure may play a primary role in the determination of specific interactions with different species of phospholipids, and the specific association not only of lipids but also of proteins may be responsible for the formation of respective specialized membrane systems.

The specific sites on the surface of cell membranes associated with either glycoprotein, glycolipid, acidic mucopolysaccharide, or glycolipoprotein complexes, may be isolated dependent upon the method of treatment for isolation.

Summary

The protein-integrated lipid bilayer model proposed by us for the basic molecular organization of cellular membranes is justified on the basis of the experimental data from biochemical and electron microscopic observations of mitochondrial and plasma membranes along with a review of some of the representative molecular models of biological membranes. This model may be considered to be a marriage of the classical lipid bilayer or unit membrane model and the original repeating unit model, as well as a compromise between the modified repeating unit model and the fluid mosaic model. In our model, the backbone of the membrane is composed of a bimolecular leaflet of phospholipid micelles and integrated globular proteins. The globular proteins of the amphipathic structure are associated with phospholipids by their hydrophobic areas and integrated into the phospholipid bilayer matrix. Their hydrophilic groups are protruded from the membrane into the aqueous phase. Integral proteins exist either separately or in a set of organized complexes as repeating units. The density of the integral proteins is diverse among respective membrane systems dependent upon the differentiation and specialization of the membranes, which may also be responsible for the diversity of widths of a variety of cellular membrane systems (50–100 Å). In addition to this, in certain species of specialized membrane systems, characteristic functional proteins (extrinsic or peripheral proteins) are bound either by stalks or without stalks directly to the surface of the membranes. The outer surface of the plasma membrane is further covered by polysaccharides and oligosaccharides associated with membrane proteins and lipids. The chemical structure of receptor sites on the surface of the cell membranes determines the specific interaction with external information and thus serves to control the metabolic processes in the cells.

REFERENCES

[1] BENSON A. A. (1966)   On the orientation of lipids in chloroplast and cell membranes.   J. Amer. Oil Chem. Soc., *43:* 265.

[2] BLASIE J. K. and DEWEY M. M. (1965)   Electron microscope and low-angle X-ray diffraction studies on outer segment membranes from the retina of the frog.   J. Mol. Biol., *14:* 143.

[3] BLASIE J. K. and WORTHINGTON C. R. (1969)   Planar liquid-like arrangement of photopigment molecules in frog retinal receptor disk membranes.   J. Mol. Biol., *39:* 417.

[4] BRANTON D. (1966)   Fracture faces of frozen membranes.   Proc. Natl. Acad. Sci. U. S., *55:* 1048.

[5] BRANTON D. and DEAMER D. W. (1972)   Protoplasmatologia.   *In:* Membrane Structure, Springer-Verlag, Wien & New York.

[6] CAIN J., SANTILLAN G. and BLASIE J. K. (1972)   Molecular motion in membranes as indicated by X-ray diffraction.   *In:* Membrane Research. Fox C. F. (ed.) pp. 3–14, Academic Press, New York & London.

[7] DANIELLI J. F. (1954)   The present position in the field of faciliated diffusion and selective active transport.   Colston Papers, *7:* 1.

[8] DANIELLI J. F. and DAVSON H. (1935)   A contribution to the theory of permeability of thin films.   J. Cell Comp. Physiol., *5:* 495.

[9] EDIDIN M. and WEISS A. (1972)   Antigen cap formation in cultured fibroblasts: A reflection of membrane fluidity and of cell motility.   Proc. Natl. Acad. Sci. U.S., *69:* 2456.

[10] ENGELMAN D. M. (1970)   X-ray diffraction studies of phase transitions in the membrane of *Mycoplasma laidlawii*.   J. Mol. Biol., *47:* 115.

[11] EYLAR E. H., SALK J., BEVERIDGE G. C. and BROWN L. V. (1969)   Experimental allergic encephalomyelitis. An encephalitogenic basic protein from bovine myelin.   Arch. Biochem. Biophys., *132:* 34.

[12] FERNÁNDEZ-MORÁN H., ODA T., BLAIR P. V. and GREEN D. E. (1964)   A macromolecular repeating unit of mitochondrial structure and function.   J. Cell Biol., *22:* 63.

[13] FRYE L. D. and EDIDIN M. (1970)   The rapid intermixing of cell surface antigens after formation of mouse human heterokaryons.   J. Cell Sci., *7:* 319.

[14] GREEN D. E. (1965)   An introduction of membrane biochemistry.   Israel Med. J., *1:* 1187.

[15] GREEN D. E. (1972)   Membrane proteins: a perspective.   Ann. N. Y. Acad. Sci., *195:* 150.

[16] GREEN D. E., JI S. and BRUCKER R. F. (1972)   Structure-function unitization model of biological membranes.   Bioenergetics, *4:* 527.

[17] GREEN D. E. and ODA T. (1961)   On the unit of mitochondrial structure and function.   J. Biochemistry, *49:* 742.

[18] GREEN D. E. and PERDUE J. F. (1966)   Membranes as expressions of repeating units.   Proc. Natl. Acad. Sci. U.S., *55:* 1295.

[19] GREEN D. E., TZAGOLOFF A. and ODA T. (1965)   Ultrastructure and function of the mitochondrion.   *In:* Intracellular Membrane Structure. SENO S. and COWDRY E. V. (eds.) pp. 127–153, Japan Soc. Cell Biol., Okayama.

[20] HATASE O. and ODA T. (1972)   Conformational changes in submitochondrial particles of beef heart. J. Biochemistry, *71:* 759.

[21] HUBBELL W. L. and MCCONNELL H. M. (1968)   Spin-label studies of the excitable membranes of nerve and muscle.   Proc. Natl. Acad. Sci. U.S., *61:* 12.

[22] KORN E. D. (1966)   Structure of biological membranes. The unit membrane theory is reevaluated in light of the data now available.   Science, *153:* 1491.

[23] MARCHESI V. T., TILLACK T. W., JACKSON R. L., SEGREST J. P. and SCOTT R. E. (1972)   Chemical characterization and surface orientation of the major glycoprotein of the erythrocyte membrane.   Proc. Natl. Acad. Sci., *69:* 1445.

[24] McConnell H. M., Deveau P. and Scandella C. (1972) Lateral diffusion and phase separations in biological membranes. *In*: Membrane Research. Fox C. F. (ed.) pp. 27–37, Academic Press, New York & London.

[25] Milchoir D. L., Morrowitz H. J., Sturtevant J. M. and Tsong T. Y. (1970) Characterization of the plasma membrane of *Mycoplasma laidlawii*. VII. Phase transitions of membrane lipids. Biochim. Biophys. Acta, *219*: 114.

[26] Nicolson G. L. (1971) Different distribution of ferritin-conjugated concanavalin A on surfaces of normal and tumor cell membranes. Nature New Biology, *233*: 244.

[27] Nicolson G. L. (1972) Topography of membrane concanavalin A sites modified by proteolysis. Nature New Biol., *239*: 193.

[28] Nicolson G. L., Hyman R. and Singer S. J. (1971) The two dimensional topographic distribution of H-2 histocompatibility alloantigens of mouse red blood cell membranes. J. Cell Biol., *50*: 905.

[29] Nicolson G. L., Masouredies S. P. and Singer S. J. (1971) Quantitative two-dimensional ultrastructural distribution of Rh_0 (D) antigenic sites on human erythrocyte membranes. Proc. Natl. Acad. Sci. U.S., *68*: 1416.

[30] Nicolson G. L. and Singer S. J. (1971) Ferritin-conjugated plant agglutinins as specific saccharide stains for electron microscopy: Application to saccharides bound to cell membranes. Proc. Natl. Acad. Sci. U.S., *68*: 942.

[31] Oda T. (1967) Molecular organization of the electron transfer and oxidative phosphorylation systems in mitochondrial membrane. Proc. 7th Internatl. Congr. Biochem., II. (Symposium), pp. 215–216.

[32] Oda T. (1968) Macromolecular structure and properties of mitochondrial cytochrome ($b+c_1$) complex, cytochrome oxidase, and ATPase. *In*: Structure and Function of Cytochromes. Okunuki K., Kamen M. D. and Sekuzu I. (eds.) pp. 500–515, Univ. Tokyo Press and Univ. Park Press.

[33] Oda T. (1970) Molecular structure and function in living matter. *In*: Profiles of Japanese Science and Scientists. Yukawa H. (ed.) pp. 107–134, Kodansha, Ltd., Tokyo.

[34] Oda T. (1972a) Role of phospholipids in the structure and function of biological membranes. *In*: Phospholipide (Proc. of Phospholipid Symp., Schlangenbat, Germany, 1969). Schettler G. (ed.) pp. 25–41, Georg Thieme Verlag, Stuttgart.

[35] Oda T. (1972b) Molecular organization of the basic structure and specific sites of biomembranes. Invited lecture, on "Specific Sites on Biomembrane," 1st. Israel Sci. Res. Conf.

[36] Oda T. and Huzisige H. (1965) Macromolecular repeating particles in the chloroplast membrane. Exptl. Cell Res., *37*: 481.

[37] Oda T. and Nishi Y. (1963) Fundamental structure and function of mitochondrial membrane. J. Electron Microscopy, *12*: 290.

[38] Oda T. and Sato R. (1964) Elementary particles of the microvilli of intestinal epithelial cells. Invited lecture, 4th Symp. Amer. Soc. Cell Biol., Cleveland.

[39] Oda T. and Seki S. (1965) Molecular structure and biochemical function of the microvillus membrane of intestinal epithelial cells with special emphasis on the elementary particles. J. Electron Microscopy, *14*: 210.

[40] Oda T. and Seki S. (1966a) Molecular organization of the energy transducing system in the mitochondrial membrane. *In*: Electron Microscopy (Proc. 6th Internatl. Congr., Tokyo). Ueda R. (ed.) Vol. 2, pp. 369–370, Maruzen Co. Ltd., Tokyo.

[41] Oda T. and Seki S. (1966b) Molecular basis of structure and function of the plasma membrane of the microvilli of intestinal epithelial cells. *In*: Electron Microscopy (Proc. 6th Internatl. Congr., Tokyo). Ueda R. (ed.) Vol. 2, pp. 387–388, Maruzen Co. Ltd., Tokyo.

[42] Oda T., Seki S. and Watanabe S. (1969) Molecular basis of structure and function of the microvillus membrane of intestinal epithelial cells. Acta Med. Okayama, *23*: 357.

[43] Oda T., Seki S., Yamamoto G., Hayashi H., Hatase O. and Wakabayashi A. (1966) Structure and function of the mitochondria with a brief note on blood cell mitochondria. Acta Haem. Jap., *29*: 108.

[44] Park R. B. (1965) Substructure of chloroplast lamellae. J. Cell Biol., *27*: 151.

[45] Park R. B. and Pon N. G. (1961) Correlation of structure with function in *Spinacea oleracea* chloroplasts. J. Mol. Biol., *3*: 1.

[46] PINTO DA SILVA P. and BRANTON D. (1970) Membrane splitting in freeze-etching. J. Cell Biol., *45:* 598.
[47] RAZIN S., MOROWITZ H. J. and TERRY T. M. (1965) Membrane subunits of *Mycoplasma laidlawii* and their assembly to membrane-like structures. Proc. Natl. Acad. Sci. U.S., *54:* 219.
[48] ROBERTSON J. D. (1959) The ultrastructure of cell membranes and their derivatives. Biochem. Soc. Symp., *16:* 3.
[49] ROSENBLITH J. Z., UKENA T. E., YIN H. H., BERLIN R. D. and KARNOVSKY M. J. (1973) A comparative evaluation of the distribution of Concanavalin A-binding sites on the surfaces of normal, virally-transformed, and protease-treated fibroblasts. Proc. Natl. Acad. Sci. U.S., *70:* 1625.
[50] SEKI S., HAYASHI H. and ODA T. (1970) Studies on cytochrome oxidase. I. Fine structure of cytochrome oxidase-rich submitochondrial membrane. Arch. Biochem. Biophys., *138:* 110.
[51] SEKI S. and ODA T. (1970) Studies on cytochrome oxidase. II. Ultrastructure of cytochrome oxidase. Arch. Biochem. Biophys., *138:* 122.
[52] SEKI S. and ODA T. (1971) Molecular structure and organization of cytochrome oxidase-rich submitochondrial membrane. J. Electron Microscopy, *20:* 232.
[53] SEKI S., OMURA S. and ODA T. (1971) Surface structure of the plasma membranes isolated from rat liver and ascites hepatoma. GANN, *62:* 89.
[54] SINGER S. J. and NICOLSON G. L. (1972) The fluid mosaic model of the structure of cell membranes. Science, *175:* 720.
[55] STEIM J. M., TOURTELLOTTE M. E., REINERT J. C., McELHUNEY R. N. and RADER R. L. (1969) Calorimetric evidence for the liquid-crystalline state of lipids in a biomembrane. Proc. Natl. Acad. Sci. U.S., *63:* 104.
[56] TOURTELLOTTE M. E., BRANTON D. and KEITH A. (1970) Membrane structure: Spin lebeling and freeze etching of *Mycoplasma laidlawii*. Proc. Natl. Acad. Sci. U.S., *66:* 909.
[57] VANDERKOOI G. (1972) Molecular architecture of biological membranes. Ann. N. Y. Acad. Sci., *195:* 6.
[58] VANDERKOOI G. and GREEN D. E. (1970) Biological membrane structure. I. The protein crystal model for membranes. Proc. Natl. Acad. Sci. U. S., *66:* 615.
[59] YAMAMOTO T., SEKI S., HIRATA S. and ODA T. (1971) Properties of ATPase of the microvillus membrane isolated from rabbit small intestinal epithelial cells. Acta Med. Okayama, *25:* 13.

Fine Structure of the Synaptic Membrane in the Central Nervous System

Koji UCHIZONO

The notion of a synapse was introduced into neurophysiology by SHERRINGTON (1897) as a putative structural correlate of a transmission in the central nervous system. He introduced the term to specify a part of the nervous system where specific nervous functions should take place. This particular site should be differentiated from the other regions of the nervous system by several kinds of characteristics; polarity of the nervous conduction—specifically one-way conduction of impulses, easy fatigability, delay in conduction time and so on. However, it was only recently that the morphological correlate of this specific nervous activity had been established, beautifully explaining almost all functional specificities of synapses.

It must be remembered that two schools of researchers had coincidentally substantiated synapses, one by electrophysiology, the other by the morphology of the synapse. DEL CASTILLO and KATZ (1954) proposed that the transmitter substance was liberated spontaneously in random fashion without impulse invasion at the neuromuscular junction of a frog. They characterized the spontaneous discharges of the miniature endplate potentials by their quantal nature. It was a mere coincidence that in the same year DE ROBERTIS and BENNETT (1954) found a new structure at the frog neuromuscular junction. It has been established that a synapse contains synaptic vesicles—small organelles in the cytoplasm of the nerve terminal which could be taken as suitable correlates of the quantal nature of spontaneous discharges at the nerve terminals. The quantal nature of spontaneous miniature potentials at the frog neuromuscular junction seemed to fit well with the discrete structure of synaptic vesicles in the terminals of neuronal axons where transmission should take place. Unfortunately, however, there was no unequivocal evidence by which the quantal nature of miniature endplate potentials was completely explained. The synapse was physiologically defined as that site where the one-way transmission of traveling nervous impulses took place and its morphological counterpart was a swelling of the nerve terminal which was characterized by its major content, synaptic vesicles. Beside the synaptic vesicles, the synapse contains various kinds of organelles such as mitochondria, glycogen granules, neurofilaments, ribonucleoprotein particles and so on.

It has been postulated that the synaptic vesicles *per se* are related to the transmitter substances. The ends of the neuronal axons swell enormously to make synaptic bags suited for containing the synaptic organelles when they make synaptic contact with other neurons or their target cells. The main function of the synapse is a synaptic transmission. However, a trophic function has been postulated for its activity for a long time, although there has been no definite evidence for this. Other higher activity of the central nervous system, for example, learning or memory, has also been postulated to be associated with the synapse, but definite evidence is still lacking. It is generally believed that the transmitter substance is synchronously released from the synaptic vesicles when the traveling impulse invades the synapse as is diagrammatically shown in Fig. 1.

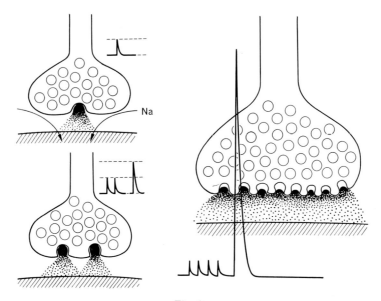

Fig. 1a

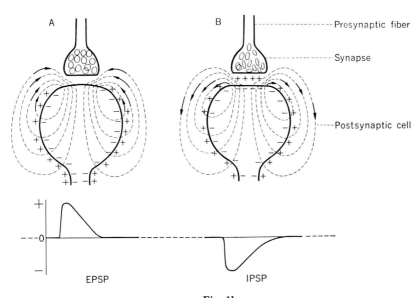

Fig. 1b

Fig. 1 a. Schematic representation of the miniature endplate potentials and action potential. In a resting state apparently irregular potentials of small amplitudes are observed, of which the quantum characteristics has been statistically established. The action potentials are thought to be produced by synchronized discharges of these small potentials.
b. A current flow in excitatory (A) and inhibitory synapse (B). In an excitatory synapse a depolarizing potential (EPSP) is produced by a release of transmitters from the presynaptic terminal, while a hyperpolarizing potential (IPSP) is produced in an inhibitory synapse (B). EPSP and IPSP have their morphological counterparts in the S-type and F-type synapse respectively.

The transmitter substance, when released through the presynaptic membrane into the synaptic cleft, makes the postsynaptic membrane permeable to specific ions, producing excitatory or inhibitory postsynaptic potentials according to the functional nature of the synapses, i. e., EPSPs or IPSPs.

General Features of Synapse

A synapse is a bulge at the end of an axon, being suitable for containing several kinds of organelles. The diameter of a synapse is varied. The most prominent characteristic of a synapse is that it contains an enormous number of synaptic vesicles. Sometimes the synapse is almost entirely filled with vesicles whose diameters are about 500 A.

The shape of the synaptic vesicle is different at functionally different synapses as is shown later, but the usual shape is a spheroid or an oval with a clear content. This type of vesicle is called a clear vesicle to differentiate it from the cored vesicle which contains a dark core inside the vesicle. A synaptic complex is defined as a complicated structure composed of a presynaptic membrane, a synaptic cleft of 200 A in width and a postsynaptic membrane. Each membrane consists of unit membranes. Thus the whole assemblage of a synaptic complex is a five layered structure. It is not unusual to see a dark homogenous substance in synaptic cleft. It is common that the postsynaptic membrane is characterized by a so-called membrane thickening, which is often conspicuous in specific types of synapses. Usually no membrane thickening is observed in the presynapitc membrane. A synaptic complex is thought to be a place where the actual transmitter release occurs, being called the synaptic site.

Sometimes it is observed that a small portion of the postsynaptic cell adjacent to the postsynaptic membrane is darkly stained in a broad area. The site was postulated by some workers to be associated with a specific type of protein (HÝDEN 1973) or a specific membrane structure which showed conformational changes according to the synaptic functions (DE ROBERTIS 1973). It is highly probable that the membranes in the synaptic complex are functionally different from the remaining parts of pre- or postsynaptic membranes. For example, DE ROBERTIS (1973) made neuropharmacological studies on the postsynaptic membrane which showed that there was a conformational change according to the type of drugs applied to it, thus elucidating membrane specificity at the molecular level. The synaptic cleft is only 200 A wide and devoid of glial invasion under normal conditions except for the degeneration of the presynaptic axon. Remaining parts of the synapse other than the synaptic complex are usually surrounded by glial membranes. It was often noticed that besides clear vesicles special types of vesicles which were surrounded by coats existed in the synapses (KANASEKI & KADOTA 1969). GRAY (1972) payed attention to the existence of a new type of component 'cytonet' in the synapse. Neurofilaments were often seen in a synapse especially in a degenerating synapse.

A few mitochondria were often observed in the synapse, but they were usually away from the synaptic site. Cored vesicles with an outer diameter of about 500 A were sometimes seen in the synapse, but they also were situated away from the synaptic site. Very few ribonucleoprotein particles were observed, consistent with the data presented by KÖNING (1958) who showed that only a small percent of the ribosomes was detected in the axonal regions of the neuron. The majority of the ribosomes were located in the neuronal soma.

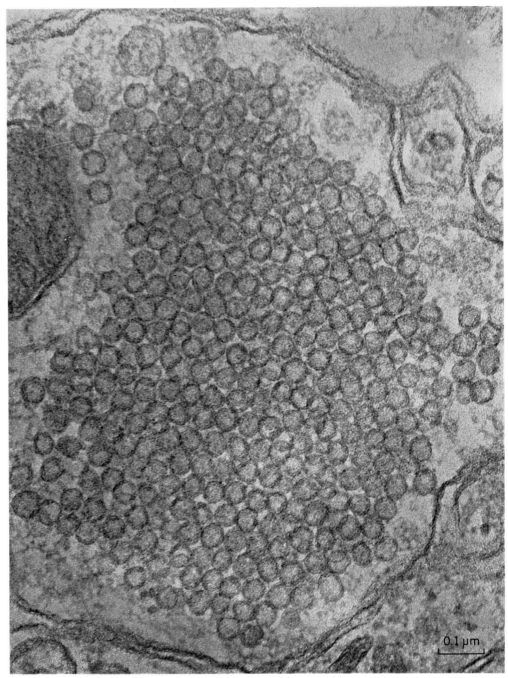

Fig. 2 Typical synapse in the central nervous system. Tremendous number of synaptic vesicles fills the synaptic bag.

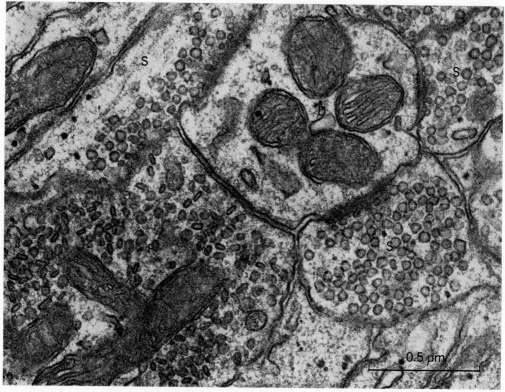

Fig. 3a

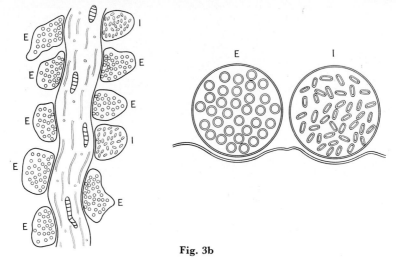

Fig. 3b

Fig. 3 a. Different types of synapses around the dendrite. Two types of synapses are observed on the surface of the dendrite (D) in the cat's cerebellum, one containing spheroidal vesicles (S-type), and the other flattened vesicles (F-type).

b. Schematic representation of two major types of synapses around the dendrite. One type (E) contains round vesicles, while the other flattened vesicles (I). In the cat's carebellum the former is excitatory, the latter inhibitory.

c. Single synaptic vesicles in functionally different synapses. The S-type synapse contains round vesicles about 500 A in diameter, while the F-type contains flattened vesicles with the major and minor diameters being 500 A and 380 A respectively.

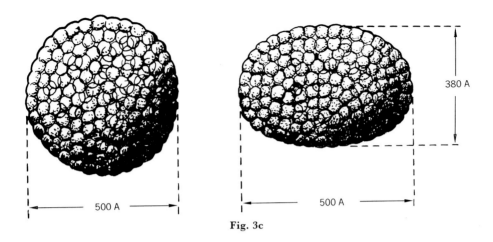

Fig. 3c

Different Types of Synapses in Various Kinds of Nervous Systems

Gray's classification of synapses into type 1 and type 2

GRAY (1961) was the first to notice the difference of synapses in the central nervous system of vertebrates. According to him the type 1 synapse was characterized by a membrane thickening of the postsynaptic element. The postsynaptic membrane thickening was usually observed in the entire area of the synaptic complex. The synaptic cleft was also characterized by the presence of a dark substance. In type 2 synapse no membrane characterization was observed. No membrane thickening in either pre- or postsynaptic membrane was recognized. In conjunction with electrophysiological events of excitatory or inhibitory postsynatpic potentials, ECCLES (1964) postulated that the type 1 synapse corresponded to the excitatory potentials, while the type 2 corresponded to the inhibitory potential. Electrophysiological investigations by ECCLES and his associates (1966) clearly indicated that the dendritic portions of the Purkinje cell in the cat's cerebellum are innervated almost exclusively by the excitatory inputs, while the soma of that cell is exclusively innervated by the inhibitory inputs. RAMÓN Y CAJAL (1911) showed that the major inputs to the Purkinje dendrite originated from the parallel fibers of the granule cells, of which the excitatory nature was established by the electrophysiological investigations of ECCLES et al. (1966). When parallel fibers were stimulated EPSPs were recorded from the soma of Purkinje cell by an intracellular microelectrode, while only IPSPs were evoked in the Purkinje cell soma when the basket cells were stimulated. These electrophysiological data conform well with this author's (1965) electron microscopic investigations which indicated that the dendrites of Purkinje cell were mostly innervated by synapses which contained clear spheroidal vesicles of about 500 A diameter, while the soma of Purkinje cell was exclusively innervated by synapses which contained flattened or elongated vesicles. The characteristic size and shape of synaptic vesicles in these cases are shown schematically in Fig. 3a, b and c.

The same characteristics of synaptic vesicles in the central nervous system was reported by BODIAN (1966) and LARRAMENDI et al. (1967). BODIAN termed the synapse containing spheroidal vesicles the S-type, and the synapse containing elongated vesicles with major diameters of 500 A and minor diameters of 380 A the F-type, respectively. Thus there was good parallelism between the shape and function of synapses as shown in Table 1.

Table 1 Morphological differentiation of synapses.

Excitatory synapse	S-type
Inhibitory synapse	F-type

S-F hypothesis

The S-F hypothesis (UCHIZONO 1974) had been proposed to denote the structural characteristics of synapses of functionally different neurons. There had been abundant evidence for this hypothesis (UCHIZONO 1974), not only in the cat's cerebellum, but also in the spinal cord and hippocampus. It had been well established by KUFFLER and EYZAGUIRRE (1955) that the dendritic region of the stretch receptor neuron of the crayfish was innervated exclusively by inhibitory nerves. Our electron microscopic investigations (1967) and those of others clearly indicated that there was only one type of synapse which was characterized by the flatness of the synaptic vesicles. IPSPs or hyperpolarization were produced in the membrane potential of a neuron by the stimulation of this inhibitory neuron. Our electron microscopic investigations (UCHIZONO 1967) elucidated without ambiguity that the inputs to the neuronal dendrite of the stretch receptor were of the F-type. The synapses contained elongated synaptic vesicle only. No S-type synapse had been observed in the dendritic region of a stretch receptor neuron. Then it was reasonable to correlate the inhibitory activity with this F-type synapse. On the other hand, inputs to the stretch receptor muscle of the crayfish were known to be exclusively excitatory, although dual innervation of excitatory and inhibitory activity in the skeletal muscles of the crayfish had been well established. Our electron microscopic investigations (KOSAKA 1966) on the stretch receptor muscle of the crayfish indicated that the majority of the synapses in this muscle were of the S-type, only a few exceptional F-type synapses were also observed in this muscle. Thus it could be concluded that the relationship between structure and function of synapses in the crayfish stretch receptor was the same as in the vertebrate cerebellum. The next table shows a parallelism of structure and function of synapses in the vertebrate cerebellum and the crayfish stretch receptor.

Table 2 Excitatory (S-type) and inhibitory (F-type) synapses are differentiated most clearly in the stretch receptor of crayfish because of the exclusive excitatory or inhibitory innervation to the specific area of the organ, without contamination from double innervation which usually occurs in other regions of this animal.

Excitatory synapse	S-type	inputs to stretch receptor muscle
Inhibitory synapse	F-type	inputs to stretch receptor neuron

Symmetrical and asymmetrical synapse

There is another way of classifying synapses in which emphasis is laid on the symmetry or asymmetry of the synaptic complex. COLONNIER (1968) proposed that in one synapse the density of the pre- and postsynaptic membrane is the same (symmetrical synapse) while in others the density of both membranes was different (asymmetrical synapse), the postsynaptic membrane being much denser and wider than the presynaptic membrane. It was noted that parallelism existed between structure and function in this classification as shown in Table 3.

Table 3 Colonnier's classification of synaptic types in the central nervous system.

Asymmetrical synapse	excitatory
Symmetrical synapse	inhibitory

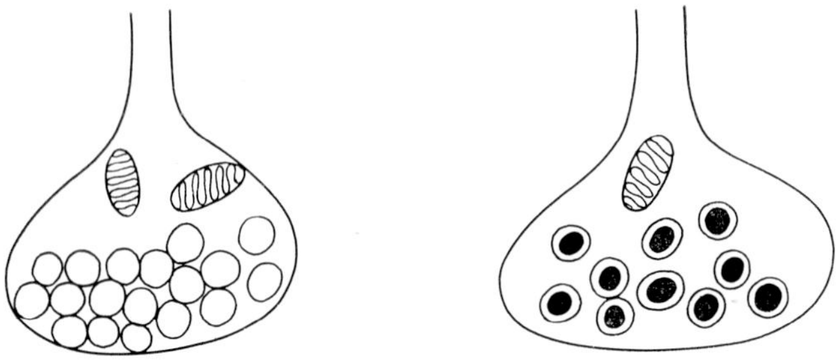

Fig. 4 Two cardinal types of synapses in the autonomic nervous system. Cholinergic and adrenergic nerve terminals are morphologically linked with the S-type and C-type synapse respectively. The latter is characterized by the presence of central cores in the vesicles.

Unfortunately, however, there are abundant intermediate instances in this classification of synapses. Owing to the ambiguity of the criteria of symmetrical and asymmetrical synapses, this type of classification is in much need for improvement.

Various kinds of synapses in the autonomic nervous system

Two major types of synapses has been known in the autonomic nervous system. It is generally expected that a different nervous system should have a different synapse. Cholinergic and adrenergic nature of the parasympathetic and sympathetic nervous system have been well established. At the skeletal neuromuscular junction of vertebrate animals only the S-type synapses are observed, conforming to the fact of monopolizing excitatory innervation of the skeletal muscle of the vertebrate. The cholinergic nature of synapses at the neuromuscular junction is represented by the spheroidal shape of the synaptic vesicles (S-type), while the adrenergic nature of synapses in the sympathetic nerves is symbolized by another type of synapses (C-type). In this system the synaptic vesicles are equipped with dense cores inside the vesicles. Cored vesicles are characteristic of the sympathetic nervous system.

A dual innervation in most of the internal organs is well represented by the presence of two types of synapses (S-type and C-type) in this system.

Occasionally a peculiar synapse is observed in a special case (BURNSTOCK 1972; UCHIZONO 1974). A large synapse which contains enormously large cored vesicles (of 2000 Å in outer diameter) as shown in Fig. 5 is sometimes observed.

The existence of a third innervation in addition to the cholinergic and adrenergic nerves, inhibitory in nature, was electrophysiologically and neuropharmacologically postulated in taenia coli. BURNSTOCK (1972) suggested that the transmitter of the third innervation was ATP or a related substance. In his opinion the synapses of this class should be called the P-type (purinergic synapse) because the putative transmitters in this system such as ATP and other purine bodies were suggested. UCHIZONO (1974) reported the existence of another type of synapse in taenia coli of the guinea pig which was similar to the F-type synapse in the somatic nervous system. In conclusion, at least four types of synapses (S-, C-, P- and F-type) exist in the autonomic nervous system, although the rate of occurrence of each of these synapses are very different from each other.

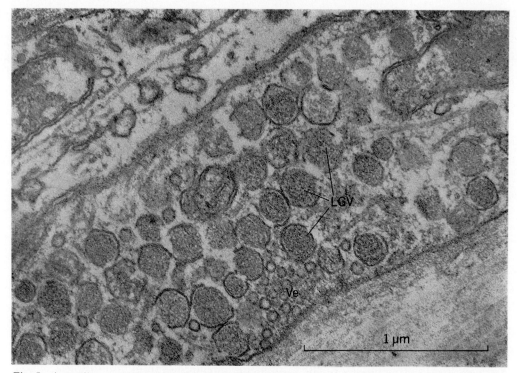

Fig. 5 A peculiar nerve terminal containing a mixture of two types of vesicles in a taenia coli.
A nerve terminal is shown which contains numerous very large granulated vesicles (LGV) together with a small number of spheroid vesicles of the usual size. Dark cores fill the vesicles to the limiting membrane.

Functional Significance of the Synaptic Difference in Various Kinds of Nervous Systems

Although the real mechanism of a release of transmitters at synapses has not yet been fully explained, the intimate relationship between synaptic transmitters and synaptic vesicles should be born in mind. Cholinergic synapses, one at the neuromuscular junction and another at the terminals of the parasympathetic nerves, are structurally very similar and belong to the S-type. It is not known whether the transmitter is contained inside the vesicles as is the case with the sympathetic nerve terminals or it is attached to the vesicles themselves. In the case of the adrenergic nervous system there is no doubt that the transmitter is contained inside the vesicles in the form of dense cores in the synapse and directly associated with catecholamine, because they completely disappear after an application of reserpine which is known to make the cores of the vesicles discharge. The recovery of the dense cores of these vesicles takes place gradually in a course of a week or so (CLEMENTI 1966).

Spheroidal clear vesicles in the synapses of both somatic and autonomic nervous systems belong to the S-type. Shape and size of the S-type synaptic vesicles in the somatic and autonomic nerves are similar to ecah other. UCHIZONO (1965, 1967) and LARRAMENDI (1967) reported that the inhibitory synapse contains elongated vesicles (500 A × 380 A) which belong to the F-type. It has been established by OBATA et al. (1967) that GABA is a strong candidate for the inhibitory transmitter in the cat's cerebellum. It is highly probable that the F-type synapses in the vertebrate cerebellum are associated with GABA.

In the spinal cord of the cat, however, WERMANN et al. (1968) proposed that glycine but

Table 4 Transmitters in the nervous systems are tentatively classified into three types on the structural basis of synapses.

S-type	Acetylcholine
	Dorsal root peptide
	Glutamate (in crayfish)
F-type	GABA
	Glycine
C-type	Catecholamine

not GABA was the inhibitory transmitter. In the case of presynaptic inhibition in the spinal cord, DAVIDOFF (1973), DE GROAT, (1973) and BARKER and NICOLL (1973) reported new findings indicating that GABA played a role in the primary afferent depolarization. UCHIZONO (1973) presented a paper concerning the synaptic structures in a presynaptic inhibition of the spinal cord. He emphasized that GABA was represented here too as the F-type synapse in spite of the depolarizing action of this substance in this special case. Usually GABA acts as an inhibitory transmitter which hyperpolarizes the membrane potential of the postsynaptic membrane. In the spinal cord it is suggested that GABA and glycine act as presynaptic and postsynaptic transmitter respectively. Both transmitters are associated with the F-type. In particular, it must be emphasized that the excitatory transmitter in the primary afferents has recently been identified by OTSUKA and his associate (1972).

The dorsal root peptide which is extracted only from the dorsal root but not from the ventral root of the cat has been established as the sensory transmitter in the primary afferent fibers. Substance P had been the candidate for the sensory transmitter for the past several decades but it lacked the definite evidence for being the sensory transmitter.

The quantity of dorsal root peptide decreases gradually and finally disappears upon sectioning the dorsal root. Extracted dorsal root peptide depolarizes the membrane potential of the ventral motoneuron in the vertebrate. A drastic change in the morphology of synapses on the surface of the motorneuron occurs following a degeneration experiment of the dorsal root. The change occurred mostly in synapses which originally contained spheroidal vesicles. RÉTHELYI (1970) determined, utilizing the CLARKE's column of the cat's spinal cord that the dorsal root terminals made giant axon terminals (GAT) on the dendrite of the relay neurons. He observed that GAT synapse contained spheroidal vesicles. It was then assumed that the S-type synapse was indicative of the excitatory nature of the neuron. He also reported that the GAT was contacted by a small synapse which contained flattened vesicles.

In conjunction with our electron microscopic investigations (UCHIZONO 1974) neuropharmacological approaches by DAVIDOFF (1973), DE GROAT (1973) and BARKER and NICOLL (1973) suggested that this small F-type synapse was associated with GABA. It had been observed by DE GROAT (1972) that the membrane of the neurons in the spinal ganglion was depolarized by the application of GABA. It is highly probable that the axon terminal of the neuron of the spinal ganglion (GAT) in the Clarke's column is also sensitive to GABA. OTSUKA and his associates (1970) reported that the area of the dorsal horn in the spinal cord of the cat contained a high concentration of GABA, which was remarkably decreased by a suppression of the blood supply by arterial occlusion. It is well known that the arterial occlusion gives rise to a degeneration of the interneurons in the spinal cord. Therefore, it is speculated that the GABA in the dorsal horn of the spinal cord is released from the axon terminals of inhibitory interneurons.

Table 4 shows the relationship between a synaptic structure and chemical nature of the transmitter in the synapse in the various kinds of neurons.

Synaptic Structure in Presynaptic Inhibition

It has been postulated that there should be two different mechanisms of inhibition existing in the spinal cord of the vertebrate. These two mechanisms of pre- and postsynaptic inhibition had been well differentiated by electrophysiological and neuropharmacological methods. As already described in the preceding chapters, there is a reasonable structural counterpart of the postsynaptic inhibition, but unfortunately there is considerable controversy on the structural correlate of a presynaptic inhibition.

The electrophysiological investigation by Eccles and his associates (1961) assumed a serial or double synapse as the mechanism of presynaptic inhibition. Since the early description of double synapse by Gray (1962) and Khattab (1968), keen attention had been focussed on the fine structure of presynaptic inhibition. Réthelyi (1970) and Kerr (1970) presented good evidence to explain the mechanism of presynaptic inhibition, but controversy still remained from the point of view of the S-F hypothesis of a synapse. Quite recently, however, Uchizono (1973) presented a new interpretation of serial or double synapse which would explain the mechanism of presynaptic inhibition both electrophysiologically and neuropharmacologically. We had been unsuccessful so far in our research to find a serial synapse in the ventral horn of the spinal cord where the mechanism of presynaptic inhibition should occur.

Conradi (1969) constructed a three-dimensional model of a serial synapse with a reconstruction of serial sections by electron microscopy and suggested a structural correlate of the presynaptic inhibition. Uchizono (1973) presented new evidence for the fine structure of presynaptic inhibition which would solve the difficulties in the interpretation of a serial synapse. Serial synapses have been rather frequently encountered in a specific area of the dorsal horn of the spinal cord, where Otsuka and his associates (1970) found a high concentration of GABA. Fig. 6 shows a serial synapse composed of a F- and a S-synapse in series, with the polarity of the synapse being always from the F-synapse to the S-synapse.

Table 5 Relationship among function, structure and chemical nature of the transmitters involved in presynaptic inhibition in the spinal cord of a cat.

Function	Transmitter	Structure
Excitatory synapse	Dorsal root peptide	S-type
Inhibitory synapse	GABA	F-type

The area is very high in GABA concentration, so it may be that the transmitter of the F-type synapse is also GABA, while the transmitter for the S-type may be the dorsal root peptide (Otsuka et al. 1972). Neuropharmacological investigations by De Groat (1972), Barker and Nicoll (1972) indicated that the GABA was the transmitter of the presynaptic inhibition which depolarized the membrane potential of the sensory axon terminal which contained an excitatory transmitter (dorsal root peptide). It is highly probable that this sensory terminal releases the depolarizing transmitter associated with the S-type. Table 5 shows the relationship of transmitter substances and the synaptic structure which is consistent with the S-F hypothesis in the somatic nervous system.

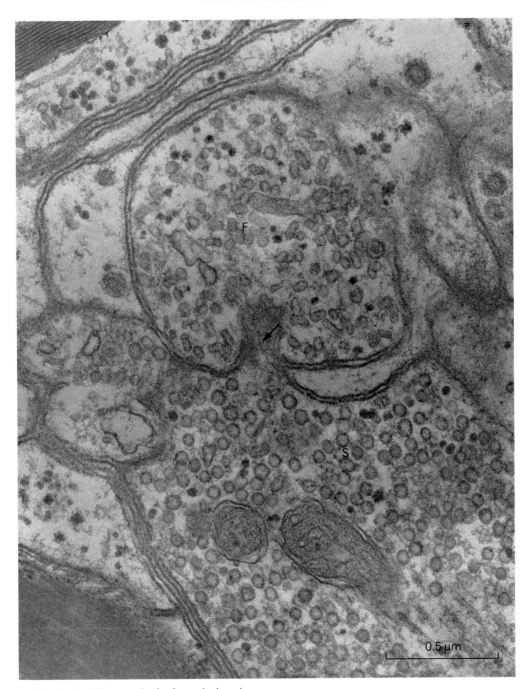

Fig. 6 Serial synapse in the frog spinal cord.
In this type of synapse the presynapse contains flattened vesicles (F-type) while the postsynapse contains spheroidal vesicles. The latter makes synaptic contact on the dendrite. Arrows show the synaptic polarity.

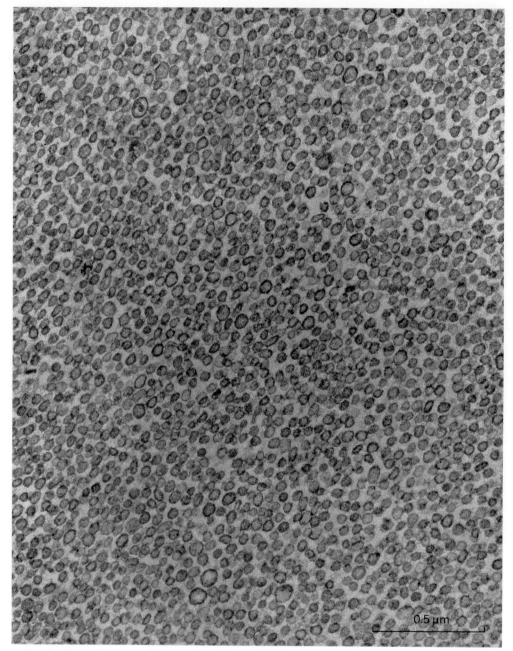

Fig. 7 Homogenate of synaptic vesicles.
Huge numbers of synaptic vesicles have been accumulated by ultracentrifugation of rat brain homogenates. The over-all characteristics of vesicles is similar to those in living specimens except for a slight deformation and shrinkage of the vesicles.

Membrane Recycle

The vesicle hypothesis of synaptic transmission which insists that the transmitter substance is firmly associated with the synaptic vesicles is more substantially supported by the discovery of a membrane recycle in the synapse. This new understanding of synaptic transmission has come about from a reappraisal of the complex of coated vesicles. Such vesicles are assumed to be produced by endocytosis. Coated vesicles (ROTH et al. 1964; NICKEL et al. 1967; KANASEKI & KADOTA 1969; NAGASAWA et al. 1970) in the nervous system consisted of smoothly walled vesicles surrounded by a shell formed of cytonet (GRAY 1972) specifically arranged in a hexagonal array. Coated vesicles fuse to form the cisternae that bud off synaptic vesicles. Every component in their scheme (HEUSER et al. 1973) had been observed quite frequently under specific experimental conditions, which were expected to be favorable for revealing metabolic changes of the nerve terminal (CLARK et al. 1970, 1972).

HEUSER et al. (1973), CECCARELLI et al. (1973) used horse radish peroxidase (HRP) or dextran to demonstrate the fate of coated vesicles in nerve terminals. A neuromuscular preparation of a frog sartorius was incubated with HRP under conditions favorable for acetylcholine release, such as bathing in elevated potassium concentrations or black widow spider venom (BWSV) application and fixation with a fixative after varying time of incubation for electron microscopy. After a short incubation period a loss of synaptic vesicles and an apparent increase in the area of the membrane lining the synaptic cleft were observed together with the appearance of HRP in the complex vesicles forming in the membrane which faced the synaptic cleft.

At later stages after fixation HRP was observed in the cisternae and in the synaptic vesicles. It was interpreted that HRP was absorbed by pinocytosis into these minute organelles. There have been sufficient evidences supporting this interpretation. An application of BWSV (CLARK et al. 1970, 1972), which is known to be potent in reducing the vesicle population, produces infoldings of the presynaptic membranes, being consistent with an increase in the membrane area at this site. This may be due to prevention of the formation of complex vesicles by BWSV, arrest of recycling of the synaptic vesicles, or reduction of the number of vesicles in a synapse.

Membraneous structures are sometimes seen in the synapse away from the synaptic sites. Synaptic vesicles seem to be produced by budding from the end of the cisternae in either the S-type or F-type synapse (UCHIZONO 1974).

Biochemical Approach to Synaptology

A pure fraction of synaptic vesicles is obtained by ultracentrifugation of a brain homogenate from a guinea pig or rat. It is rather amazing that the vesicles are only slightly damaged in spite of the drastic physicochemical processes in the ultracentrifugation as shown in Fig. 7. We cannot recognize differences in the shape of the vesicles except for a slight shrinkage of the vesicles obtained by biochemical procedures by merely inspecting the electron micrographs of synapses obtained by the different methods. The membrane structure of single vesicles is well maintained in both cases. There has been a little success in attempts to obtain a different fraction of synapses from brain homogenate indicating a structural difference of the S-type and F-type synaptic vesicles (unpublished).

REFERENCES

[1] BARKER J. L. and NICOLL R. A. (1972)   GABA: Role in primary afferent depolarization. Science, *176:* 1043–1045.

[2] BODIAN D. (1966)   Electron microscopy: tow major synaptic types on spinal motoneurons. Science, *151:* 1093–1094.

[3] BURNSTOCK G. (1972)   Purinergic nerves.   Pharmacol. Rev., *24:* 509–581.

[4] CICCARELLI B., HURLBUT W. P. and MAURO A. (1973)   Turnover of transmitter and synaptic vesicles at the neuromuscular junction.   J. Cell Biol., *57:* 499–524.

[5] CLARK A. W., HURLBUT W. P. and MAURO A. (1972)   Changes in the fine structure of the neuromuscular junction of the frog caused by black widow spider venom.   J. Cell Biol., *52:* 1–14.

[6] CLARK A. W., MAURO A., LONGENECKER H. E. and HURLBUT W. P. (1970)   Effect on the fine structure of the frog neuro-muscular junction.   Nature, *225:* 703–705.

[7] CLEMENTI F., MANTEGAZZA P. and BOTTURI M. (1966)   Pharamcological and morphologic study on the nature of the dense-core granules present in the presynaptic endings of sympathetic ganglia.   Int. J. Neuropharmacol., *5:* 281–285.

[8] COLONNIER M. (1968)   Synaptic patterns on different cell types in the different laminae of the cat visual cortex. An electron microscope study.   Brain Res., *9:* 268–287.

[9] CONRADI S. (1969)   Ultrastructure of dorsal root boutons on lumbosacral motoneurons of the adult cat, as revealed by dorsal root section.   Acta Physiol. Scand., Suppl., *332:* 85–115.

[10] DAVIDOFF R. A. (1972)   GABA antagonism and presynaptic inhibition in the frog spinal cord.   Science, *175:* 331–333.

[11] DE GROAT W. C. (1972)   GABA-depolarization of a sensory ganglion: Antagonism by picrotoxin and bicuculline.   Brain Res., *38:* 429–432.

[12] DE ROBERTIS E. D. P. and BENNETT H. S. (1954)   Submicroscopic vesiclar component in the synapse.   Fed. Proc., *13:* 35.

[13] DE ROBERTIS E. D. P. (1973)   Molecular biology of chemical synaptic transmission.   Abstract ISN Tokyo Meeting p. 10.

[14] DEL CASTILLO J. and KATZ B. (1954)   Quantal components of the endplate potential.   J. Physiol., *124:* 560–573.

[15] ECCLES J. C., ECCLES R. M. and MAGNI F. (1961)   Central inhibitory action attributable to presynaptic depolarization produced by muscle afferent volleys.   J. Physiol., *159:* 147–166.

[16] ECCLES J. C. (1964)   The Physiology of Synapses. p. 19, Springer-Verlag, Berlin-Göttingen-Heidelberg.

[17] ECCLES J. C., LLINAS R. and SASAKI K. (1966)   The action of antidromic impulses on the cerebellar Purkinje cells.   Exp. Brain Res., *1:* 161–183.

[18] GRAY E. G. (1961)   Ultrastructure of synapses of the cerebral cortex and of certain specializations of neuroglial membrane.   *In:* Electron Microscopy in Anatomy. BOYD E. S. et al. (eds.) pp. 54–73, Edward Arnold, London.

[19] GRAY E. G. (1962)   A morphological basis for presynaptic inhibition.   Nature, *193:* 82–83.

[20] GRAY E. G. (1972)   Are the coats of coated vesicles artefacts?   J. Neurocytol., *1:* 363–382.

[21] HEUSER J. E. and REESE T. S. (1973)   Evidence of recycling of synaptic vesicle membrane during transmitter release at the frog neuromuscular junction.   J. Cell Biol., *57:* 315–344.

[22] HÝDEN H. (1973)   Organized complexities and protein response of brain cell.   Abstract ISN Meeting, Tokyo p. 8.

[23] KANASEKI T. and KADOTA K. (1969)   The vesicle in a basket.   J. Cell Biol., *42:* 202–220.

[24] KERR F. W. L. (1970)   The organization of primary afferent in the subnucleus caudalis of the trigeminal: A light and electron microscope study of degeneration.   Brain Res., *23:* 147–165.

[25] KHATTAB F. I. (1968)   A complex synaptic apparatus in spinal cords of cats.   Experientia, *24:* 690–691.

[26] KÖNIG H. (1958)   An autoradiographic study of nucleic acid and protein turnover in the mammlian neuraxis.   J. biophys. biochem. Cytol., *4:* 785–792.

[27] Kuffler S. W. and Eyzaguirre C. (1955) Synaptic inhibition in an isolated nerve cell. J. gen. Physiol., *39:* 155–184.

[28] Larramendi L. M. H., Fickenscher L. and Lemkey-Johnston N. (1967) Synaptic vesicles of inhibitory and exciattory terminals in the cerebellum. Science, *156:* 967.

[29] Nagasawa J., Douglass W. W. and Schulz R. A. (1970) Ultrastructural evidence of secretion by exocytosis and of a synaptic vesicle formation in posterior pituitary gland. Nature, *227:* 407–409.

[30] Nickel E., Vogel A. and Waser, P. G. (1967) "Coated vesicles" in der Umgebung der neuromuskularen Synapsen. Z. Zellforsch. Mikro. Anat., *78:* 261–266.

[31] Obata K., Ito M., Ochi R. and Sato N. (1967) Pharmacological properties of the postsynaptic inhibition by Purkinje cell axons and the actino of GABA on Deiters' neurones. Exp. Brain Res., *4:* 43–57.

[32] Otsuka M. et al. (1972) The presence of a motor-neuron depolarizing peptide in bovine dorsal roots of spinal nerves. Proc. Jap. Acad., *48:* 342–346.

[33] Otsuka M., Obata K., Miyata Y. and Tanaka Y. (1970) The meassurement of GABA in isolated nerve in cat central nervous system. J. Neurochem., *18:* 287–295.

[34] Ramón y Cajal S. (1911) Histologie du système nerveux de l'homme et des vertébrés. Maloine, Paris.

[35] Réthelyi M. (1970) Ultrastructural synaptology of Clarke's column. Exp. Brain Res., *11:* 159–174.

[36] Roth T. F. and Porter K. R. (1964) York protein uptake in the oocyte of the mosquitto Aedes aegipti. J. Cell Biol., *20:* 313–332.

[37] Sherrington C.S. (1897) The central nervous system. *In:* A Text-Book of Physiology, 7th ed., Foster M. (ed.) vol. 3, Macmillan, London.

[38] Uchizono K. (1965) Characteristics of excitatory and inhibitory synapses in the central nervous system of the cat. Nature, *207:* 642–643.

[39] Uchizono K. (1967) Inhibitory synapses on the stretch receptor neurone of the crayfish. Nature, *214:* 833–834.

[40] Uchizono K. (1973) Structural and chemical consideration on the presynaptic inhibitory synapses. Proc. Jap. Acad., *49:* 569–574.

[41] Uchizono K. (1975) Excitation and Inhibition: Synaptic Morphology. Igaku Shoin Ltd., Tokyo.

[42] Werman R. and Aprison M. H. (1968) Glycine: the search for a spinal cord inhibitory transmitter. *In:* Structure and Functions of Inhibitory Neuronal Mechanisms. von Euler C., Skoglund S. and Söderberg U. (eds.) pp. 473–486, Pergamon Press, Oxford.

2. FILAMENTS

Fine Structure of Myofilaments in Chicken Gizzard Smooth Muscle

Yoshiaki NONOMURA

As regards the structure and function of vertebrate smooth muscle, the often used phrase, "headache muscle" seems to be continuing alive even at the present time when remarkable progress is believed to have been made by the technical improvement in electron microscopical and electrophysiological investigations.

There has long been an unproductive controversy regarding the problem of the mode of existence of myofilaments, particularly the thick filament. During the early period from 1967 to 1971, it was disputed whetehr the essential myofilaments under the physiological conditions consisted of one filament (PANNER & HONIG 1967, 1970; COOKE et al. 1970; ROSENBLUTH 1971) or two filaments (NONOMURA 1968; RICE et al. 1970; HEUMANN 1971; DEVINE & SOMLYO 1971). After the two filaments' theory was widely accepted, a new confusing problem surrounding the presence of a ribbon structure as a essential structure of the thick filament was presented by LOWY's group on the basis of electron microscopical and X-ray diffractional studies (LOWY & SMALL 1970; LOWY et al. 1970, 1973; SMALL & SQUIRE 1972; SMALL & SOBIESZEK 1973; SOBIESZEK & SMALL 1973). The structure of the ribbon supposed by LOWY's group was a very long and wide filament, 3 μm in length and 50 to 80 nm in width, and consisted of the lent filaments (usually named as intermediate filaments) as a core and myosin molecules with polarity attached to the core surface. Recently the ribbon observed in a transverse section has been explained as an artifact product of fixation (SOMYLO et al. 1971, 1973) or the result of a free movement of thick filaments in the spacing of the cytoplasm by stretching the fiber (NONOMURA 1975a). What could we understand about the structure of myofilaments in vertebrate smooth muscle during these seven years? We should perceive that the morphological situation of the smooth muscle field has now been in a state comparable to the early part of the 1950's in the striated muscle filed where the existence of two filaments was established, but there was no unanimous theory on the contraction mechanism.

This article is intended to observe the fine structure of spearated myofilaments in the homogenates or the crude fractions by negative staining and to clarify the filament structure in relation to the sectioned images.

Experimentals

Chicken gizzard (CG) was dissected in small pieces and placed directly into 3% glutaraldehyde buffered with 0.125 M cacodylate (pH 7.2). The tissue was washed with a cacodylate buffer, teased into much smaller pieces in the buffer solution and fixed in 1% osmium tetroxide buffered with cacodylate. After the tissue was washed repeatedly with distilled water, block staining was carried out in 1% uranyl acetate solutions, and dehydrated with an alcohol series and acetone: The tissue was then embedded in Epon according to the usual method. Materials were sectioned with a Porter-Blum MT-II type microtome and stained with lead citrate.

For negatively stained specimens, the tissue was cut into small pieces in Locke solution, transferred into a modified relaxing medium which contained 0.15 M KCl, 25 mM tris-maleate buffer (pH 7.0), 5 mM $MgCl_2$ and 10 mM or more ATP, and homogenized with a Polytron homogenizer (Kinematica GMBH Luzern, Switzerland) for 15 to 30 sec.

A crude fraction of thin filaments was prepared by the following procedure. The homogenate mentioned above was centrifuged at 7,000 rpm for 30 min. The resultant supernatant was recentrifuged at 30,000 rpm for 2 hr. The sediment was dissolved in 0.1 M KCl and 20 mM tris-maleate buffer (pH 7.0).

The fraction, including dense bodies and intermediate filaments (Fraction D-I), was prepared by some modification of the method of COOKE and FAY (1972). After homogenizing the tissue in the relaxing medium as mentioned above, the material was centrifuged at 18,000 rpm for 1 hr. The sediment was suspended in 0.6 M KCl and 10 mM tris-maleate buffer (pH) and the resultant suspension was centrifuged at 5,000 rpm for 30 min. The sediment was dissolved in 0.1 M KCl and 10 mM tris-maleate buffer. This fraction contained many dense bodies, intermediate filaments and mitochondria.

Actomyosin of CG was prepared by the following procedure: The tissue was homogenized with a Polytron homogenizer in 0.5 M KCl and 30 mM tris-maleate buffer (pH 7.0) for a few minutes. After standing at room temperature for 30 min, the solution was homogenized once more for a few minutes and the homogenate was centrifuged at 3,000 rpm for 10 min. The supernatant was diluted with 10 mM tris-maleate buffer into a final concentration of 0.1 M KCl. This solution was centrifuged at 8,000 rpm for 30 min and the resultant sediment was dissolved in 0.6 M KCl buffered with 10 mM tris-maleate. The dissolution and centrifuging procedure was repeated twice.

Negative staining was carried out with 4% uranyl acetate as follows: One drop of the sample was placed on a mesh with a carbon-coated collodion film. Immediately several drops of stain were placed upon the specimen and the stain was promptly rinsed and dried with a piece of filter paper. A JEM 100 B-4 type electron microscope was used with an accelerating voltage of 80 KV and with a 30 μm objective aperture.

Results and Discussion

Images in the sections

In a transverse section, smooth muscle of CG clearly shows two kinds of myofilaments, thick and thin filaments as demonstrated in Fig. 1. The thick filament reveals an irregular profile rather than the round shape usually observed in striated muscle. In many places the thick filament appears as a ribbon-like structure; narrow in width and long in length. The origin of the ribbon in CG was explained from the observations of a longitudinal section (Fig. 2) that thick filaments ran obliquely to the long axis of the fiber. On the other hand, the ribbon produced by an intense stretching of the guinea pig's taenia coli was considered to be the result of the close apposition of thick filaments (NONOMURA 1975a). Thus, the ribbon was produced by various means. Thin filaments with a diameter of 6 to 7 nm were randomly distributed inbetween the thick filaments. Ratio of the number of thin to thick

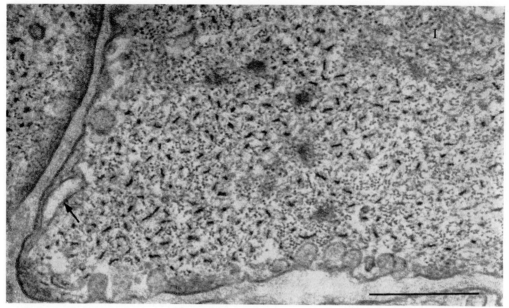

Fig. 1 Transverse section of a chicken gizzard.
Thick and thin filaments are randomly distributed and dense bodies appear with their clear outlines. In the area marked I, intermediate filaments are gathered. An arrow shows a sarcoplasmic reticulum under the subcytoplasmic membrane. Many micro-inpocketings or surface vesicles are visible as invaginations of membranes. (Inserted bar indicates the scale of 0.5 μm)

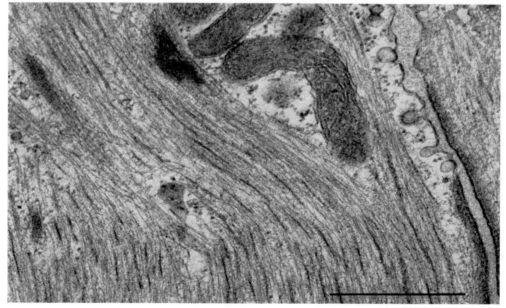

Fig. 2 Longitudinal section of a chicken gizzard.
Thick and thin filaments run in a disorderly arrangement along the fiber axis. Dense bodies and the dense area under the subcytoplasmic membrane are recognizable with their clear outlines. Intermediate filaments are identified by their special wavy profiles. Mitochondria and micro-inpocketings are visible. (Scale indicates 1 μm)

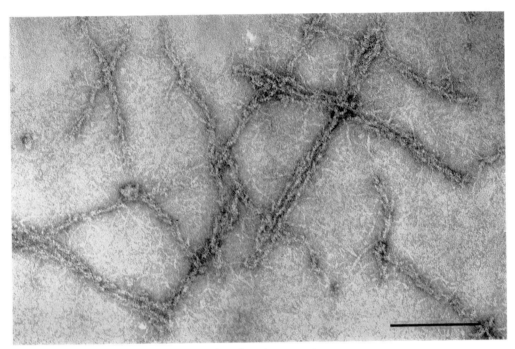

Fig. 3 Negatively stained image of chicken gizzard actomyosin.
Actomyosin was dissolved in 0.6 M KCl buffered with bicarbonate. Notice the arrowhead structure on all the filaments. Two or three filaments are aligned in parallel. Negatively stained with 4% uranyl acetate. (Scale indicates 0.25 μm)

filaments was calculated to be nearly 8 in the transverse section and this ratio was much less than the ratio of 15 in the guinea pig's teania coli which was described in details in other paper (NONOMURA 1975a). This is meant to include the higher number of the thick filament in CG than in GTC. Dense bodies with clear outlines were distributed in the transverse section. In the longitudinal section, the dense body appeared fusiform and a bundle of thin filaments was seen to enter into the end of the dense body. Intermediate filaments were distinguished by their special size and shape: in a transverse section by their round profile with a diameter of nearly 10 nm and in a longitudinal section by their long wavy profile.

Negatively stained image of actomyosin

Generally a preparation of actomyosin in smooth muscle is difficult because of the small amounts of myosin. For this reason CG is the preferable material for an actomyosin preparation in smooth muscle because of its relatively high contents of myosin molecules. A negatively stained image of CG actomyosin exhibits a typical structure of the arrowhead resembling actomyosin when it is prepared from skeletal muscle as shown in Fig. 3. In some instances two or three actomyosin filaments are observed to be aligned in parallel.

When relative amounts of myosin to actin in the actomyosin filaments were small, the arrowhead formation was poorly developed. In relation to this evidence, we were reminded of some reports about ten years ago that a protein was extracted at low ionic strength from arterial (LASZT & HAMOIR 1961) and uterine (HUYS 1961) smooth muscle. LASZT and HAMOIR suggested that this protein, named "tonoactomyosin", was a special form of actomyosin which coexisted with normal actomyosin. At present, tonoactomyosin

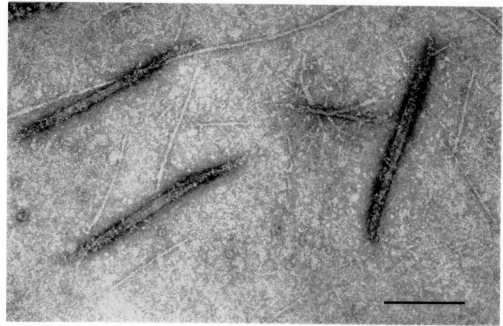

Fig. 4 Negatively stained image of separated myofilaments.
Thick and thin filaments are clearly visible. In the thick filaments, notice the central bare zone, surface projections and tail ends. (Scale indicates 0.25 μm)

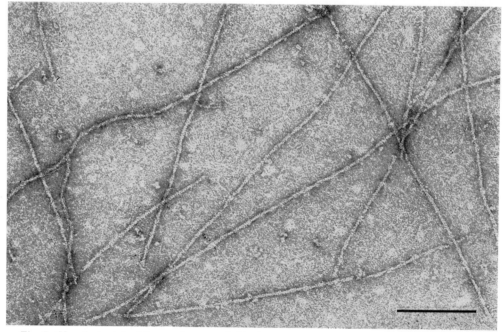

Fig. 5 Negatively stained image of separated thin filaments.
A pitch of 36 to 37 nm is clearly visible on the filament. The length of some filaments is 2 μm or more. (Scale indicates 0.25 μm)

is understood to be only a special state of actomyosin in which the ratio of myosin to actin is very low. A negatively stained image of tonoactomyosin showed an actin-like structure because of the poorly developed arrowhead formation (SHOENBERG et al. 1966).

Separated myofilaments—negatively stained image of thick filaments

To obtain natural forms of myofilaments from fresh muscle, high concentrations of ATP were necessary in the relaxing medium. Similar situation was observed in separated myofilaments of molluscan catch muscle in the presence of high concentrations of ATP (NONOMURA 1974). If 3 to 4 mM of ATP were used in the case of CG, the interaction of thick to thin filaments occurred promptly and the thick filament was surrounded by many thin filaments like the superprecipitated state of the actomyosin system (NONOMURA & EBASHI 1975). Thus, the natural profile of the thick filament was not apparent in the electron microscope. When actomyosin was used as the starting material to obtain separated thick and thin filaments in the dissociated state, ATP was not needed in such high concentrations as with the fresh homogenate. The negatively stained images of separated thick filaments reported until now were considered to be obtained by the addition of ATP to actomyosin (RICE et al. 1966; KELLY & RICE 1968; SHOENBERG 1969).

Fig. 4 shows the separated myofilaments obtained from a homogenate of fresh muscle in the presence of high concentrations of ATP. Separated thick and thin filaments are visible. Profiles of the thick filament resemble the thick filament or the synthetic myosin filament from striated muscle with the presence of a central bare zone, surface projections of myosin heads and both ends of a tail. In spite of such a resemblance in structure, its dimension was different from that of striated muscle. The length of the filament was nearly 0.7 μm, shorter than the 1.5 μm in striated muscle and the width was about 14 to 20 nm, broader than the 12 to 15 nm in striated muscle. The result is almost the same as that in the guinea pig's taenia coli (NONOMURA 1975a).

Possibility of the presence of a ribbon structure in CG was carefully checked with negative staining of the whole muscle homogenate. As described above, two kinds of filaments with the usual types of profiles were recognizable and other organellae, mitochondria, dense bodies, intermediate filaments, microsomes and collagen were often visible. However, the negatively stained image of the very long filament with a broader width expected from the structure of the ribbon reported by SMALL and SOBIESZEK (1973) and SOBIESZEK and SMALL (1973) could not be observed in the whole samples of homogenate.

Negatively stained image of a thin filament

A crude fraction of separated thin filaments exhibited a double stranded structure consisting of beaded units like the structure of an actin filament (Fig. 5). Sometimes longer filaments around 1.5 μm were visible, but we could not conclude from this result that the length of the natural thin filament in smooth muscle was over 1.0 μm because of the possibility of new actin polymerization on the thin filaments. From the evidences that an extremely purified actin filament could not form such a double stranded structure under the conditions (NONOMURA, KATAYAMA & EBASHI 1975) and that tropomyosin-troponin complex was obtained from CG (EBASHI et al. 1966), the thin filament of smooth muscle might be constructed by an actin-tropomyosin-troponin complex as well as that of striated muscle (EBASHI et al. 1969).

Negatively stained image of a dense body and intermediate filament

Fig. 6 shows an image of Fraction D-I. The dense body reveals a fusiform profile as well as in the longitudinal section. It has a dimension of 0.4 to 0.8 μm in length and 40 to 60

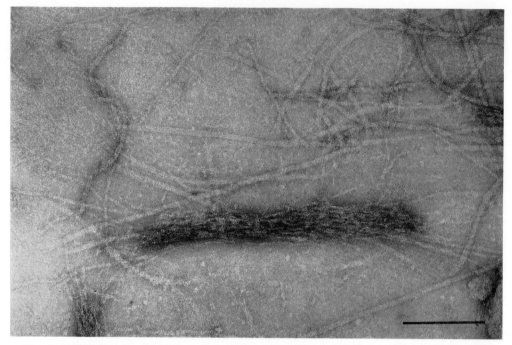

Fig. 6 Negatively stained image of a dense body and intermediate filaments. Some thin filaments are recognizable as entering into the dense body at both ends. Notice the difference of shape and size between thin filaments and intermediate filaments. (Scale indicates 0.25 μm)

nm in width. A bundle of thin filaments enters into the dense body at both ends, but in the middle portion these thin filaments cannot be followed. Instead, very much thinner filaments and amorphous dense materials are observable in the middle portion. Although many intermediate filaments are suspended around the dense body a direct structural relationship between the dense body and the intermediate filaments is not observed.

The intermediate filament, which was pointed out be ISHIKAWA et al. (1968) for the first time in a developmental state of the smooth muscle, was distinguished by its unique structure and different dimension from other myofilaments. The diameter was nearly 8 to 10 nm and the length was too long to determine its entire length on the mesh, perhaps 3 μm or more. The structure revealed a rope-like feature consisting of three or four long twisted chains with a diameter of 2 to 3 nm. This structure was definitely different from the double stranded feature of the thin filament.

Smooth muscle contraction on a structural basis

Generally a proviso for arguing the sliding mechanism is considered to fulfill the following conditions: the existence of two kinds of myofilaments, the thick and thin filament, and the presence of polarity on both filaments. This work shows that the smooth muscle of CG satisfies the former condition by presenting two kinds of filaments (Figs. 1 and 4) and also the latter condition by demonstrating the arrowhead structure of the actomyosin filament (Fig. 2) and a surface polarity in the separated thick filament. Thus, the sliding mechanism of the smooth muscle contraction is easily explained by researchers of smooth muscle (PANNER & HONIG 1967; SMALL & SQUIRE 1972).

However, the randomly distributed myofilaments in the cytoplasm (Figs. 1 and 2) indicate to us the difficulty of inducing an orderly process of sliding. Furthermore due to a deficiency or lack of any holding device of myofilaments like the Z-band and the M-line in striated muscle except for a small number of dense bodies, the usual sliding mechanism cannot work as efficiently in smooth muscle. Such a difficulty increases in the smooth muscle of the guinea pig's taenia coli in which the dense body is completely unrecoginzable in sections and in negatively stained samples of whole homogenates (SMALL & SQUIRE 1972; NONOMURA 1975b).

Before we hastily reach a conclusion on the contraction mechanism of smooth muscle, we should accumulate more information regarding the fine structure of myofilaments and their arrangement in the relaxed and contracted state. Contraction of smooth muscle would have various features except for the sliding mechanism which is a close apposition of thick and thin filaments like the image obtained in the superprecipitation process and/or partial folding of myofilaments particularly in the intense contracted state.

Conclusion

The fine structure of myofilaments in vertebrate smooth muscle was investigated mainly by the negative staining method using chicken gizzard as the material. Existence of two kinds of filaments, the thick and thin filaments was confirmed in the homogenate of fresh muscle. Presence of high concentrations of ATP in the relaxing medium was necessary to keep the natural profiles of the thick filament. Although the shape of the thick filament resembled that of skeletal muscle, its dimensions were slightly different from that of skeletal muscle; shorter in length and broader in width. The structure and dimensions of the thin filament were almost the same as that of skeletal muscle. A dense body was recognized as consisting of thinner filaments and amorphous materials with a bundle of thin filaments at both ends. The intermediate filaments exhibited their unique long profiles independent of other myofilaments and the dense body. While the dense body and the intermediate filaments were unique constituents characteristic of vertebrate smooth muscle, the thick and thin filaments essentially had the same molecular structure as that of skeletal muscle. However, myofilaments in smooth muscle cells were distributed in a disorderly arrangement because of the lack of a holding device for the myofilaments like the Z-band and the M-line in skeletal muscle. Further structural investigation should be aimed at the conversion of myofilament arrangements between the relaxed and contracted state. The contraction mechanism of smooth muscle should be discussed on the basis of accumulating much more informations regarding the fine structure and arrangement of myofilaments in smooth muscle.

REFERENCES

[1] COOKE P. H., CHASE R. H. and CORTES J. M. (1970) Thick filaments resembling myosin in electrophoretically-extracted vertebrate smooth muscle. Exp. Cell Res., *60:* 237.

[2] COOKE P. H. and FAY F. S. (1972) Correlation between fiber length, ultrastructure and length-tension relationship of mammalian smooth muscle. J. Cell Biol., *52:* 105.

[3] DEVINE C. E. and SOMLYO A. P. (1971) Thick filaments in vascular smooth muscle. J. Cell Biol., *49:* 636.

[4] EBASHI S., IWAKURA H., NAKAJIMA H., NAKAMURA R. and OOI Y. (1966) New structural proteins from dog heart and chicken gizzard. Biochem. Z., *345:* 201.

[5] EBASHI S., ENDO M. and OHTSUKI I. (1969) Control of muscle contraction. Quart. Rev. Biophys., *2:* 351.

[6] HEUMANN H. G. (1971) Über die Funktionsweise glatter Muskelfasern Elektroneumikroskopische Untersuchungen an der Darmmuskulatur der Hausmaus. Cytobiol., *3:* 259.

[7] Huys J. (1961)　Isolement et propriétés de la tonoactomyosine d'utérus de vache.　Arch. Internat. Physiol., *69:* 677.

[8] Ishikawa H., Bischoft R. and Holtzer H. (1968)　Mitosis and intermediate-sized filaments in developing skeletal muscle.　J. Cell Biol., *38:* 538.

[9] Kelly R. E. and Rice R. V. (1968)　Localization of myosin filaments in smooth muscle.　J. Cell Biol., *37:* 105.

[10] Laszt L. and Hamoir G. (1961)　Études par électrophorèse et ultracentrifugation de la composition protéinique de la couche musculaire des carotides de bovide.　Biochim. biophys. Acta, *50:* 430.

[11] Lowy J. and Small J. V. (1970)　The organization of myosin and actin in vertebrate smooth muscle.　Nature, *227:* 46.

[12] Lowy J., Poulsen F. R. and Vibert P. J. (1970)　Myosin filaments in vertebrate smooth muscle.　Nature, *225:* 1053.

[13] Lowy J., Vibert P. J., Haselgrove J. C. and Poulsen F. R. (1973)　The structure of the myosin elements in vertebrate smooth muscle.　Phil. Trans. Roy. Soc. Lond., *B265:* 191.

[14] Nonomura Y. (1968)　Myofilaments in smooth muscle of guinea pig taenia coli.　J. Cell Biol., *39:* 741.

[15] Nonomura Y. (1974)　Fine structure of the thick filament in molluscan catch muscle.　J. Mol. Biol., *88:* 445.

[16] Nonomura Y. (1975a)　Basic structure of vertebrate smooth muscle. I. The mode of existence of myofilaments. submitted to J. Cell Biol.

[17] Nonomura Y. (1975b)　Basic structure of vertebrate smooth muscle. II. The structure of dense body and intermediate filament. submitted to J. Cell Biol.

[18] Nonomura Y. and Ebashi S. (1975)　Isolation and identification of smooth muscle contractile proteins.　*In:* Methods in Pharmacology. Daniel E. E. (ed.) Vol. III, p. 141, Plenum Press, New York.

[19] Nonomura Y., Katayama E. and Ebashi S.(1975)　The effect of phosphates on the structure of the actin filament.　J. Biochem., in press.

[20] Panner B. J. and Honig C. R. (1967)　Filament ultrastructure and organization in vertebrate smooth muscle.　J. Cell Biol., *35:* 303.

[21] Panner B. J. and Honig C. R. (1970)　Locus and state of aggregation of myosin in tissue sections of vertebrate smooth muscle.　J. Cell Biol., *44:* 52.

[22] Rice R., Brady A. C., Dupue R. H. and Kelly R. E. (1966)　Morphology of individual macro-molecules and their ordered aggregates by electron microscopy.　Biochem. Z., *345:* 370.

[23] Rice R. V., Moses J. A., McManus G. M., Brady A. C. and Blasik L. M. (1970)　The organization of contractile filaments in a mammalian smooth muscle.　J. Cell Biol., *47:* 183.

[24] Rosenbluth J. (1971)　Myosin-like aggregates in trypsin-treated smooth muscle cells.　J. Cell Biol., *48:* 174.

[25] Shoenberg C. F. (1969)　An electron microscope study of the influence of divalent ions on myosin filament formation in chicken gizzard extracts and homogenates.　Tissue and Cell, *1:* 83.

[26] Shoenberg C. F., Rüegg J. C., Needham D. M., Schirmer R. H. and Nemetchek-Gansler H. (1966)　A biochemical and electron microscope study of the contractile proteins in vertebrate smooth muscle.　Biochem. Z., *345:* 255.

[27] Small J. V. and Sobieszek A. (1973)　The core component of the myosin-containing elements of vertebrate smooth muscle.　Cold Spr. Harb. Symp., *37:* 439.

[28] Small J. V. and Squire J. M. (1972)　Structural basis of contraction in vertebrate smooth muscle.　J. Mol. Biol., *67:* 117.

[29] Sobieszek A. and Small J. A. (1973)　The assembly of ribbon-shaped structures in low ionic strength extracts obtained from vertebrate smooth muscle.　Phil. Trans. Roy. Soc. Lond., *B265:* 203.

[30] Somlyo A. P., Devine C. E., Somlyo A. V. and Rice R. V. (1973)　Filament organization in vertebrate smooth muscle.　Phil. Trans. Roy. Soc. Lond., *B265:* 223.

[31] Somlyo A. P., Somlyo A. V., Devine C. E. and Rice R. V. (1971)　Aggregation of thick filaments into ribbons in mammalian smooth muscle.　Nature New Biology, *231:* 243.

Degeneration and Regeneration of Striated Muscle Fibers: Fine Structural Changes of the Rat Skeletal Muscle in Response to a Myotoxic Drug, Bupivacaine Hydrochloride

Atsushi ICHIKAWA *and* Misao ICHIKAWA

Degeneration and regeneration of skeletal muscles have been intensively studied by many investigators in various aminals of either diseased or experimentally injured conditions. There are many reports that mammalian skeletal muscle fibers are capable of regeneration following partial destruction induced by various forms of injuries such as the applications of heat (Shafiq & Gorycki 1965), cold (Price et al. 1964a, b; Reznik 1969a; Baloh et al. 1972), ischemia (Stenger et al. 1962; Reznik 1967, 1969a, b; Kasper et al. 1969), crushing (Sloper & Pegnum 1967; Church et al. 1969; Church 1969, 1970a, b; Teräväinen 1970), cutting (Bintliff & Walker 1960; Walker 1963, 1972) and myotoxic agents (Price et al. 1962; D'Agostino 1963; Libelius et al. 1970; Jirmanová & Thesleff 1972). Though the degenerative changes of damaged muscle fibers vary in appearance from fiber to fiber depending upon the degree of injury, muscle degeneration more or less involves disarrangement and disorganization of myofibrils and fine structural changes in the sarcoplasmic organelles such as the mitochondria and the sarcoplasmic reticulum. When muscle fibers are severely injured, the sarcolemma disappears and all components in the sarcoplasm are destroyed resulting in necrosis of the muscle fibers. Within a few days after injury, necrotic muscle fibers are infiltrated with phagocytic cells which ingest the damaged sarcoplasm. About that time, mononuclear cells with poorly developed organelles and small amount of cytoplasm are also distinguishable within the basement lamina of degenerating muscle fibers. These cells appear to progressively increase in number and to differentiate into myogenic cells, which are usually called the presumptive myoblasts to form myotube by fusing with each other. Such histological and cytological events in muscle regeneration have recently been reviewed by Carlson (1973) and Hudgson and Field (1973).

As Carlson pointed out in his review, however, despite the general agreement that the presumptive myoblasts of freshly injured mammalian muscle go on to form muscle, their origin had not been satisfactorily determined. Since Mauro (1961) had first described the satellite cells in the frog muscle, these cells have been found in a wide variety of vertebrates (Muir 1970) and were considered to play an important role as precursors of myoblasts in muscle regeneration after injury (Walker 1963; Shafiq & Gorycki 1965; Allbrook et al. 1966; Church et al. 1966; Church 1969; Jirmanová & Thesleff 1972) or in muscle diseases (Laguens 1963; Howes et al. 1966; Shafiq et al. 1967; Aloisi 1970; Conen & Bell 1970; Mastsglia et al. 1970; Baloh et al. 1972) as well as in myogenesis during postnatal development (Chiakulas & Pauly 1965; Muir et al. 1965; Ishikawa 1966; Shafiq et al. 1968), growth (Moss & Leblond 1970, 1971) and hypertrophy (Reger & Craig 1968). Some investigators contended that the myonuclei of damaged muscle fibers break off from degenerating sarcoplasm and were surrounded by thin rims of cytoplasm invested with plasma membranes, giving rise to the satellite cells which were capable of differentiation into myoblasts (WALKER 1963; LEE 1965; REZNIK 1969a, b, 1970; HESS & ROSNER 1970; ONTELL 1974). Another opinion that cells other than these two varieties of cells such as the mesenchymal cells in the connective tissue or some blood cells could be included in the reserve populations of myogenic cells in muscle regeneration has not become indisputable (BATESON et al. 1967).

The present study was designed to determine the origin of myogenic cells in muscle regeneration by analyzing morphological changes in the skeletal muscle which was totally damaged by the administration of a myotoxic drug, bupivacaine hydrochloride. The lumbrical muscle of rat hind limb was used for this purpose. Because a single subcutaneous injection of the drug into the planter region produced total damage of the muscle. This muscle is composed of four small muscles arising from the ventral surface of the tendon of flexor digitorum longus to insert on the proximal ends of the first phalanx of digits two, three, four and five. Since dimensions of each muscle are as small as 3.0 to 3.5 mm long and less than 1.5 mm in maximum diameter, it is not difficult to observe the entire muscle with serial sections both at light and electron microscopic levels.

Materials and Methods

Albino rats of both sexes ranging from 250 to 380 g in body weight were used in the experiments. The animals were given a single injection of 0.25 ml of 0.5% bupivacaine hydrochloride* subcutaneously into the planter region of the left hind limb and the same amount of Ringer's solution into the opposite side for controls. In some cases, two injections at a three hour interval were given because occasionally a single injection was insufficient to cause complete degeneration throughout the muscle. The results obtained in both cases were, however, primarily the same with respect to histological and cytological events in degeneration and regeneration of muscle fibers. At various time intervals after injection, such as 1, 3, 6, 12, 24, 36, 48, 60, 72, 96 and 120 hours, each experimental animal was anesthetized with an intraperitoneal injection of Nembutal and the toes of the hind limbs were fixed on a plastic board to keep the muscles in a stretched position and perfused into the abdominal aorta with a 2% glutaraldehyde solution buffered to pH 7.2 to 7.4 with a cacodylate buffer. Ten to fifteen minutes later, the lumbrical muscles were dissected out *en toto* under a binocular microscope, immersed into cold glutaraldehyde for 2 hours, washed in buffer and postfixed in buffered 1% osmium tetroxide. After dehydration with increasing concentrations of cold ethanol, the tissues were embedded in a mixture of Epon-Araldite within an aluminum scale. Thin sections were mounted on uncoated copper grids, stained with uranyl acetate and lead citrate and examined with Hitachi HS-8 and JEOL 100C electron microscopes. Thick sections representing the entire surfaces of each block in the middle of both normal and injured muscles were cut, stained with toluidine blue and examined by light microscopy to count the numbers of muscle fibers, satellite cell nuclei and myonuclei. To assess the accuracy with which the satellite cells were identified by light microscopy, thin sections adjacent to the thick sections were cut from materials of normal muscles and mounted on one-hole meshes (about 1.0 mm in diameter) coated with collodion film for electron microscopy. A montage of electron micrographs (taken at $\times 2,000$) of a thin section which represented approximately 3/4 of the entire face of a block was prepared.

* 1-n-butyl-DL-piperidine-2-carboxylic acid-2, 6-dimethyl-anilide-hydrochloride; Marcaine, manufactured by AB Boforos, Nobel-Pharma, Molndal, Sweden.

Observations

Light microscopy

In cross-sectional regions in the middle of a normal muscle, muscle fibers were closely packed with less connective tissue containing smaller numbers of fibroblasts (Fig. 1a, b). The number of muscle fibers composing the most lateral lumbrical muscle was 695 on the average from three different animals. Myofibrils of the muscle fibers were homogeneously distributed with narrow interfibrillar spaces of sarcoplasm. The nuclei closely associated with the surface of the muscle fibers were oval or slightly flattened in shape. Many of them had one or two prominent nucleoli within a pale matrix with a moderately abundant chromatin apposed to the nuclear envelope, while small numbers of nuclei were distinguishable by their coarser chromatin pattern in a relatively dark matrix. In correlative electron microscopic examination of thin sections, the latter were identified as being those of the satellite cells and their proportion to muscle nuclei was estimated to be approximately 7% with light microscopy, although their electron microscopic estimation was slightly higher and was about 8% as shown in Table 1. This was due to the fact that with light microscopy some nuclei were difficult to identify as to their type and were counted into the number of myonuclei or excluded from the count.

Table 1 Changes in the numbers of muscle fibers, myonuclei and satellite cell nuclei after the administration of bupivacaine.

Case examined (time interval after injury)	Number of muscle fibers (A)	Number of myonuclei (B)	Number of SC nuclei (C)	$\frac{C}{B+C}$ (%)	$\frac{C}{A}$ (%)
normal control*	699**	343	29	7.1	4.2
1 hour	657	280	34	10.8	5.1
3 hours	505	190	37	16.3	7.3

* From electron microscopy of a thin section, serial to the thick section and mounted on a one-hole mesh, the numbers of myonuclei and satellite cell nuclei of 486 muscle fibers were 243 and 23 respectively. Accordingly, $\frac{C}{B+C} = 8.6\%$ and $\frac{C}{A} = 4.7\%$.

** The intrafusal fibers of muscle spindles were not calculated in the number of muscle fibers. Change in materials later than 3 hours after injury could no be estimated because of extensive alterations.

One hour after a bupivacaine administration, the muscle fibers were loosely disposed with edematously distended interstitial tissue which contained a few mast cells and fibroblasts. Though the degenerative reaction of the muscle fibers was not always uniform throughout the muscle, most of the muscle fibers were stained lightly and their myofibrils were more or less disorganized and coarsely scattered throughout the sarcoplasm. The mitochondria were granular in profile and tended to aggregate into clusters especially in the subsarcolemmal region. The myonuclei which had one or two condensed nucleoli and coarse chromatin in a pale matrix were frequently displaced into the degenerating sarcoplasm far from the sarcolemma, while the satellite cells with relatively dark nuclei appeared to be intact. The latter became more easily distinguishable from the myonuclei, because their nuclei had the same features as that in the normal control and were surrounded with thin rims of cytoplasm exhibiting a clear contrast to the degenerating sarcoplasm. Changes in the number of muscle fibers were not evident at this stage and the satellite cells seemed to be unchanged in occurrence (Table 1). In materials 3 hours after injury, almost all of the muscle fibers fell into a degeneration process exhibiting

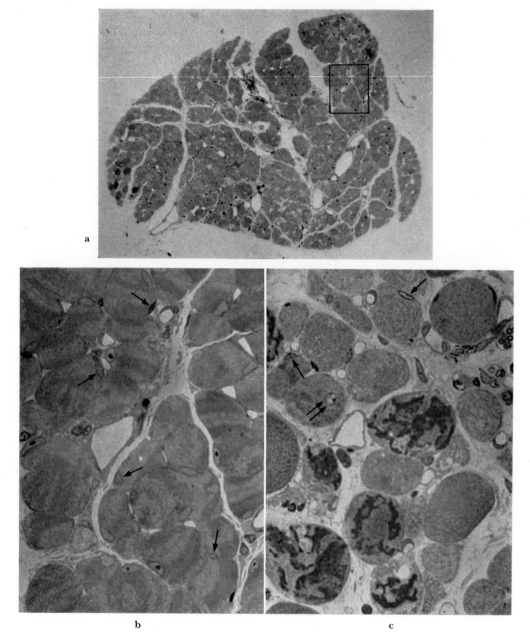

Fig. 1 a. A cross sectional region in the middle of a normal rat lumbrical muscle. ($\times 90$) **b.** Framed area in Fig. 1a at high magnification. The satellite cells with dark nuclei are distinguishable (arrows). ($\times 600$) **c.** A cross section of a muscle 3 hours after bupivacaine administration. Most of the muscle fibers degenerate resulting in necrosis and are surrounded with edematously distended connective tissue. The satellite cells become more clearly distinguishable (arrows), but the myonuclei are hardly detectable except for a displaced one (double arrow). ($\times 600$)

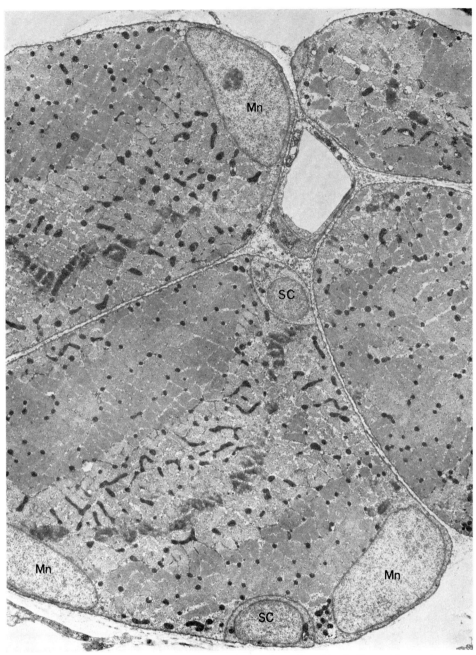

Fig. 2 An electron micrograph of a cross section of normal muscle. Muscle fibers with smooth contour are closely apposed to each other with a small amount of endomysial connective tissue. The satellite cells (SC) are clearly distinguishable from the myonuclei (Mn). ($\times 5,000$)

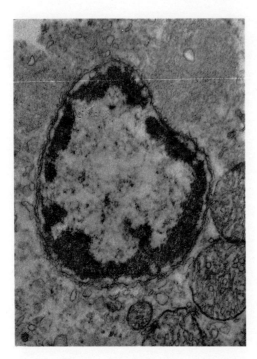

Fig. 3 A pyknotic myonucleus of a degenerating muscle fiber 1 hour after injury. It appears that the heterochrom atinaggregates along the nuclear envelope during an early stage of pyknosis and then, is detached from it becoming finely granulated. (×24,000)

various degrees of alterations (Fig. 1c). Some of them were quite irregular in contour appearing to be shrunken while possessing pyknotic myonuclei within lightly stained sarcoplasm containing debris of myofibrils. And, others were clearly invested with a sarcolemma with a smooth contour and darkly stained but cross striations of their myofibrils were occasionally smeared in longitudinal sections. The invasion of a number of capillaries into the endomysial connective tissue was noticeable, but the infiltration of mononuclear cells into the interstitial tissue space was not obvious. Few leukocytes with lobated nuclei and numbers of mononuclear cells were seen to infiltrate into the connective tissue for 6 to 12 hours following injury. Some of them were attached to the surface of the muscle fibers but no figures showing their invasion into the degenerating sarcoplasm were detectable. Some mononuclear cells had light spherical nuclei in light, finely granulated cytoplasm while others had dark nuclei in small amounts of slightly darkened cytoplasm. In the 24 to 36 hour period, a number of irregularly shaped cells with dark coarse granules in the cytoplasm which were presumed to be macrophages and some flattened cells with basophilic cytoplasm were seen within the basement membrane of degenerating fibers. About 48 to 60 hours after injury, almost all of the degenerated muscle fibers were transformed into cell cords which were composed of macrophages, basophilic mononuclear cells and polymorphonuclear leukocytes. The first appearance of a new formation of myofibrils was recognizable in the cytoplasm of the mononuclear cells in stages later than 72 hours. Myotubes which were composed of cells with centrally placed nuclei surrounded by thin bundles of myofibrils were detectable in materials obtained later than 120 hours after injury.

Electron microscopy

Changes in the muscle fibers were varied in appearance from fiber to fiber within a few

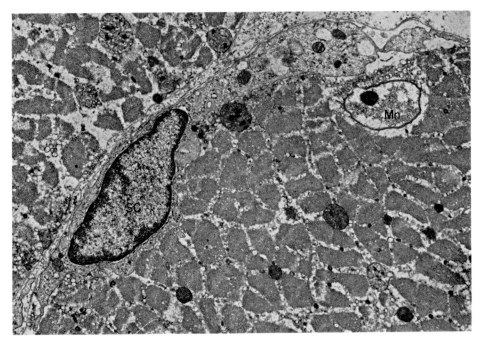

Fig. 4

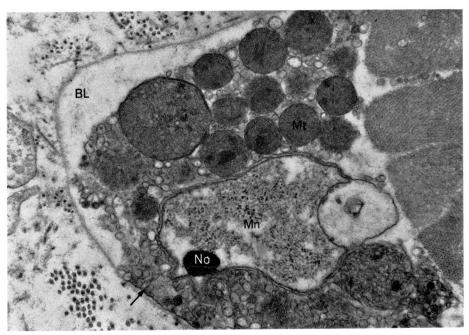

Fig. 5

Figs. 4 and 5 A part of several necrotic muscle fibers 15 hours after injury. The sarcolemma is fragmented (arrow in Fig. 5) and invisible in part. The myonuclei (Mn) which are severely destroyed and identified only by their nuclear envelope with pores contain finely granulated chromatin debris and condensed nucleoli (No); whereas the satellite cell (SC) appears intact and is elongated along the basement lamina. Note the swollen mitochondria with tubular or vesicular cristae and the accumulation of dense material into the matrix and numerous small vesicles probably originating from fragmentation of the sarcoplasmic reticulum. (Fig. 4; ×7,000, Fig. 5; ×18,000)

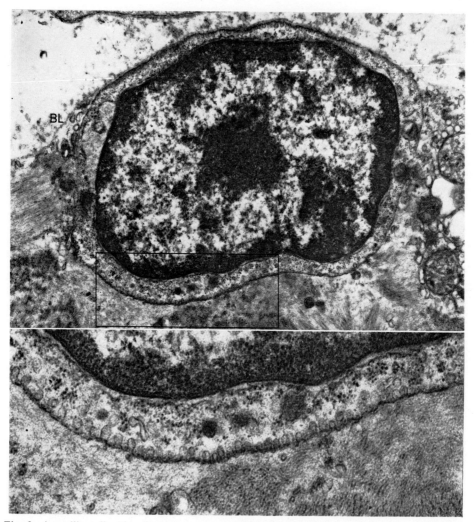

Fig. 6 A satellite cell within the basement lamina (BL) of a degenerating muscle fiber 6 hours after injury. It contains a few small mitochondria and numerous free ribosomes in its cytoplasm. Pinocytotic vesicles associated with the plasma membrane appear to be more numerous than in the normal control. (upper ×12,000; lower ×40,000)

hours after injury, during the early stage of degeneration. Some of them appeared to be intact but in many muscle fibers the plasma membrane was fragmented at many sites or disappeared completely so that the sarcoplasm was invested only by the basement lamina (Figs. 4 and 5). Myonuclei were frequently displaced into the degenerating sarcoplasm far from the sarcolemma and contained one or two prominent nucleoli and heterochromatin accumulations along the inner surface of the nuclear envelope (Fig. 3) The mitochondria of the granular profiles appeared to be swollen and displayed abundant tubular or vesicular cristae with occasional accumulations of dense material in the matrix (Fig. 5). The myofibrils were broken up into pieces with the disappearance of the Z lines and their debris were scattered throughout the sarcoplasm. Numerous small vesicles which were probably derived from fragmentation of the sarcoplasmic reticulum gathered in clusters around the degenerating myofibrils.

Fig. 7 A longitudinal section of a slightly damaged muscle fiber 15 hours after injury. The sarcolemma and the myonucleus appear intact, but the Z lines of the myofibrils are disorganized appearing zig-zag. Some mitochondria (framed in the upper right) show a degenerative change in their cristae as shown in Fig. 8. ($\times 8,000$)

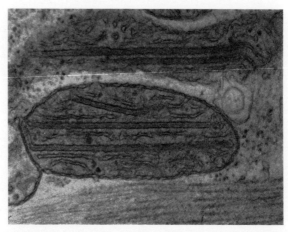

Fig. 8 The mitochondria of a degenerating muscle fiber shown in Fig. 7, at a higher magnification. Note an accumulation of moderately dense material into the intracristal space. (×52,000)

It was quite peculiar that the satellite cells associated with these muscle fibers resisted the drug thus appearing to be intact and to have a number of pinocytotic vesicles associated with the plasma membrane, appearing to be more numerous than in those of the normal muscle fibers (Figs. 4 and 6). Their cytoplasm progressively increased in volume and contained a few mitochondria and a number of free ribosomes. This might suggest that the satellite cells were not only undamaged by the myotoxic drug but were activated in response to the injury. In lightly damaged muscle fibers which were still invested with the plasma membrane and contained a normal configuration of myofibrils with narrow interfibrillar spaces, some mitochondria in the subsarcolemmal region showed an accumulation of moderately dense material into the intracristal space (Fig. 8). Such a feature of the mitochondria could be seen even in muscle fibers which appeared intact in other components, so it might be one of the initial changes in response of the muscle fibers to bupivacaine administration. Disarrangement of the myofibrils was not obvious in these fibers but their Z lines were frequently disorganized and appeared zig-zag in longitudinal sections to the long axis of the muscle fibers (Fig. 7). No remarkable changes in the myonuclei and other cellular components as well as the associated satellite cells were observed in these fibers in the following stages of the experiment.

In a later stage of degeneration, from 6 to 12 hours after injury, the myonuclei in the necrotic muscle fibers became irregular in shape appearing to be shrunken and contained finely granulated chromatin debris and condensed nucleoli in a pale matrix (Fig. 5). Some of the satellite cells associated with these fibers remained at their original site and became elongated along the basement lamina. They had a flattened nucleus and a moderate amount of cytoplasm containing a few flattened saccules of the granular endoplasmic reticulum, a few mitochondria and numerous free ribosomes. While, the others appeared to protrude toward the interstitial space lifting up the basement lamina, and were occasionally seen to be invested in part with the basement lamina (Figs. 9 and 10). They had round nuclei with somewhat undulated surface and peripherally accumulated heterochromatin in dark matrices and intracytoplasmic organelles such as a few mitochondria and numerous free ribosomes.

Free cells with quite similar cytological features but without the basement lamina were seen in the endomysial and perivascular connective tissues, and mitotic figures could

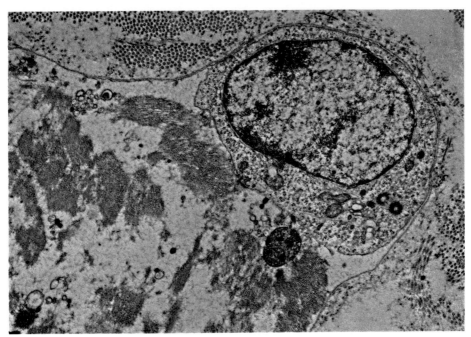

Fig. 9 A satellite cell protruding toward the interstitial space. 15 hours after injury. (×10,000)

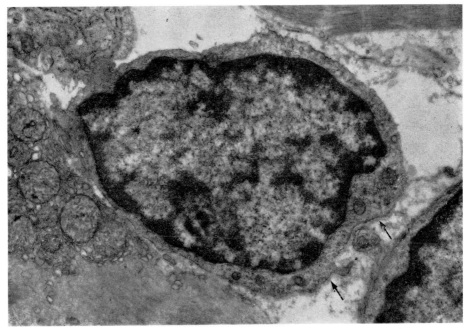

Fig. 10 A satellite cell migrating into the interstitial tissue. Note that it is partly invested with the basement lamina (arrow). 6 hours after injury. (×18,000)

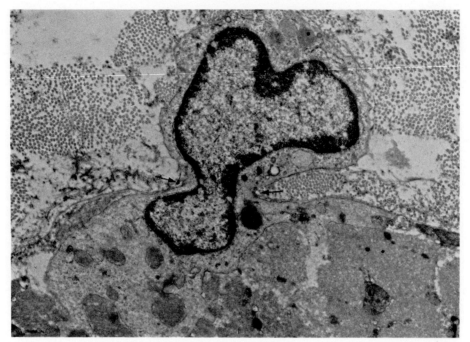

Fig. 11 Invasion of a macrophage into a degenerating muscle fiber passing through the basement lamina (arrow). 15 hours after injury. (×10,000)

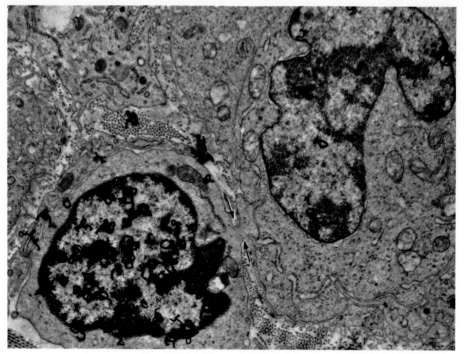

Fig. 12 An electron microscopic autoradiograph showing the invasion of a ^{3}H-thymidine-labeled free cell into a cell cord passing through the basement lamina (arrows). 65 hours after injury. (×9,000)

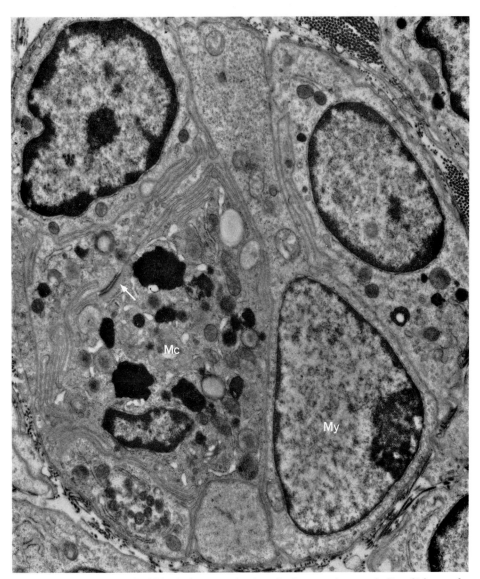

Fig. 13 A cell cord formed within the basement lamina of a degenerating muscle fiber 65 hours after injury. It is composed of macrophages (Mc), presumptive myoblasts (My) and occasional leukocytes which are interdigitated with each other and occasionally bound by desmosomes (arrow). (×11,000)

be detected among them. This suggested that possibly some activated satellite cells wandered away from their original site into the interstitial tissue space, became free cells and increased in number by mitosis. In fact, our recent radioautographic studies which will be reported elsewhere revealed that many of the satellite cells labeled with ^{3}H-thymidine migrated into the interstitial space and became free cells. A few heterophilic leukocytes with lobated nuclei and round or ovoid specific granules in their cytoplasm as well as presumed macrophages with a number of primary lysosomes and irregular contours of cell bodies were also detectable in the connective tissue at this stage. During 24 to 36 hours, macrophages were observed within the basement lamina of the degenerating

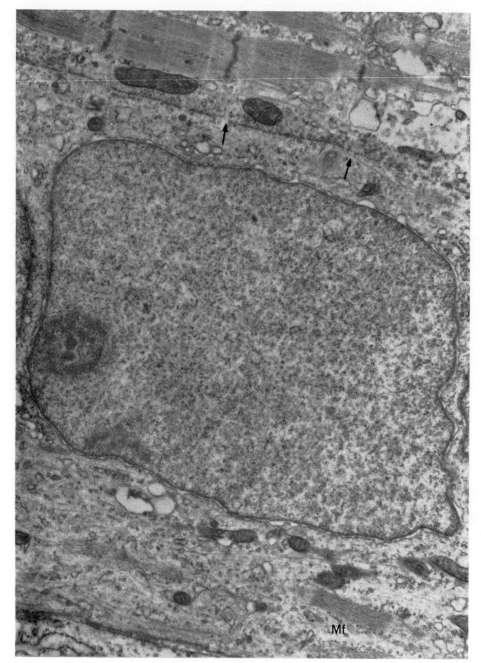

Fig. 14 A part of a longitudinal section of a myotube 120 hours following injury. Newly formed myofibrils are scattered in the cytoplasm around a centrally-placed nucleus. Arrows show portions of the fusion between the myotube and a newly formed muscle fiber. (×12,000)

muscle fibers, and the debris of disorganized myofibrils and other components appeared to be surrounded with their cytoplasmic processes (Fig. 11).

The following stages showed that the sarcoplasmic areas of degenerating muscle fibers were replaced by the cytoplasm of those cells which were numerous macrophages of

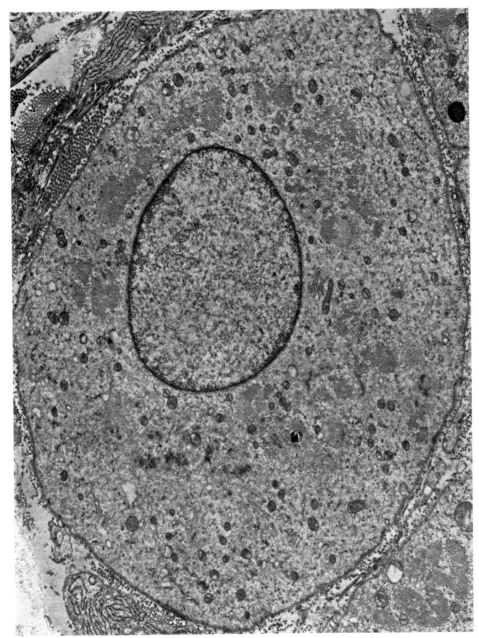

Fig. 15 A cross section of a myotube in the same stage of regeneration as Fig. 14. (×9,000)

various size and form of phagosomes, the slightly flattened satellite cells standing along the basement lamina and occasionally a few heterophilic leukocytes. Mononuclear cells in the interstitial tissue which had a few mitochondria and numerous free ribosomes in the cytoplasm were occasionally seen to insert their cytoplasmic processes into the cell cord within the basement lamina of degenerating fibers (Fig. 12).

Thus, necrotic muscle fibers are progressively transformed into the cell cords which were composed of several kinds of cells interdigitating each other in complexity (Fig. 13).

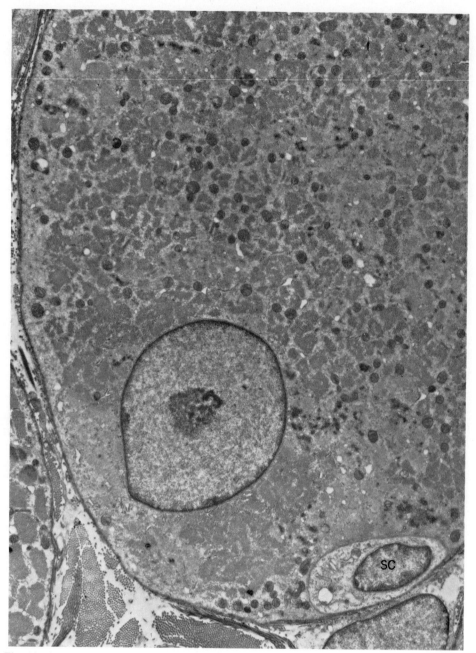

Fig. 16 A cross section of a newly formed muscle fiber. A spherical myonucleus is apposed to the surface of the fiber which is attached to a satellite cell (SC). 120 hours following injury. (×8,000)

Mitotic figures were seen in cells with numerous free ribosomes. Later than 72 to 96 hours, thin bundles of myofilaments arose in the cytoplasm of presumptive myoblasts in the cell cord. The myotubes with centrally-placed nuclei surrounded by several thin myofibrils were then formed by fusion of the myoblasts (Figs. 14 and 15). In newly formed muscle fibers which were observable in specimens later than 120 hours, a number of round or ovoid nuclei were closely associated with the sarcolemma and thin myofibrils were regularly arranged with relatively wide interfibrillar spaces. The satellite cells with moderate amount of cytoplasm were frequently observed within the basement lamina of these fibers (Fig. 16).

Discussion

Since Sokoll et al. (1968) had reported that a long-acting local anesthetic, bupivacaine hydrochloride, had a myotoxic action giving rise electrophysiological changes in muscle fibers similar to those following surgical denervation or botulinum poisoning, several investigators reported morphological changes in response to the drug on rat skeletal muscles both at the light and electron microscopic levels (Benoit & Belt 1970; Libelius et al. 1970; Jirmanová & Thesleff 1972). Jirmanová and Thesleff (1972) studied the rat extensor digitorum longus muscle treated with methylbupivacaine and reported that a single subcutaneous injection of the drug caused a uniform, complete and irreversible destruction of the superficial layers of the muscle fibers followed by phagocytosis and complete regeneration. The degeneration of muscle fibers described in this paper mostly agreed with their data. But the present study dealing with a quantitative analysis of the changes in the numbers of muscle fibers, myonuclei and satellite cell nuclei following injury clearly showed that bupivacaine administration produced a complete destruction of almost all of the muscle fibers, in which the myonuclei were selectively and completely destroyed along the entire length of the necrotic fibers whereas the associated satellite cells remained intact.

Because the degenerative change in the myonuclei proceeded so rapidly as within a few hours after injury, it could only be confirmed if the sampling was started immediately after injury and carried out at short time intervals. The attitude of the satellite cells in the degeneration process that some of them remained at their original site within the basement lamina while others migrated from their parent fibers and became free cells was also detectable at a considerably early stage. The mononuclear free cells with dark nuclei, numerous free ribosomes and poorly developed organelles in their cytoplasm were quite similar in appearance to the presumptive myoblasts in the cell cord which was formed during the later stage of regeneration. They appeared to increase in number by mitosis in the interstitial tissue.

Hess and Rosner (1970) have mentioned that the satellite cells do not wander away from their parent fibers in the degeneration process of damaged muscle, whereas Terävainen (1970) has suggested the possibility that some of the newly formed satellite cells detach themselves from the muscle fibers to proliferate in the interstitial tissue space. The present study supports the latter opinion. The satellite cells within the basement lamina also show mitoic figures in the regeneration process but it occurs later in the stage of cell cord formation.

The proportion of satellite cell nuclei to muscle nuclei has been variously given as 10% in the web muscle of the fruit bat (Muir et al. 1965), 4.8 to 5.8% in the mouse levator ani muscle (Venable 1966), approximately 1% in the rat gastrocnemius (Aloisi 1970) and 1.6% in young rat extensor digitorum longus (Ontell 1974). The present result (about 8%) is considerably higher than the data in the rat reported previously. Whether this depends upon the age of the animals used or a difference in the muscle examined is unknown, although it seems that the population of the satellite cells in the adult muscle is larger than that generally accepted.

With respect to experimental models of mammalian skeletal muscle regeneration, previous studies have dealt mainly with alterative changes in partially damaged muscle fibers. It was believed that any possible precursors of the myoblastic cell originated from those cells in or near the intact segment of the injured muscle fibers. However, the observation that bupivacaine administration produces complete destruction of muscle fibers along their entire length and selectively destroys the myonuclei is of quite interesting and should favor the determination of the origin of myoblastic cells in muscle regeneration. The present study at least ruled out the opinion that the presumptive myoblasts in muscle regeneration originated from pre-existing myonuclei pinched off from damaged muscle fibers. And, it is also confirmed that even if almost all of the muscle fibers are damaged resulting in necrosis following injury, the muscle is capable of regeneration unless the associated satellite cells are destroyed and the blood supply is interrupted. Whether or not the mesenchymal cells in the connective tissue or any circulating blood cells participate in muscle regeneration as precursors of myoblasts could not be definitely determined in this study. However, it appears that the mononuclear cells with numerous free ribosomes in the cytoplasm are detectable in the interstitial tissue slightly earlier than during the stage of regeneration in which infiltration of macrophages and leukocytes occurs. A radioautographic study in which any possible precursor cells of the myoblast are selectively labeled with tracer, is needed for resolving this problem.

Summary

Degeneration and regeneration of mammalian skeletal muscle were studied in the rat lumbrical muscles treated with a myotoxic drug, bupivacaine hydrochloride. The results obtained are summarized as follows:

1) Most muscle fibers of the muscle examined underwent a characteristic necrosis extending over their entire length following injury, in which the myonuclei were completely destroyed while the associated satellite cells resisted the drug action appearing to be intact and were activated to differentiate into myogenic cells in the regeneration process.

2) Some of the activated satellite cells remained attached to their parent fibers and were elongated along the basement lamina during the degeneration process, while others seem to migrate into the interstitial tissue to become free cells which frequently exhibited mitotic figures in the regeneration process.

3) In the later stage of regeneration, degenerated muscle fibers were replaced with cell cords which were composed of those cells which were elongated satellite cells, phagocytic cells and mononuclear cells invading from the interstitial tissue. Subsequently, these cell cords transformed into myotubes by fusion of the myoblastic cells originating from the satellite cells which might or might not have passed through a phase of free cells.

Thus, it may be concluded that the muscle is capable of regeneration even when many of the muscle fibers are destroyed and lose their myonuclei along their entire length, and also that the activated satellite cells associated with degenerating muscle fibers may play an important role in muscle regeneration by differentiating into myoblastic cells.

REFERENCES

[1] ALLBROOK D. (1962) An electron microscopic study of regenerating skeletal muscle. J. Anat., *96*: 137.

[2] ALLBROOK D., BAKER, W. DE C. and KIRKALDY-WILLIS W. H. (1966) Muscle regeneration in experimental animals and in man: The cycle of tissue change that follows trauma in the injured limb syndrome. J. Bone & Joint Surg., *48-B*: 153.

3. Crystalloids

Crystalloid Structure in Granules of Human Mast Cells

Shiro MATSUURA *and* Yutaka TASHIRO

In mast cells histamine and heparin are localized in basophile granules where they exist as histamine-heparin-protein complexes (LAGUNOFF et al. 1964; UVNÄS et al. 1970). In response to various stimuli, for example antigenic stimulus in the case of sensitized cells, the granules are extruded out of the cells by exocytosis (RÖHLICH et al. 1971) and this degranulation process triggers a series of important biological responses such as inflammatory or allergic reactions (OSLER et al. 1968; ISHIZAKA et al. 1969).

The ultrastructure of human mast cell granules has been investigated by various authors (STOECKENIUS 1956; BENDITT 1960; HIBBS et al. 1960; PHILLIPIS et al. 1960; MORAT & FERNANDO 1962; THIERY 1963; FEDORKO & HIRSH 1965; SELYE 1965; HASHIMOTO et al. 1966, 1967; ORFANOS 1966; WEINSTOCK & ALBRIGHT 1967; BRINKMAN 1968; KOBAYASHI et al. 1968; FUJITA et al. 1969; HIRAKAWA 1969a, b), and a number of structures have been described; homogeneous structure, reticular or granular structure, scroll-like or cylindrical lamellae frequently running in pairs, parallel crystalline pattern with ordered array of dense and less dense lines, and a hexagonal or honeycomb pattern. It is not clear, however, whether these crystalloids do represent different structures or they are simply different views of one structure.

We have investigated the fine structure of the mast cell granules with an electron microscope equipped with a tilting stage. It is concluded that the hexagonal and parallel crystalline patterns seen in the mast cell granules could be interpreted as arising from different views of one structure, which consists of spherical particles, ~ 30 Å in radius. It is further suggested that the scroll-cylindrical structure is also composed of similar spherical particles. It is not clear whether the latter structure is a precursor form of the former structure.

Materials and Methods

Mast cells were obtained by a sternal puncture from human bone marrow. They were immediately fixed in cold 2% glutaraldehyde buffered at pH 7.2 with phosphate buffer (SABATINI et al. 1963), post fixed in 1% OsO_4 after repeated washing with the same buffer, dehydrated with ethanol and finally embedded in Epon 812 (LUFT 1961). Sections were cutw ith a diamond knife on a Porter-Blum MT-2 microtome and stained successively with uranyl acetate and lead citrate. They were examined in a Hitachi HU-12 electron microscope equipped with a Hitachi tilting goniometer stage of the HK-5 type.

Results

General views of the mast cell granules

Figs. 1–3 show electron micrographs of the human mast cell granules. In accordance with description by previous authors (STOECKENIUS 1956; BENDITT 1960; HIBBS et al. 1960; PHILLIPIS et al. 1960; MORAT & FERNANDO 1962; ORFANOS 1966) the homogeneous structure, reticular structure, scroll-like structure, cylindrical lamellae usually running in pairs, parallel crystalline pattern, and the hexagonal or honeycomb pattern can be identified. All the these structures are not in direct contact with the cytoplasm but are enclosed within a smooth membrane. Detailed analyses of each of these patterns will be described in the following sections.

Parallel crystalline pattern

Fig. 4 is a mast cell granule with a typical parallel crystalline pattern. Usually dense and less dense lines run straight throughout the granules as shown in this figure. Sometimes several parallel crystalline patterns crossing at various angles coexist within a granule. The average distance between the lamellae is 114 ± 4 A (Table 1). In some part of the granules, especially at the peripheral portions of the granules, the lamellae appear to consist of round, dense particles regularly arranged in a linear fashion with an average periodicity of 64 ± 2 A (Fig. 4, arrow).

Table 1 Periodicity in A of the repeating structures found in the mast cell granules. Periodicity found in the hexagonal-parallel crystalline structure and in the scroll-cylindrical structure are given in the table. Figures in the bracket show the number of granules used for the measurement.

Structure	Distance (A)
Parallel crystalline structure	
line-line	114+4 (9)
particle-particle	64+2 (10)
Hexagonal structure	
line-line	114+8 (9)
Scroll-cylindrical structure	
particle-particle	66+4 (9)
line-line	68+3 (4)

Hexagonal pattern

Fig. 5 shows a typical hexagonal or honeycomb pattern with an identical periodicity which is composed of an electron-opaque margin surrounding a less dense core. The average distance between the electron-opaque lamellae is 114 ± 8 A, and they cross each other at an angle of about 60°.

Granules with the hexagonal pattern vary in contour and their edges are irregular. In favorable cases, however, such granules are polygonal in shape and the shape is probably a hexagonal prism or its derivatives.

Scroll-like or cylindrical pattern

Figs. 1–3 show that a number of granules contain scroll-like lamellae and cylindrical structure. They exist in many granules independent of whether they contain parallel crystalline structures or not. When the scroll structure and crystalline structure coexist in the same granule, the former is found usually at the peripheral portion of the granule. The lamellae in some granules appear to be in continuity with the parallel crystalline or

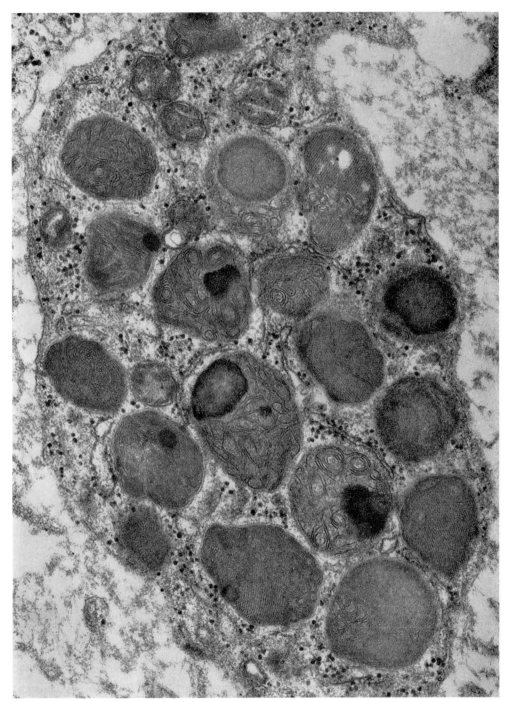

Fig. 1

Figs. 1–3 General views of the specific granules of human mast cells. Homogeneous structures with various densities (Figs. 1–3), reticular structure (Fig. 2), scroll-like and cylindrical structure (Figs. 1–3), parallel crystalline pattern (Figs. 1–3) and hexagonal pattern (Figs. 2 and 3) are observed. (Fig. 1; ×52,000, Figs. 2 & 3; ×73,000)

Fig. 2

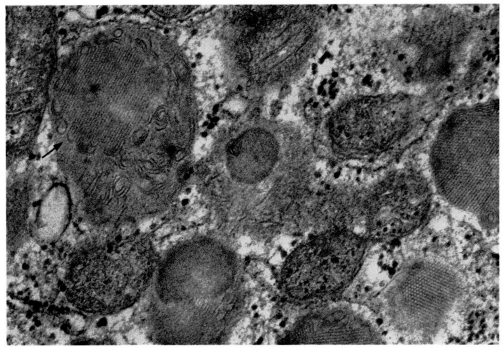

Fig. 3

hexagonal structure, suggesting an intimate relationship of both structures (Fig. 3, arrow). The lamellae frequently run in pairs and each lamella appears to consist of round dense particles regularly arranged in a linear fashion as pointed out by WEINSTOCK and ALBRIGHT (1967).

Longitudinal sections of the scrolls show that they are cylindrical (Figs. 1, 2, 6 and 7). When sectioning is done at a grazing angle to the plane of the lamellae, the lamellae seem to be composed of parallel and closely packed dense lines separated by spaces of lesser densities (Figs. 6 and 7, shown by arrows) with periodicities of 68 ± 3 A (Table 1). The direction of these lines is not perpendicular but appears to be tilted $\sim 60°$ from the long axis of the cylinder. The walls of the cylinder also appear to be made up of round dense particles regularly arranged in a linear fashion with an average periodicity of 66 ± 4 A (Fig. 7, shown by stars). Apparently most of the cylindrical structure seen in Figs. 6 and 7 exist freely in the cytoplasm. This is probably because the membrane which is surrounding the cylindrical structure is sectioned tangentially.

Homogeneous pattern

Figs. 1–3 show that a number of the granules have homogeneous or amorphous materials with variable electron densities. In some granules such homogeneous regions coexist with crystalline as well as scroll-cylindrical structure. There seem to be two kinds of homogeneous pattern. In one case the homogeneous region can be clearly distinguished from the surrounding region by the difference in density (true homogeneous structure) but in the other the crystalline region is gradually transformed into a homogeneous pattern without a clear boundary, suggesting that such a pattern is only apparent (apparently homogeneous structure) as described in page 75.

Reticular structure

Reticular structure is the other type of structure which is found in the mast cell granules.

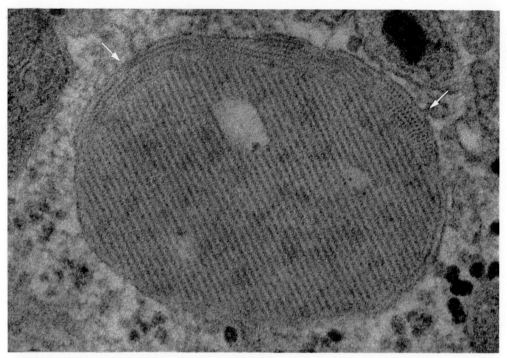

Fig. 4 A mast cell granule with a parallel crystalline pattern. Round dense particles are observed (arrow) at the periphery of the granule regularly arranged in a linear fashion with a periodicity of ~64 A. (×192,000)

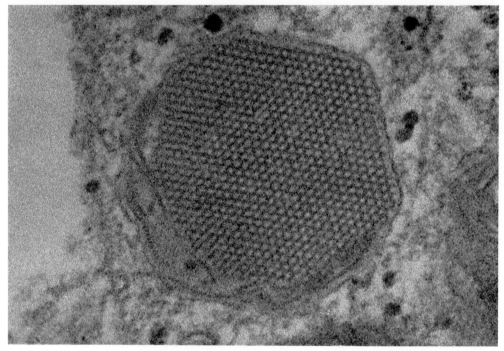

Fig. 5 A mast cell granule with a hexagonal honyecomb pattern. The contour of this crystal suggests that it is a hexagonal prism in shape. (×200,000)

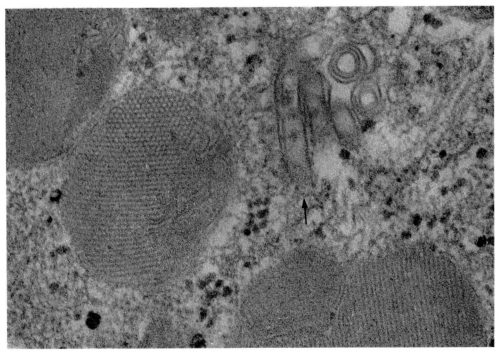

Fig. 6 Several scroll-like and cylindrical lamellae are observed at the upper right portion of the figure. The one pointed out by an arrow shows that the lamellae are composed of parallel dense lines which are separated by spaces of lesser densities and tilted $\sim 60°$ from the long axis of the cylinder. The granule located at the left half of the same figure indicates that the hexagonal pattern is gradually transformed to a parallel crystalline pattern. ($\times 120,000$)

Sometimes this structure also coexists with various other types of structure as shown in Fig. 2. Some of the reticular structures are loose while the others are so compact and dense that they cannot be distinguished from a true homogeneous structure, suggesting their formation by condensation of some of the true homogeneous structures from the reticular structures (Fig. 2).

The relationship between the parallel crystalline, hexagonal and homogeneous patterns

The existence of parallel crystalline and hexagonal patterns in human mast cell granules have been described by various authors (FEDORKO & HIRSH 1965; WEINSTOCK & ALBRIGHT 1967; KOBAYASHI et al. 1968). The identical periodicity of the two patterns has suggested that the honeycomb pattern probably represents the parallel crystalline structure, sectioned 90 degrees to the plane of sectioning (WEINSTOCK & ALBRIGHT 1967). No experimental evidence, however, has so far been published to support this suggestion.

In Figs. 3 and 6 are shown some mast cell granules with the hexagonal honeycomb pattern at some portions of the granules. Since the hexagonal pattern is gradually transformed to the parallel crystalline pattern in other portions of the granules, it is very probable that they are simply different views of the same structure.

In order to confirm this conclusion, electron microhgraps of the mast cell granules taken before and after tilting (at $\sim 40°$ in this case) were compared (Figs. 8 and 9). In one granule, shown by an arrow, a hexagonal honeycomb pattern is converted into a parallel crystalline pattern while in the other granules the parallel crystalline pattern either disappears or the direction of the crystalline pattern changes by about $60°$ by the tilting.

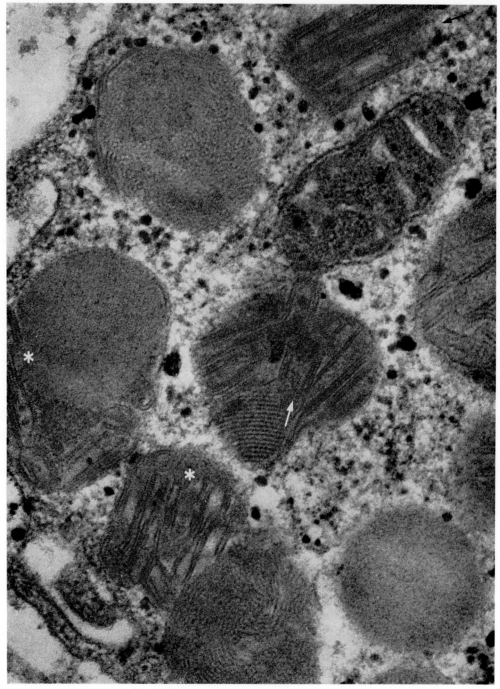

Fig. 7 Several mast cell granules with cylindrical pattern. The cylindrical structure indicated by the arrow shows the substructure with parallel dense and less dense lines. Some of the cylindrical lamellae (star mark) suggest that they are composed of round dense particles, regularly arranged in a linear fashion with a periodicity of ∼66 Å. (×112,000)

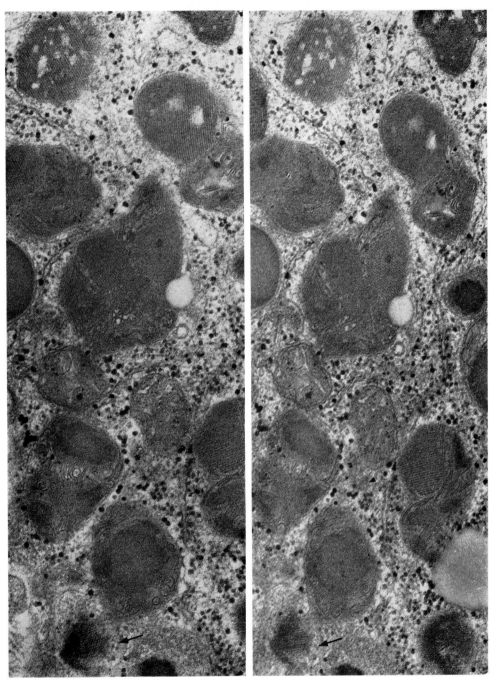

Fig. 8   **Fig. 9**

Figs. 8 and 9 A pair of tilted electron micrographs showing several mast cell granules. The section was tilted at $\sim 40°$ (Fig. 9). In one granule shown by an arrow, the hexagonal pattern was converted to a parallel crystalline pattern, while in the other granules parallel crystalline pattern either disappeared or the direction of crystalline pattern changed $\sim 60°$ after tilting. It can also be seen that in some granules the homogeneous pattern is convertible to the parallel crystalline pattern, indicating that some of the granules with the homogeneous appearances are merely apparent (apparent homogeneous structure). (Figs. 8 & 9; ×48,000)

We could safely conclude, therefore, that the parallel crystalline pattern and the hexagonal honeycomb pattern are interconvertible and that they are simply different views of the same structure.

A comparison of Figs. 8 and 9 also enables us to conclude that some of the homogeneous structures exhibit the crystalline structure appearance when observed from the proper direction, indicating that the apparently homogeneous structure is simply a different view of the crystalline structure.

Discussion

A model for the hexagonal and parallel crystalline structure

From the electron micrographs we have constructed a tentative model which seems to be consistent with these observations (Figs. 10 and 11). Each particle represents a building unit which is probably composed of a heparin-protein complex, and is in close contact with the surface of the six identical particles.

When sectioned perpendicular to the long axis, a hexagonal honeycomb pattern will be observed as shown in Fig. 10. When sectioned parallel to the long axis and observed from the directions shown by arrows (Fig. 11), each crossing at an angle of $60°$, parallel crystalline patterns will be observed. When observed from any intermediate angle, homogeneous structure will be observed (apparent homogeneous structure). If the specimens are cut parallel to the long axis and contain only one or two layers of the lamellae at the peripheral portion of the granules, dense particles regularly arranged in a linear fashion will be observed.

Thus all the electron micrographs of parallel crystalline and hexagonal structure and the particulate structure regularly arranged in a linear fashion seems to be explained by this model.

Assuming that this model is correct, the radius of the sphere is calculated from the distances between the parallel crystalline lines as 33 ± 1 A, twice this value is in good agreement with the particle-particle distance (64 ± 2 A) observed at peripheral portions of the parallel crystalline structure. We could safely conclude, therefore, that the building unit is a spherical particle ~33 A in radius.

A model for the scroll-cylindrical lamellar structure

It has been suggested by BRINKMAN (1968) that the scroll structure and cylindrical structure is a transverse and longitudinal section of the same structure, respectively. This conclusion is also supported by the present electron microscopic observations: all the intermediate patterns between the scroll structure and cylindrical structure have been observed.

A tentative model for the scroll-cylindrical lamella is shown in Fig. 12, which seems to be consistent with all the electron micrographs of the structure. For a matter of convenience, the model for the cylindrical lamella is flattened and photographed from the top of the model. The long axis of the cylinder is shown by an arrow. If this model is accepted, the radius of each particle is calculated as 38 A from a line-to-line distance of the cylindrical lamella (68 A, Table 1). This means that the building unit of the scroll-cylindrical structure is slightly larger than that of the crystalline structure.

Maturation process of the mast cell granules

Electron microscopic studies on the maturation process of mast cell granules have been reported by COMBS (1966) and FUJITA et al. (1969). Unfortunately, however, no suggestion has been proposed for the formation of the crystalline structure.

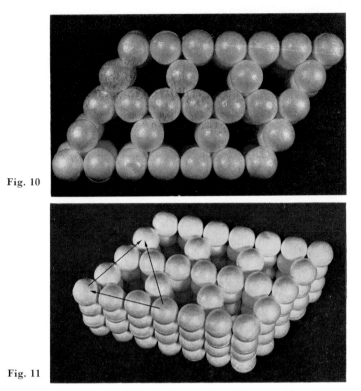

Figs. 10 and 11 A model for the hexagonal-parallel crystalline structure. The top view (Fig. 10) and side view (Fig. 11) of the model. It is suggested that the building unit of the crystalline structure is a spherical particle, ~33 Å in radius. When sectioned perpendicular to the long axis of the model, a hexagonal honeycomb pattern will be observed as shown in Fig. 10. When sectioned parallel to the long axis and viewed from the direction shown by the arrows of Fig. 11, parallel crystalline structure will be observed.

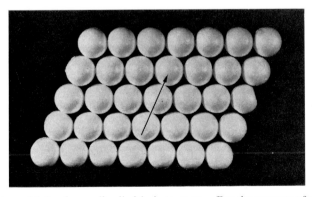

Fig. 12 A model for the scroll-cylindrical structure. For the purpose of simplicity, the cylindrical lamella is flattened on a plane and viewed from the top of the model. The long axis of the cylinder is shown by an arrow.

We have suggested that both the scroll-cylindrical structure and the hexagonal parallel crystalline structure are composed of spherical particles similar in size. This fact suggests the possibility that the latter structure is a precursor of the scroll-cylindrical structure. A comparison of the two models, however, shows that the arrangement and diameter of the spherical particles in both structures are probably different. At the present moment, therefore, we prefer another possibility, that they are chemically different and formed independently.

We do not know whether the reticular structure is formed by a condensation of true homogeneous structures or vice versa. It is very probable, however, that the chemical composition of the latter two structures is quite different from that of the former crystalline and scroll-cylindrical structures.

Estimation of the particle weight of the building unit in the hexagonal-parallel crystalline structure

Since the radius of the spherical particles in the hexagonal-parallel crystalline structure is 33 A, the volume of the building unit is calculated as 1.51×10^{-19} ml. If the partial specific volume of the heparin-protein complex is assumed to be 0.72, the particle weight of the building unit may be 1.26×10^5 daltons. It has been reported by LAGUNOFF et al. (1964) that heparin and basic protein constitute about 30% and 35% of the dry weight of the granules, respectively. The molecular weight of heparin is reported by MATHEWS (1966) to be $7\text{-}16 \times 10^3$ and a somewhat higher value of $6\text{-}20 \times 10^3$ has been reported by GOODMAN and GILMAN (1970). From these data it is suggested that the minimum molecular weight of the heparin-protein complex is $\sim 2.5 \times 10^4$. This value is about one fifth of the particle weight of the building unit. If the particles are 20% hydrated, the particle weight of the building unit may be $\sim 10 \times 10^4$ and the particles are probably composed of four molecules of the heparin-protein complex.

It is not clear what kind of material exists in the less dense cores of the hexagonal structure. One possibility is that some low molecular weight substances such as water and/or histamine exist in these holes, and that they are lost from the granules during the fixation and dehydration procedures. This kind of localization of histamine, if it is true, may explain the rapid release of histamine from a heparin-protein complex during the degranulation process of a mast cell.

Summary

Specific granules of a human mast cell have been investigated by electron microscopy. It has been concluded that the hexagonal honeycomb pattern is another view of the parallel crystalline structure, and a model was proposed which consists of spherical particles, ~ 30 A in radius and $\sim 10^5$ in molecular weight. The scroll-cylindrical structure seems to be composed of a similar but a slightly larger spherical particle. It is not clear, however, whether the latter structure is a precursor of the former structure or if they are formed independently.

ACKNOWLEDGEMENTS

We thank Dr. JAMES A. LAKE of the New York University, School of Medicine, N.Y. for helpful advice.

This work was supported by a grant from the Ministry of Education of Japan (Grant No. 844020, 1973).

REFERENCES

[1] BENDITT E. P. (1960) 5-Hydroxytryptamine and 5-Hydroxytryptophan decarboxylase in rat mast cells. In 5-Hydroxytryptamine. Ann. N. Y. Aca. Sci., *73:* 204.
[2] BRINKMAN G. L. (1968) The mast cell in normal human bronchus and lung. J. Ultrast. Res., *23:* 115.

[3] COMBS J. W. (1966)   Maturation of rat mast cells. An electron microscope study.   J. Cell Biol., *31:* 563.

[4] FEDORKO M. F. and HIRSH J. C. (1965)   Crystalloid structure in granules of guinea pig basophils and human mast cells.   J. Cell Biol., *26:* 973.

[5] FUJITA H., ASAGAMI C., MINOZUMI S., YAMAMOTO K. and KINOSHITA K. (1969)   Electron microscopic studies of mast cells of human fetal skins.   J. Ultrast. Res., *28:* 353.

[6] GOODMAN L. S. and GILMAN A. (1970)   The pharmacological basis of therapeutics. 4th edition, p. 1447, Macmillan Comp., London & Toronto.

[7] HASHIMOTO K., CROSS B. G. and LEVER W. F. (1966)   An electron microscopic study of the degranulation of mast cell granules in urticaria pigmentosa.   J. Invest. Dermatol., *46:* 139.

[8] HASHIMOTO K., TARNOWSKI W. M. and LEVER W. F. (1967)   Reifung und Degranulierung der Mastzellen in der menschlichen Haut.   Hautarzt., *18:* 318.

[9] HIBBS R. C., BURCH G. H. and PHILLIPS J. H. (1960)   Electron microscopic observations on the human mast cell.   Am. Heart J., *60:* 121.

[10] HIRAKAWA K. (1969a)   Electron microscopic studies on the mast cells from human bone marrow.   J. Kansai Med. School, *21:* 257.

[11] HIRAKAWA K. (1969b)   Fine structure of the specific granules of the mast cells from human bone marrow.   J. Kansai Med. School, *21:* 269.

[12] ISHIZAKA T., ISHIZAKA K., JOHANSON S. G. O. and BENNICH H. (1969)   Histamine release from human leukocytes by anti-rE antibodies.   J. Immunol., *102:* 884.

[13] KOBAYASHI T., MIDTGÅRD K. and ASBOE-HANSEN G. (1968)   Ultrastructure of human mast-cell granules.   J. Ultrast. Res., *23:* 153.

[14] LAGUNOFF D., PHILLIPS M. T., ISERI O. A. and BENDITT E. P. (1964)   Isolation and preliminary characterization of rat mast cell granules.   Lab. Invest., *13:* 1331.

[15] LUFT J. H. (1961)   Improvements in epoxy resin embedding methods.   J. Biophys. Biochem. Cytol., *9:* 409.

[16] MATHEWS M. B. (1966)   Clin. Orthop. *48:* 267, cited from "Handbook of Biochemistry", SOBER H. A. (ed.) p. D-69, The Chemical Rubber Co., Ohio.

[17] MOVAT H. Z. and FERNANDO N. V. (1962)   The fine structure of connective tissue. I. The fibroblast.   Exptl. Mol. Pathol., *1:* 509.

[18] ORFANOS C. (1966)   Mastzelle und mastzelldegranulation.   Klin. Wochschr., *44:* 1177.

[19] ORFANOS C. (1966)   Elektronenmikroskopische Beobachtungen zur Mastzelldegranulation bei der deffiusen Mastozytose des Menschen.   Arch. Klin. Exptl. Dermatol., *214:* 521.

[20] OSLER A. G., LICHTENSTEIN L. M. and LEVY O. A. (1968)   In vitro studies of human reaginic allergy.   Adv. Immunol., *8:* 183.

[21] PHILLIPS J. H., BURCH G. H. and HIBBS R. C. (1960)   Significance of tissue chromaffin cells and mast cells in man.   Circulation Res., *8:* 692.

[22] RÖHLICH P., ANDERSON P. and UVNÄS B. (1971)   Electron microscope observations on compound 48/80—induced degranulation in rat mast cells.   J. Cell Biol., *51:* 465.

[23] SABATINI D. D., BENSCH K. and BARRNETT R. J. (1963)   Cytochemistry and electron microscopy. The preservation of cellular ultrastructure and enzymatic activity by aldehyde fixation. J. Cell Biol., *17:* 19.

[24] SELYE H. (1965)   The Mast Cells. p. 51, Butterworth, London.

[25] STOECKENIUS W. (1956)   Zur Feinstruktur der Granula menschlicher Gewebsmastzellen. Exp. Cell Res., *11:* 656.

[26] THIERY J.P.(1963)   Étude au microscope électronigue de la maturation et de l'excrétion des granules des mastocytes.   J. Microscopic., *2:* 549.

[27] UVNÄS B., ÅBORG C. H. and BERGENDORFF A. (1970)   Storage of histamine in mast cells. Evidence for an ionic binding of histamine to protein carboxyls in the granule heparin-protein complex.   Acta Physiol. Scand. Suppl., *78:* 1.

[28] WEINSTOCK A. and ALBRIGHT J. T. (1967)   The fine structure of mast cells in normal human gingiva.   J. Ultrast. Res., *17:* 245.

Crystalloid Inclusions in Human Testicular Cells

Toshio NAGANO

It has been shown by light microscopy that human testicular cells contain regular, periodic structures called crystalloid or crystal inclusions. According to Stieve (1930), in the human seminiferous tubule, the crystalloid of Lubarsch is in the spermatogonium and is 20 μm in length, 1.5 to 3 μm in thickness with a spindle or needle shape. In the Sertoli cell, the crystalloid of Charcot-Böttcher and the smaller inclusion called Spangaro's crystalloid were described. He also summarized two kinds of inclusion bodies in the interstitial cell: the crystal of Reinke (1896) and the rice-form body of Winiwarter (1912). It seems that these crystalloids have been found in men (Gatenby & Beams 1935; Ito & Oinuma 1939; Ito & Hioki 1940) but never in animals. However, recently a few exceptional cases have been observed in some animals (see below). Since Fawcett and Burgos (1956, 1960) reported the fine structure of the Reinke's crystal by means of electron microscopy, several papers dealing with these inclusions have been published (Bawa 1963; Yamada 1965; Nagano 1966, 1967, 1968, 1969; DeKretser 1967a, b, 1968; Sisson & Fahrenbach 1967; Yasuzumi et al. 1967; Burgos et al. 1970; Nagano & Ohtsuki 1971; Sohval et al. 1971). This communication will contain a review of these inclusions with some findings by the author.

Materials presented here were obtained from patients of various ages. Castrated testes from prostate cancer patients were fixed with glutaraldehyde by vascular perfusion. Testicular tissues from biopsies for diagnostic purposes were also studied. These patients showed that the number of sperms in their semen ranged between "normal" and "oligospermic". The tissues from aspermatic patients were eliminated from this study.

Crystalloid of Lubarsch

In human spermatogonia, a needle shape structure was reported in the cytoplasm by Lubarsch (1896). The crystalloid consists of a bundle of fine filaments running parallel to the long axis when observed by electron microscopy (Nagano 1967, 1969; DeKretser 1968; Sohval et al. 1971). The filament is about 100 A in diameter and appears to be tubular (Figs. 1 and 2). The filaments are connected to each other by fine dense strands and embedded in dense material (Fig. 3). Near the ends of each crystalloid, dense granules similar to ribosomes are found between the filaments (Fig. 2). The nature of the crystalloid is not known. The digestion test with RNase failed to be demonstrated (Nagano 1969). It is interesting to note that a bundle of filaments was reported in the cytoplasm of the rooster spermatogonium (Nagano 1969).

It is recognized that the two types of spermatogonia in mammals including men are identified by histologic and radioautographic methods (Clermont 1963, 1972). The type A spermatogonium can increase in number of the same type of the spermatogonium

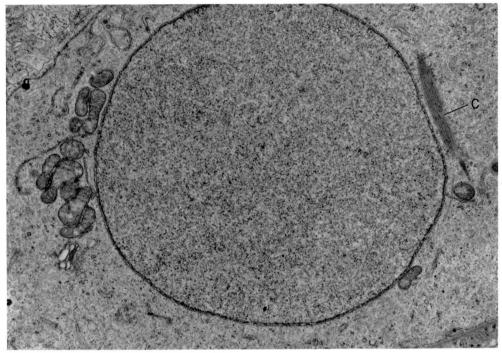

Fig. 1 A human spermatogonium contains a spindle-form inclusion (C) of Lubarsch. Mitochondria in group and Golgi complex of small size can be seen. Chromatin appears to be homogeneous. This spermatogonium is possibly the A type. (×14,000)

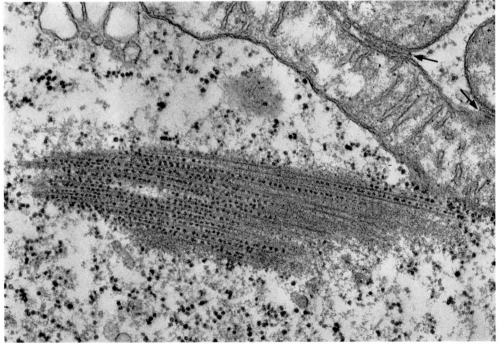

Fig. 2 Higher magnification view of the crystalloid of Lubarsch in the human spermatogonium. The crystalloid consists of filaments, granules and matrix. The mitochondria are closely associated with dense material (arrows). (×51,000)

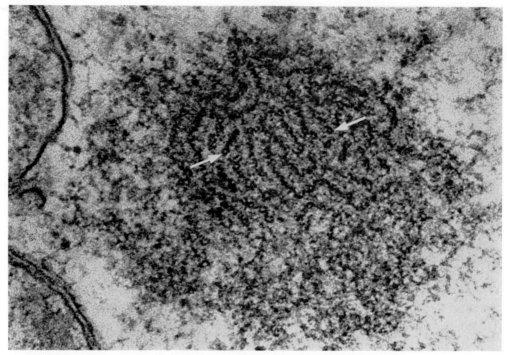

Fig. 3 Cross section of the crystalloid of Lubarsch. The tubular appearance (arrows) of the filaments and interconnection among the filaments can be seen. (×120,000)

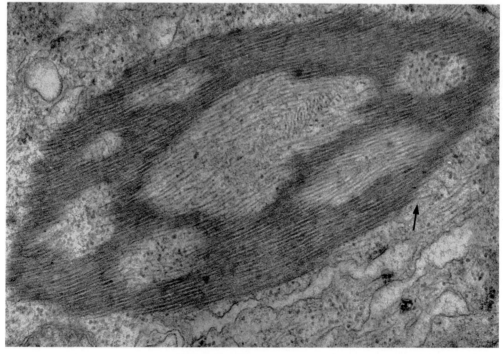

Fig. 4 Crystalloid of Charcot-Böttcher in the human Sertoli cell. It consists of filaments with tubular appearance. The less dense area within the crystalloid can be seen. Some filaments distributed in the cytoplasm are connected with the crystalloid filaments (arrows). (×50,000)

by mitosis. Some of them transfer into an intermediate type spermatogonium. The type B spermatogonium derived from the intermediate type has the capacity to differentiate into a primary spermatocyte in which meiosis takes place. It seems that both types of human spermatogonia contain the crystalloid of Lubarsch (SOHVAL et al. 1971). The spermatogonium with the cytoplasmic bridge connecting with another spermatogonium contains the crystalloid (unpublished). Furthermore, the crystalloid was observed in the early spermatocyte identified by its size and position in the tubule (NAGANO 1969). A quantitative analysis of existence of this crystalloid remains to be studied. The biotransformation of the crystalloid as spermatogenesis proceeds is not known.

If the granular material in the crystalloid is ribonucleoprotein, the crystalloid of Lubarsch is similar to the granule-lamella complex found in the kidney tubule (BULGER 1968) and many other cells (HOSHINO 1969; KRISHAN 1970). The biological significance of this complex has yet to be explained.

Crystalloid of Charcot-Böttcher

In the cytoplasm of the Sertoli cell, large crystalloids can be observed frequently. According to STIEVE (1930), the crystalloid is needle shape and 15 to 25 μm in length. Sometimes, it shows ramification. The crystalloid consists of a bundle of fine filaments similar to that of in the spermatogonium (BAWA 1963; NAGANO 1966; SOHVAL et al. 1971). The diameter of the filaments ranges between 100 and 180 A. Sometimes they are tubular in appearance (Fig. 4). SOHVAL et al. (1971) reported that the difference between spermatogonial and Sertoli cell crystalloids is the absence of granules among the filaments of the Sertoli crystalloid. The Sertoli cell crystalloid also exhibits less dense area which is not present in the crystalloid of Lubarsch.

Regarding the cytoplasmic filaments in the Sertoli cell, two kinds of filaments were reported (NAGANO 1971). One is 50 A and the other is 100 to 150 A in diameter. The former is found throughout the cytoplasm, while the latter is distributed in the peripheral portion of the cytoplasm (Figs. 5 and 6). When the nucleus of the maturing spermatids elongates, a specialized junctional area can be seen in the Sertoli cell (DYM & FAWCETT 1970; FAWCETT 1973). Cisterns of the smooth membrane are arranged in a row at a constant distance from the cell membrane of the Sertoli cell along the spermatid nucleus. Between the Sertoli cell membrane and the cistern, the filaments are arranged along the head of the spermatid (Fig. 7) (FLICKINGER & FAWCETT 1967; DYM & FAWCETT 1970). This special arrangement of the filaments is explained as a special role of elongation, holding and release of the spermatid during differentiation. In the swine Sertoli cell, structures quite similar to the Sertoli crystalloid of the human are found. Filaments in the junctional area between the spermatid and the Sertoli cell of the swine develop much more than those in men, and are arranged in a hexagonal fashion (TOYAMA 1975).

Crystalloids of SPANGARO (1902) have not been identified by electron microscopy.

The functional significance of the Sertoli crystalloid is unknown. However, filaments consisting of the crystalloid are morphologically quite similar to those in the specialized junctional areas facing the maturing spermatid, and between two Sertoli cells (Fig. 6). Therefore, if the filaments are reutilized in the Sertoli cytoplasm, the crystalloid may be a reservoir of the filaments closely packed together. In the human Sertoli cell, the microtubules are not as developed as those of the Sertoli cell of the guinea pig (FAWCETT 1966). However, it is not likely to be assumed that the filaments are transformed from the microtubules either in the living state or during the preparation processes.

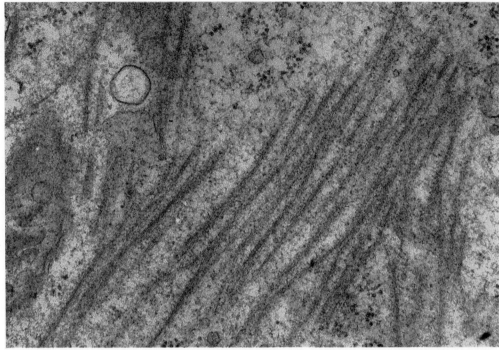

Fig. 5 In the peripheral part of the Sertoli cell cytoplasm, the filaments about 100 A in diameter distributed in random directions are illustrated. (×50,000) (From NAGANO T. (1966) Z. Zellforsch., 73: 89)

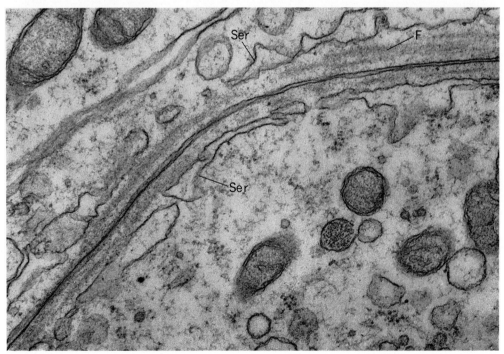

Fig. 6 In a specialized junction between two Sertoli cells of the human testis, the filaments (F) sectioned longitudinally are illustrated. The cisterns of the smooth endoplasmic reticulum (Ser) are arranged along the junction. (×55,000)

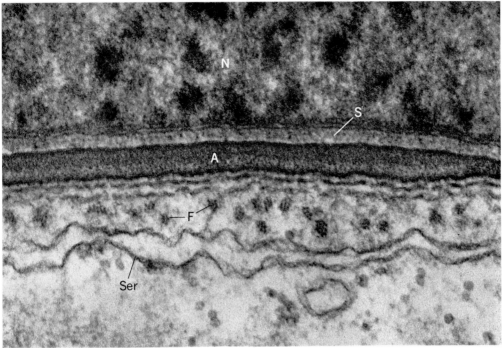

Fig. 7 A junctional area between the Sertoli cell and the developing spermatid. Cross section of the filaments (F) in the Sertoli cytoplasm between the cell membrane of the Sertoli cell and the cistern of the smooth membrane (Ser) can be seen. N: nucleus of the spermatid, A: acrosome, S: subacrosomal space. (×100,000)

Reinke's Crystal

The crystal of Reinke is easily found in human testicular interstitial cells using light and electron microscopy. REINKE (1896) described that this crystal was constantly found in the interstitial cells of all actively spermatogenic testes from 15 to 65 years of age. FAWCETT and BURGOS (1956, 1960) published the first electron micrographs of Reinke's crystals and noted that they were composed of globular macromolecules, 150 Å in diameter, uniformly spaced, about 190 Å apart along two axes at right angles to each other. A regular hexagonal pattern consisting of 200 Å tubular hexagons and filaments regarding to the formation of the crystal were demonstrated by YAMADA (1965). SISSON and FAHRENBACH (1967) reported that the crystal was composed of 50 Å filaments. The fine structure regarding to tri-dimensional analysis was studied (NAGANO & OHTSUKI 1971). The crystal is a hexagonal prism or its derivatives (Figs. 8–11). An optical diffraction analysis was also made on the electron micrographs (Figs. 13–15). Three types of internal patterns are obtained in relation to the main planes of the crystal. In a section through the perpendicular plane to the long axis, the crystal exhibits a hexagonal surface outline and a hexagonal honeycomb pattern in the interior of it (Fig. 9). When a section was cut through a plane parallel to the axis, two kinds of internal patterns can be observed. One pattern is a periodic repeart of a set of three parallel lines (each is 50 Å in diameter) perpendicular to the long axis. The first and the second are straight lines, while the third is a row of dense dots 300 Å apart (Fig. 10). Another pattern shows parallel lines perpendicular to the long axis. These lines are arranged at 150 Å intervals (Fig. 11). These two kinds of internal patterns are

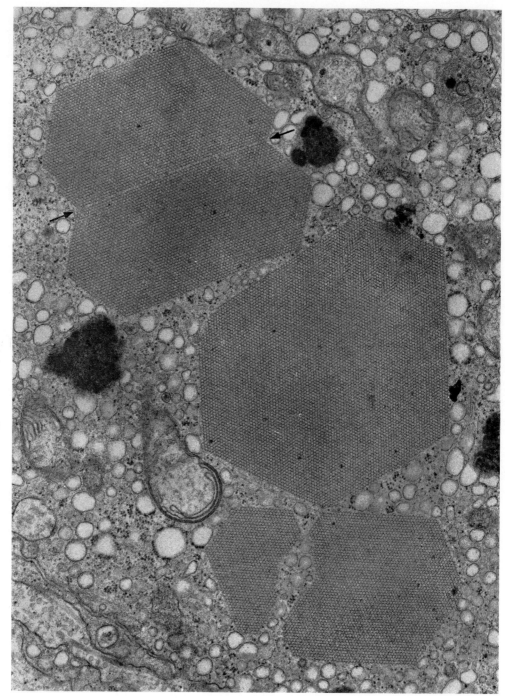

Fig. 8 Crystals of Reinke in the human interstitial cell. A granular reticulum is vesicular in form due to the soaked fixation, but the crystals are well preserved. They are polygonal in shape and their edges are sharp. Note that each side is parallel to the corresponding opposite side. The arrows indicate dislocation of the crystal. (×22,000) (From NAGANO T. & OHTSUKI I. (1971) J. Cell Biol., *51:* 148)

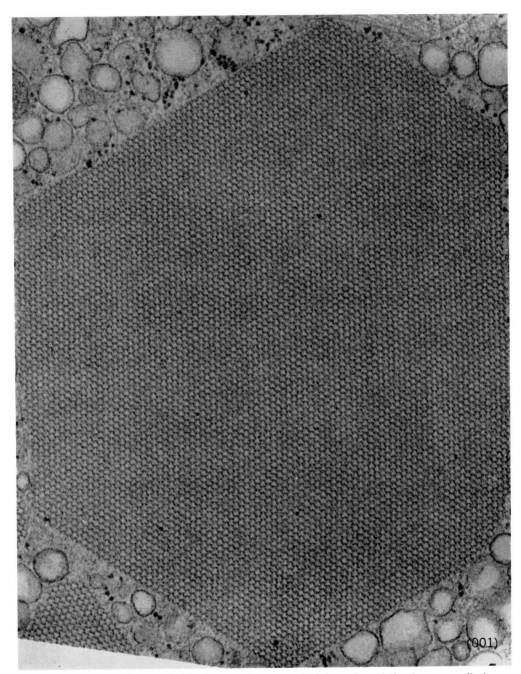

Fig. 9 A section of the crystal with a hexagonal contour. This is cut through the plane perpendicular to the long axis, parallel to the (001) plane. The internal pattern shows a honeycomb structure consisting of hexagons with 200 A on each side. (×55,000) (From NAGANO T. & OHTSUKI I. (1971) J. Cell Biol., *51*: 148)

Fig. 10 The crystal is sectioned through the plane parallel to the long axis (bar) and to the (110) plane. The lines, 50 A in diameter, at right angles to the axis are situated at 150 A intervals. The first (1) and the second (2) lines are straight, and the third (3) is seen as a row of dense dots. (×90,000) (From NAGANO T. & OHTSUKI I. (1971) J. Cell Biol., *51:* 148)

Fig. 11 The crystal shows parallel lines at 150 A intervals perpendicular to the axis (bar), and is cut parallel to the (100) plane. (×35,000) (From NAGANO T. & OHTSUKI I. (1971) J. Cell Biol., *51:* 148)

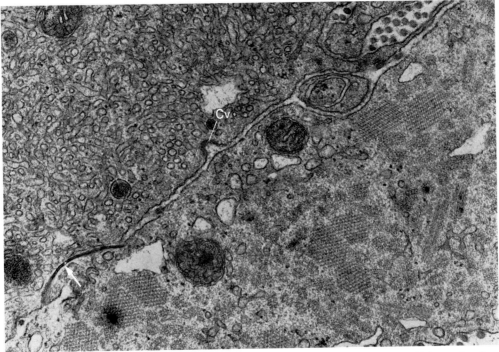

Fig. 12 Showing parts of two interstitial cells. In the cytoplasm, tiny crystals which consist of filaments are sectioned through various planes. In the other cell, the cytoplasm is largely occupied by the tubular agranular reticulum. The close junction (arrow) between the cells and the caveole (Cv) with some fuzz material can be seen. ($\times 40,000$)

interconvertible when the section is tilted about 30 degrees along the long axis. Specimen tilting and optical transform methods revealed that the crystal consists of 50 A thick filaments and has a trigonal (three-fold screw) axis, $a=300$ A, $c=450$ A, along the long axis (Figs. 13–15). A three-dimensional model is proposed (Fig. 16). Dislocation of the crystal that has been demonstrated by FAWCETT and BURGOS (1960) is also analyzed (Fig. 8) (NAGANO and OHTSUKI 1971). In the dislocation, one out of three of the filaments in each set is lacking.

The Reinke's crystal is sometimes found in the nucleus of the human testicular interstitial cell (YAMADA 1965; YASUZUMI et al. 1967). Some interstitial cells without crystal of Reinke contain filaments similar to those of the Reinke's crystal but not as closely packed (Fig. 12). The rice-shaped body of WINIWARTER (1912) may correspond with these filaments. A transitional structure between Reinke's crystal and the rice shaped body has been described (NAGANO and OHTSUKI 1971).

Although the crystal of Reinke has been studied in considerable detail with the thin-section method, nobody except Reinke has been succeeded in its isolation and chemical analysis. He has isolated the crystal using a primitive procedure and concluded that the crystal is protein in nature (REINKE 1896).

Similar crystals were said to exist in the hilus cell of the human ovary (WATZKA 1957), though ovarian crystals remain to be studied further. It is interesting to note that crystal containing cells in the adrenal cortex of the human male have been reported (MAGALHÃES 1972). The crystal appears to possess essentially the same internal structure as the Reinke's crystal in the interstitial cell of the testis. She reports that the crystal containing cell in the

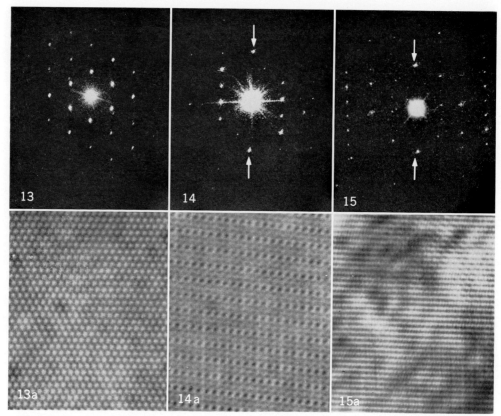

Figs. 13–15 Optical diffraction patterns of the electron micrographs of Figs. 9–11, respectively. Fig. 13 shows the hexagonal pattern. In Figs. 14 and 15, the relative strength of the third-order meridional reflection (arrows) indicates the presence of a three-fold screw axis along the long axis. Figs. 13a–15a: Optical filtering images of Figs. 9–11. The repeated patterns are more evident. (From NAGANO T. & OHTSUKI I. (1971) J. Cell Biol., *51*: 148)

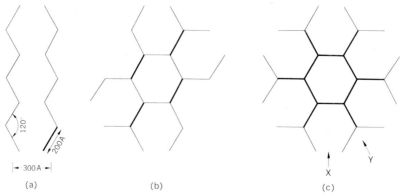

Fig. 16 A diagramatic drawing of the crystal of Reinke viewed perpendicular to the long axis showing the arrangement of 50 Å filaments. Each set of filaments folded 120° alternately at 200 Å intervals is related to its neighbor by a two-fold axis running parallel to and halfway between the folded filaments (a). A similar second set of the filaments is related to the next by a rotation of 120°, forming a orthohexagon of 200 Å on each side (b). The third set is also arranged by a similar way (c). The sets are actually arranged at 150 Å intervals along the long axis. In (c), the sectional plane perpendicular to (X), i.e. parallel to the (110) plane, shows the pattern of Fig. 10, while the plane perpendicular to (Y), parallel to the (100) plane shows that of Fig. 11.

adrenal cortex appears to be exclusive to the human male and it may be related to the adrenal androgen secretion. Since the crystal of Reinke is only found in the human male, a presumptive function of the crystal may be a role in androgenic synthesis for the human.

In the testicualr interstitial cells of the squirrel monkey, fibrillar patterned structures showing the honeycomb array have been reported during seasonal regression (BELT & CAVAZOS 1971). They reproted the patterned structures not similar to the Reinke's crystal nor the crystalloid of Charcot-Böttcher.

In summary, the ultrastructure of the crystalloid inclusions in the spermatogenic, Sertoli and interstitial cells in the human testis were described. Although the functional evidence of these inclusions is not available, their significance is suggested. Biophysical and biochemical studies on the isolated inclusions as well as histochemical studies are desired.

REFERENCES

[1] BAWA S. R. (1963) Fine structure of the Sertoli cell of the human testis. J. Ultrastr. Res., 9: 459.

[2] BELT W. D. and CAVAZOS L. F. (1971) Fine structure of the interstitial cells in the squirrel monkey during seasonal regression. Anat. Rec., 169: 115.

[3] BULGER R. E. (1968) Granule-lamella complex in monkey renal proximal tubular cells. J. Ultrastr. Res., 24: 150.

[4] BURGOS M. H., VITAL-CALPE R. and AOKI A. (1970) Fine structure of the testis and its functional significance. In: The testis. JOHNSON A. D., GOMES W. R. and VANDEMARK N. L. (eds.) Vol. 1, p. 551, Academic Press Inc., New York.

[5] CLERMONT Y. (1963) The cycle of the seminiferous epithelium in man. Amer. J. Anat., 112: 35.

[6] CLERMONT Y. (1972) Kinetics of spermatogenesis in mammals: seminiferous epithelium cycle and spermatogonial renewal. Physiol. Rev., 52: 198.

[7] DEKRETSER D. M. (1967a) The fine structure of the testicular interstitial cells in men of normal androgenic status. Z. Zellforsch., 80: 594.

[8] DEKRETSER D. M. (1967b) Changes in the fine structure of the human testicular interstitial cells after treatment with human gonadotrophins. Z. Zellforsch., 83: 344.

[9] DEKRETSER D. M. (1968) The fine structure of the immature human testis in hypogonadotrophic hypogonadism. Virshows Arch. B. Zellpath., 1: 283.

[10] DYM M. and FAWCETT D. W. (1970) The blood-testis barrier in the rat and the physiological compartmentation of the seminiferous epithelium. Biol. Reprod., 3: 308.

[11] FAWCETT D. W. (1966) The Cell: Its Organelles and Inclusions. W. B. Saunders, Philadelphia.

[12] FAWCETT D. W. (1973) Observations on the organization of the interstitial tissue of the testis and on the occuding cell junctions in the seminiferous epithelium. In: Advances in the Biosciences, 10. Schering Workshop on Contraceptions: The Masculine Gender. RASPE G. and BERNHARD S. (eds.) Pergamon Press, London.

[13] FAWCETT D. W. and BURGOS M. H. (1956) Observations on the cytomorphology of the germinal and interstitial cells of the human testis. In: Ciba Foundation Colloquia on Aging. WOLSTENHOLME G. E. W. and MILLAR E. C. P. (eds.) Vol. 2, p. 86, Little, Brown and Co., Boston.

[14] FAWCETT D. W. and BURGOS M. H. (1960) Studies on the fine structure of the mammalian testis. II. The human interstitial tissue. Amer. J. Anat., 107: 203.

[15] FLICKINGER C. and FAWCETT D. W. (1967) The junctional specializations of Sertoli cells in the seminiferous epithelium. Anat. Rec., 158: 207.

[16] GATENBY J. B. and BEAMS H. W. (1935) The cytoplasmic inclusions in the spermatogenesis of man. Quart. J. micr. Sci., 78: 1.

[17] HOSHINO M. (1969) "Polysome-lamellae complex" in the adrenal cells of the human adrenal cortex. J. Ultrastr. Res., *27:* 205.

[18] ITO T. and OINUMA SH. (1939) Zytologische Untersuchungen über die Hodenzwischenzellen des Menschen. Fol. anat. jap., *18:* 497.

[19] ITO T. and HIOKI K. (1940) Zur Zytologie der Sertolizellen im menschlichen Hoden. Fol. anat. jap., *19:* 301.

[20] KRISHAN A. (1970) Ribosome-granular material complexes in human leukemic lymphoblasts exposed to Vinblastine sulfate. J. Ultrastr. Res., *31:* 272.

[21] LUBARSCH O. (1896) Über das Vorkommen krystallinischer und krystalloider Bildungen in den Zellen des menschlichen Hodens. Virchows Arch. path. Anat., *145:* 316.

[22] MAGALHÃES M. C. (1972) A new crystal-containing cell in human adrenal cortex. J. Cell Biol., *55:* 126.

[23] NAGANO T. (1966) Some observations on the fine structure of the Sertoli cell in the human testis. Z. Zellforsch., *73:* 89.

[24] NAGANO T. (1967) Spermatogonium. *In:* Fine Structure of Cells and Tissues. Vol. III. Urogenital Organ. YAMADA E., UCHIZONO K. and WATANABE Y. (eds.) pp. 66–67, Igaku Shoin Ltd., Tokyo. (in Japanese)

[25] NAGANO T. (1968) Fine structural relation between the Sertoli cell and the differentiating spermatid in the human testis. Z. Zellforsch., *89:* 38.

[26] NAGANO T. (1969) The crystalloid of Lubarsch in the human spermatogonium. Z. Zellforsch., *97:* 491.

[27] NAGANO T. and OHTSUKI I. (1971) Reinvestigation of the fine structure of Reinke's crystal in the human testicular interstitial cells. J. Cell Biol., *51:* 148.

[28] REINKE F. (1896) Beiträge zur Histologie des Menschen. Arch. mikr. Anat., *47:* 34.

[29] SISSON J. K. and FAHRENBACH W. H. (1967) Fine structure of steroidogenic cells of a primate cutaneous organ. Amer. J. Anat., *121:* 337.

[30] SOHVAL A. R., SUZUKI Y., GABRILOVE J. L. and CHURG J. (1971) Ultrastructure of crystalloids in spermatogonia and Sertoli cells of normal human testis. J. Ultrastr. Res., *34:* 83.

[31] SPANGARO S. (1902) Über die histologischen Veränderungen des Hodens, Nebenhodens und Samenleiters von Geburt und bis zum Greisenalter, mit besonderer Berücksichtigung der Hodenatrophie, des elastischen Gewebes und des Vorkommens von Krystallen im Hoden. Anat. Heft., *18:* 593.

[32] STIEVE H. (1930) Harn- und Geschlechtsapparat. *In:* Handbuch mikr. Anat. Mensch.,

[33] TOYAMA Y. (1975) Ultrastructural study of crystalloids in Sertoli cells of the normal, intersex and experimental cryptorchid swine. Cell Tiss., *158:* 205.
MÖLLENDORFF W. (ed.) Vol. 7/2, Verlag von Julius Springer, Berlin.

[34] WATZKA M. (1957) Das Ovarium. *In:* Handbuch mikr. Anat. Mensch., MÖLLENDORFF W. and BARGMANN W. (eds.) Vol. 7/3, Springer Verlag, Berlin.

[35] WINIWARTER H. (1912) Observations cytologiques sur les cellules du testicule humain. Anat. Anz., *41:* 309.

[36] YAMADA E. (1965) Some observations on the fine structure of the interstitial cell in the human testis as revealed by electron microscopy. Gunma Symp. Endocrinol., *2:* 1.

[37] YASUZUMI G., NAKAI Y., TSUBO I., YASUDA M. and SUGIOKA T. (1967) The fine structure of nuclei as revealed by electron microscopy. IV. The intranuclear inclusion formation of Leydig cells of aging human testes. Exp. Cell Res., *45:* 261.

Crystal Structures in Amphibian Yolk Platelets

Ryohei HONJIN

A yolk platelet is the most prominent component in the cytoplasm of eggs and embryonic cells and is also the most important material to be utilized during embryonic development of all animals. HOLTFRETER (1946a, b, c) suggested the presence of an oriented ultrastructure in amphibian yolk platelets by polarization microscopy though his reports were wanting in deatil. Recently much attention has been focused on this subject and a periodic structure indicating the existence of a crystalline lattice has been demonstrated by electron microscopy (KEMP 1956; EAKIN & LEAMANN 1957; WISCHNITZER 1957, 1964; HONJIN & TSUDA 1958; KARASAKI & KOMODA 1958; TSUDA & TAKAHASHI 1958; KARASAKI 1959, 1962, 1963a, b; LANZAVECCHIA 1960; TSUDA 1961; RIGLE & GROSS 1962; WALD 1962; WARTENBERG 1962; WALLACE & KARASAKI 1963; MASSOVER 1970, 1971; ARMSTRONG 1972; LEONARD et al. 1972). Several models for the crystalline structure of a platelet have been proposed, based upon electron microscopic and biochemical information (HONJIN & TSUDA 1958; KARASAKI 1959, 1963a; TSUDA 1961; WARTENBERG 1962; WALLACE 1963; LEONARD et al. 1972). However, a reliable determination of the accurate shape and size of a unit cell of the crystalline lattice cannot be obtained, as long as investigations rely only upon electron microscopic images. Recent X-ray diffraction analyses combined with electron microscopic investigations have pointed out the presence of a hexagonal system regarding the crystalline lattice of the main body of a yolk platelet (HONJIN 1961, 1966, 1967, 1968a, b, 1969; HONJIN & NAKAMURA 1961, 1969; HONJIN & SHIMASAKI 1962, 1965; HONJIN et al. 1965; SHIMASAKI 1965).

The present paper gives an account of the structural unit of the three-dimensional macromolecular organization in the main body of several amphibian yolk platelets obtained from polarization microscope, electron microscope and X-ray diffraction analyses.

Materials and Methods

Observations were made on yolk platelets in early embryos and mature eggs of *Rhacopholus schlegelii* var. *arborea* and *Triturus pyrrhogaster* (Boie). The developmental stages of the embryos were estimated by ICHIKAWA's (1931) and OKADA and ICHIKAWA's methods (1947) respectively.

Polarizing microscopic studies were carried out on fresh yolk platelets isolated from early embryos or mature eggs. Both orthoscopic and conoscopic examinations were performed on platelets dispersed in a 0.64% saline solution or distilled water by means of a POM type polarizing microscope (Olympus Optical Co., Ltd.).

Electron microscopic studies were performed on slices of presumptive ectodermal cells or isolated yolk platelets prepared by the ESSNER's method (1954). Materials were fixed with 1% osmium tetroxide solution (pH 7.25), embedded in plastics, sectioned with ultramicrotomes, stained with lead acetate or uranyl acetate, and examined in an HU-11 type electron microscope (Hitcahi, Ltd.).

X-ray diffraction studies were carried out on isolated fresh yolk platelets obtained from embryos or eggs. Small-angle X-ray diffraction patterns were obtained with a vacuum camera having a pinhole collimating system. A densitometric tracing of the small-angle diffraction photographs was performed by a microphotometer which was connected with an electronic circuit panel to automatically record the diffraction pattern on a chart (Rigaku-Denki Co., Ltd.). Detailed preparation and condition for the X-ray diffraction studies were identical to those described earlier (HONJIN et al. 1965; HONJIN & NAKAMURA 1969).

Experimental Results

Ordinary light microscopic observations

Yolk platelets in *Triturus* appear as round or flattened oval bodies, while those in *Rhacopholus* are somewhat angular. Some of the platelets are 10 μm in diameter, but the smaller ones are near or below the limits of resolution of the light microscope.

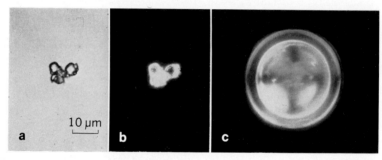

Fig. 1 a. Yolk platelets obtained from a *Rhacopholus* embryo (stage 9), under parallel nicols. They were dispersed in distilled water. **b.** Orthoscopic picture of the same yolk platelets as seen in Fig. 1a. The platelets indicated a weak birefringence under crossed nicols. **c.** A conoscopic figure of a single yolk platelet obtained from a *Rhacopholus* embryo (stage 9). There can be seen many concentric rings split by a dark cross.

Polarization optic inspections

Yolk platelets under parallel nicols are seen in Fig. 1a. Using the orthoscope with crossed nicols, the same platelets indicate a weak birefringence, as shown in Fig. 1b. This result indicates that the yolk platelet is an anisotropic crystalline substance. The conoscopic figure of a single yolk platelet is shown in Fig. 1c, which exhibits many concentric rigs split by a dark cross. This conoscopic figure bleongs to the interference figure of a uniaxial crystal system. This suggests that the yolk platelet is a uniaxial crystal.

Electron microscopic studies

Yolk platelets of amphibia are composed of a central main body displaying a crystalline lattice structure, a superficial layer of fine particles surrounding the main body, and a limiting membrane of approximately 70 A thickness enclosing the entire structure, as shown in Fig. 2. The structural feature of yolk platelets is essentially the same in the two species studied, both in mature eggs and embryos at different stages of development. In isolated yolk platelets, only a few particles have been observed adhering to the outer surface of the

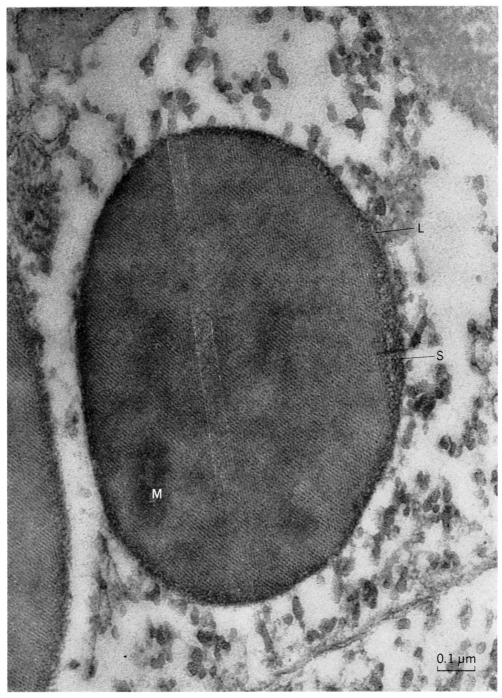

Fig. 2 Part of a presumptive ectodermal cell in a *Triturus* embryo (stage 12). The yolk platelet consisted of a central main body (M) and an outer superficial layer (S), both enclosed within a single limiting membrane (L). The entire main body displayed a hexagonal array of dots approximately 50 A in diameter. The superficial layer consisted of fine particles which were much smaller than the cytoplasmic ribonucleoprotein. ($\times$ 120,000)

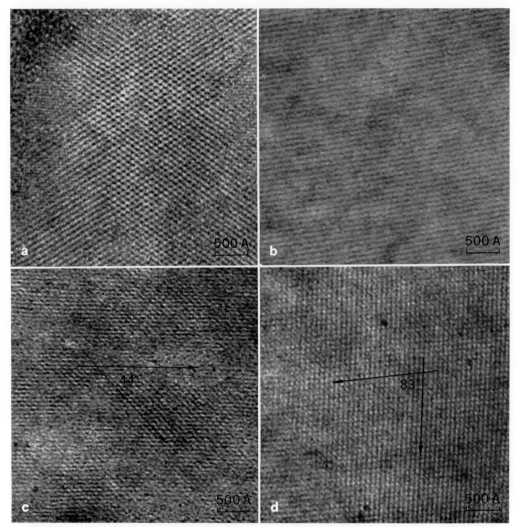

Fig. 3 a. Part of a yolk platelet in the neural crest cell of a *Triturus* embryo (stage 18). The main body displayed a hexagonal array of dots approximately 50 A in diameter. ($\times$180,000) **b.** Part of a yolk platelet isolated from a *Rhacopholus* embryo (stage 12). The alternating dense and less dense parallel lines had a spacing of 85 A. ($\times$180,000) **c.** Part of a yolk platelet in the ectodermal cell of a *Rhacopholus* embryo (stage 9). The main body displayed an intersecting line pattern with an angle of 44 degrees. ($\times$180,000) **d.** Part of a yolk platelet in the neural crest cell of a *Rhacopholus* embryo (stage 18). Two parallel line systems intersected with each other at an angle of 83 degrees. ($\times$180,000)

main body owing to the destruction of the superficial layer and the limiting membrane during the isolation procedure.

The appearance of the main bodies revealed in close-to-focus pictures of many thin sections could be divided into the following three types: dot, line, and homogeneous patterns.

The *dot pattern* consisted of electron-dense, roundish or slightly ellipticed dots packed in an almost hexagonal array and separated from one another on all sides by a less dense, intermediate substance as shown in Fig. 3a. The diameter of the dots was estimated to be 50 A. Occasionally the dot pattern appeared as a pseudohexagonal array. The dots were arranged in straight, parallel rows in three directions in the photograph. The three directions of the rows of dots intersected one another at angles ranging from 42 to 73 degrees. The rows of dots running in the same dicretion stood in a periodic pattern with a parallel arary and uniform spacing. The center-to-center distance between the adjacent rows of dots ranged from 65 to 90 A with the majority of the values falling between 70 and 85 A. The dot pattern could be recognized in unstained thin sections, but the dots increased in density when the section was stained with lead or uranium.

The *line pattern* consisted of straight, parallel alternately dense and less dense lines. The center-to-center distance between the parallel dense lines ranged from 55 to 95 A with the majority of the values falling between 65 and 85 A. The line pattern could be subdivided into two kinds: *a simple line pattern* and *an intersecting one*. The former consisted of parallel lines in one direction, as shown in Fig. 3b. The latter was composed of two systems of lines intersecting each other at angles of 30 to 90 degrees, as shown in Figs. 3c, d.

The *homogeneous pattern* appeared entirely homogeneous, even though the sections were cut very thin and correct focusing was carefully performed.

Of the 523 close-to-focus pictures of yolk platelets, only 44 cases revealed the dot pattern (8%). Two hundred ten cases displayed the simple line pattern (40%). One hundred seventy cases displayed the intersecting line pattern (33%). The remaining 99 cases revealed the homogeneous pattern (19%). All grades of transition between the dot and line patterns could be seen, as well as between the line and homogeneous pattern. In fact, all possible variations in the type of pattern, in the spacing of the dots and lines, and in the angles in the intersecting lines, were often observed in different parts of a single micrographic plate. These variations in electron microscopic images were probably due to difference in the angles of sectioning of the main bodies with respect to their crystalline lattice. It would be very difficult to determine the accurate shape and size of the unit cell of the crystalline lattice when based only upon the electron microscopic images, because electron microscopic studies offer no information regarding the angles of sectioning of the main bodies.

X-Ray Diffraction Analyses

Small-angle X-ray diffraction data from fresh yolk platelets

The powder X-ray diffraction photograph obtained from fresh yolk platelets with the pinhole collimating system revealed a typical diffraction pattern (Fig. 4). The densitometric tracing of this diffraction photograph by a microphotometer is shown in Fig. 5, which displays nineteen peaks of reflections. In Fig. 5, a weak peak is seen between the peaks labeled 5 and 6. This peak, however, is generally obscure, so that it was excluded from the measurement. The mean values of observed interplanar spacings for these nineteen reflections averaged on ten specimens were filled up in Fig. 4. As shown in Figs. 4 and 5, the second, fourth, fifth and eighth reflections are strong in relative intensity. The fourth was the strongest. The characteristic of the diffraction spacings of the yolk platelets showed no significant difference among the mature egg and the different stages of the embryos of the two species examined, though the two experimental animals actually belonged to two fairly unrelated groups, the anurans and the urodeles, respectively.

Crystallographic analysis of the diffraction data

In order to estimate the size and shape of the unit cell of the crystalline structure in the yolk platelets, the diffraction data were investigated by means of the reciprocal lattice

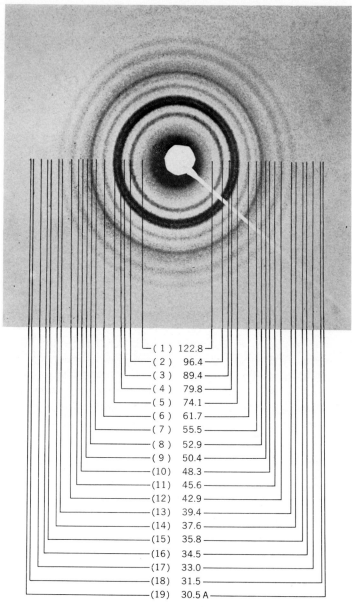

Fig. 4 A small-angle X-ray diffraction pattern of isolated fresh yolk platelets of *Rhacopholus* embryo (stage 9) taken with a pinhole collimating system and a high power X-ray source. Note many circular line profiles of the reflections. The interplanar spacings of the reflections were filled up according to the values of the observed interplanar spacings obtained from a densitometric tracing of the X-ray diffraction pattern shown in Fig. 5. (From Honjin R. & Nakamura T. (1969) J. Ultrastruct. Res., *20:* 400)

method, $\sin^2\theta$ ratio method, slide rule method (Guinier 1956), or the graphical procedure with Bjurström chart (Bjurström 1931) or Hull-Davey chart (Hull & Davey 1921). For determining the Miller indices agreeing with the powder diffraction pattern, the identification operation was performed by applying these methods to two of the seven crystal systems, tetragonal and hexagonal. The other five crystal systems were excluded from the objective of the study. Reasons for this treatment were as follows: (a) it had

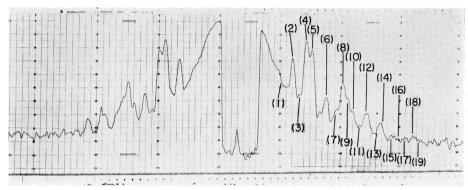

Fig. 5 A chart of the densitometric tracing of Fig. 4. Note the nineteen peaks of reflections. (From HONJIN R. & NAKAMURA T. (1969) J. Ultrastruct., *20:* 400)

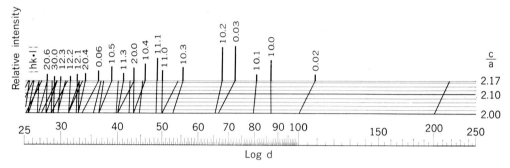

Fig. 6 An identification operation on the Hull-Davey chart for the hexagonal crystal system. The fitting position could be found on an axial ratio, $c/a = 2.17$. Note the Miller indices $(hk \cdot l)$ and the relative intensity of reflections shown by a set of proportional scales. (From HONJIN R. & NAKAMURA T. (1969) J. Ultrastruct. Res., *20:* 400)

already been pointed out on the basis of observations with the polarizing microscope that the fresh yolk platelet of *Amphibia* was a uniaxial crystal (HONJIN 1967, 1968a, b), as seen in Fig. 1 of the present paper; (b) it was also known that the cubic system showed no birefringence under crossed nicols, and both the tetragonal and hexagonal crystal systems assumed an interference figure of a uniaxial crystal in polarizing optical images, while the orthorhombic, monoclinic, and triclinic systems assumed that of a biaxial crystal; (c) the rhombohedral crystal could simply be described in terms of hexagonal axes.

Although the tetragonal system was already excluded from the range of possibility for the crystal form of the yolk platelet in our previous studies (HONJIN et al. 1965; HONJIN & SHIMASAKI 1965; HONJIN 1966, 1967, 1968a, b; HONJIN & NAKAMURA 1969), the author's identification test with regard to the tetragonal system had been tried again in the present study, by way of precaution. But no satisfactory result could be obtained. This result confirmed the early conclusion that the tetragonal system was beyond the range of possibility.

Spacings for the reflections in the diffraction pattern were examined closely by means of the Hull-Davey chart for the possibility of the hexagonal system. The fitting position was obtained at an axial ratio, $c/a = 2.17$ on the chart, where a and c were the axial lengths of the unit cell of the crystal (Fig. 6). Miller indices $(hk \cdot l)$, for each reflection are given in Table 1. The relative intensity of each reflection shown in Table 1 has calculated from the height of each peak of reflection in Fig. 5. In Fig. 6 the relative intensity was shown by a set of proportional scales.

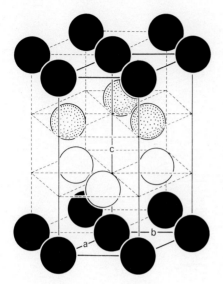

Fig. 7 A model for the arrangements of yolk protein (phosvitin molecules) within the main crystal body of the yolk platelets. An arrangement of phosvitin molecules packed in a hexagonal crystalline lattice ($a=110.5\pm1$A, $c=239.8\pm2$A, and $c/a=2.17$) is shown. The black spheres represented the phosvitin molecules in the first and fourth planes which were perpendicular to the c axis. The white and dotted spheres represented those in the second and third planes, respectively. All planes were arranged parallel to one another with a uniform spacing of 79.9 A. A unit which was shown by the solid lines connecting the molecules represented a unit cell of the hexagonal crystalline lattice. It contained three chemical component units. (Modified from HONJIN R. & NAKAMURA T. (1969) J. Ultrastruct. Res., *20:* 400)

One of the relationships between the parameters involved in coherent X-ray diffraction from the hexagonal structure is:

$$d = \frac{1}{\sqrt{\frac{4(h^2+k^2+hk)}{3a^2} + \frac{l^2}{c^2}}} \quad (1)$$

which can be rearranged into:

$$d = \frac{a}{\sqrt{\frac{4(h^2+k^2+hk)}{3} + \left(\frac{l}{c/a}\right)^2}} \quad (2)$$

where h, k, and l are Miller indices, d is the spacing between the planes of a given set of $hk \cdot l$ indices, and a and c are the axial lengths of the unit cell. From Eq. (2), the axial lengths, a and c were obtained. Then the calculated spacings for each reflection were obtained (Table 1). There was a distinct similarity between the observed spacings and the spacings calculated for the hexagonal lattice. This suggested that the results obtained by the present identification operation for determination of the crystal system and the axial ratio of the crystalline lattice of the yolk platelet were proper. This relationship was more accurately assured by the $\sin^2\theta$ test, in which a remarkable resemblance was seen between the observed $\sin^2\theta$ and the calculated $\sin^2\theta$ (Table 2). These results indicated that the yolk platelet crystal was a hexagonal system with $a=110.5\pm1$ A, $c=239.8\pm2$ A, and the axial ratio=2.17. Among the nineteen reflections, the (00·3) reflection was the strongest, and the (10·2), (11·1), and (10·0) reflections were ranked next in intensity. This indicates that each spacing of one period on the c axis contained three parallel equidistant planes prominent for each reflection, which were perpendicular to the c axis. Fig. 7 represents a model for the crystalline lattice of the yolk platelet.

Discussion

Based upon the electron microscopic observations of the periodic line pattern in amphibian yolk platelets, KARASAKI and KOMODA (1958) and KARASAKI (1959) supposed that

Table 1 X-ray data from fresh yolk platelets. (Modified from Honjin R. & Nakamura T. (1969) J. Ultrastrct. Res., *20*: 400.)

Line	$\sqrt{\frac{4}{3}(h^2+k^2+hk)+\left(\frac{l}{c/a}\right)^2}$		Length of a axis (A)	Calculated spacing (A)	Observed spacing (A)	Relative intensity of line*
	$(hk \cdot l)$	$(c/a=2.17)$				
1	00·2	0.922	113.2	119.9	122.8	0.2
2	10·0	1.155	111.3	95.7	96.4	0.6
3	10·1	1.243	111.1	88.9	89.4	0.2
4	00·3	1.383	110.4	79.9	79.8	1.0
5	10·2	1.477	109.5	74.8	74.1	0.9
6	10·3	1.801	111.1	61.4	61.7	0.4
7	11·0	2.000	111.0	55.3	55.5	0.3
8	11·1	2.052	108.6	53.8	52.9	0.7
9	10·4	2.175	109.6	50.8	50.4	0.5
10	20·0	2.309	111.5	47.9	48.3	0.4
11	11·3	2.431	110.9	45.5	45.6	0.2
12	10·5	2.577	110.6	42.9	42.9	0.4
13	00·6	2.765	108.9	40.0	39.4	0.3
14	20·4	2.955	111.1	37.4	37.6	0.2
15	12·1	3.090	110.6	35.8	35.8	0.1
16	12·2	3.198	110.3	34.6	34.5	0.1
17	12·3	3.353	110.7	33.0	33.0	0.1
18	30·0	3.464	109.1	31.9	31.5	0.1
19	20·6	3.603	109.9	30.7	30.5	0.1
Mean			110.5±1			

* The relative intensity of each reflection was calculated from the height of each peak of reflections shown in Fig. 5. The highest fourth line was set up for a standard of comparison of intensity.

Table 2 Comparison between $\sin^2\theta$ (Obs.) and $\sin^2\theta$ (Cal.).* (Modified from Honjin R. & Nakamura T. (1969) J. Ultrastruct. Res., *20*: 400.)

Line	$(hk \cdot l)$	$\sin^2\theta$ (Obs.)	$\sin^2\theta$ (Cal.)**
1	00·2	0.000039	0.000041
2	10·0	0.000064	0.000065
3	10·1	0.000074	0.000075
4	00·3	0.000094	0.000093
5	10·2	0.000108	0.000106
6	10·3	0.000156	0.000158
7	11·0	0.000193	0.000195
8	11·1	0.000212	0.000205
9	10·4	0.000234	0.000230
10	20·0	0.000254	0.000260
11	11·3	0.000286	0.000288
12	10·5	0.000324	0.000324
13	00·6	0.000383	0.000372
14	20·3	0.000420	0.000425
15	12·1	0.000465	0.000465
16	12·2	0.000500	0.000496
17	12·3	0.000546	0.000547
18	30·0	0.000599	0.000584
19	20·6	0.000640	0.000632

* The $\sin^2\theta$ was given by the formula: $\sin^2\theta = \frac{\lambda^2}{4}\left(\frac{4}{3}\frac{(h^2+k^2+hk)}{a^2}+\frac{l^2}{c^2}\right)$

** Calculation of the calculated $\sin^2\theta$ was based on $a=110.5$A and $c=239.8$A.

the line pattern was the image of *a fringe system* whose existence they hypothetically postulated. About the same time, Honjin and Tsuda (Honjin & Tsuda 1958; Tsuda & Takahashi 1958; Tsuda 1961) observed both the line and dot patterns and suggested the presence of *a close-packed structure of particles* representing a crystalline lattice built up by macromolecules. After a while, Wartenberg (1962) postulated that the variation in the appearance of dots or lines seemed to depend upon the angle of sectioning of long, thread-like micelles arranged in a parallel array. He presented *a model of long cylindrical micelles* with a diameter of 35 A. However, the plausibility of his model could be ruled out at a glance of the X-ray diffration data obtained in the present study, which gave the length of the c axis of a hexagonal lattice.

It was easy to geometrically anticipate the high probability that the sectioning was oblique to the axis of the hexagonal lattice. Each section which was not quite parallel to the axes would give an image of superimposed units, and thus be observed as parallel lines or a homogeneous state when the thickness of the section was 200 to 500 A. This satisfactorily explained the appearance of three types of pattern, with all grades of transition between them, and the large range of spacings between the rows of dots and between the periodic lines as well as the large range of intersecting angles in the paired line pattern. A reliable determination of the accurate shape and size of the unit cell of the crystalline lattice, however, could not be obtained as long as investigators relied only upon the electron microscopic images, because the angle of sectioning of the thin section to the axis of the crystalline lattice could not be determined and the accurate values of the parameters of the crystalline lattice could not be measured in the electron microscopic profiles.

After a while, Karasaki (1963a) and his colleague Wallace (1963) proposed a new model for the crystalline lattice of the yolk platelet, which they termed *a simple hexagonal lattice*. It was proposed because the micrographic dot images obtained in their later work (Karasaki 1962, 1963a) could not be explained by their former fringe system model. It was found that the particles in the main body could be visualized with an electron microscope in formalin-fixed, unstained sections (Wald 1962) and that the component protein macromolecules were found sufficiently large and discrete so as to be resolved and visualized with the electron microscope (Wallace 1963b). The X-ray diffraction study, however, made the model of a simple hexagonal lattice postulated by Karasaki and Wallace unacceptable, because of an axial ratio, $c/a=1$, calculated from their model based upon the electron microscopic images. This differed markedly from the value of the axial ratio obtained from the present X-ray diffraction analysis. According to the model shown by Wallace (1963b), all the protein macromolecules lay in a straight line along the c axis, and there was no closely packed organization of molecules in an orientation of the crystalline lattice parallel to the c axis. This kind of crystalline lattice should be very unstable, and it was very rarely met with in nature. From this point of view also, their model hardly seemed to represent the fundamental ultrastructure of the main body of the yolk platelet, which was actually stable and was preserved for a relatively long period of embryonic development, though it showed a gradual decomposition.

The hexagonal crystalline lattice postulated in the present X-ray diffraction study belonged to the uniaxial system. This result was completely in accord with the findings in the polarizing microscopic observations which indicated that the yolk platelet was a uniaxial crystal. Another kind of crystal lattice differeng in its state of packing, termed *hexagonal close-packed structure*, was present in nature. However, this crystal structure should show a conspicuous reflection on the surface having Miller indices (00·2). This differed markedly from the results obtained in the present study; hence the hexagonal close-packed structure was ruled out from consideration.

According to the biochemical analysis of yolk (BURLEY & COOK 1961; ITO 1963; WALLACE 1963a, b; BARMAN et al. 1964), the crystalline main body was composed of two kinds of proteins: a highly phosphorylated phosphoprotein, which was designated as amphibian phosvitin (component S), and a kind of lipoprotein, termed amphibian lipovitellin (component F). Two phosvitin molecules were associated with one lipovitellin molecule, and the molecular weight of these two proteins were approximately 32,000 and 420,000, respectively. Their spherical diameters had been calculated to be 40 A for phosvitin and 101 A for lipovitellin while the density of the phosvitin was greater than that of the lipovitellin. As already reported (KARASAKI 1963a; HONJIN et al. 1965; HONJIN 1967) and corroborated by the present study, the contrast of the dot or line pattern was considerably increased by uranium staining, which preferentially stained such high phosphorus-containing substances as nucleic acids (HUXLEY & ZUBAY 1961; STOCKENIUS 1961; ZOBEL & BEER 1961), and it might be assumed that the uranium would become bound to the phosphate group of the phosvitin to effect the observed increase in contrast. These observations indicated that the phosvitin molecule represented the electron-dense particles observed by electron microscopy as arranged in a hexagonal array.

The volume of the unit cell of the hexagonal lattice obtained in the present study was calculated to be 2535.7×10^{-21} ml. Since this unit cell contained three component protein units (or units of chemical substances), the volume for every chemical component unit was calculated to be 845.2×10^{-21} ml. According to biochemical studies (WALLACE 1963a, b) the volume of two phosvitin molecules and one lipovitellin molecule is 602×10^{-21} ml, and the intact yolk platelet contained 30% water by weight. If the chemical component unit of the crystalline lattice of the main body was composed of two phosvitin molecules and one lipovitellin molecule as suggested in the biochemical analysis (WALLACE 1963b), a unit cell provided in our model had 71.2% of its space accounted for by the biochemically measured volumes of the component proteins. If the component protein was spherical in shape, consideration must be given to the packing space between the molecules which was usually calculated to be about 26% of the total volume in this crystal form. Part of the remaining 28.8% of the space of the unit cell might be attributed to water and other chemical substances.

According to WALLACE (1963b) the unit cell of a yolk crystalline lattice was composed of a lipovitellin molecule which appeared as a somewhat long cylindrical rod (164 A in length) and contained two detached phosvitin molecules in itself. He was very emphatic that two phosvitin molecules in such a unit cell (one lipovitellin molecule) would be about 81 A apart from each other and also a similar distance apart from phosvitin molecules in neighboring units. Although the positioning of lipovitellin molecules within the hexagonal system of phosvitin molecules was based upon a completely arbitrary decision, if it was permissible to suppose that the lipovitellin molecule contained two phosvitin molecules in itself, it seemed natural that two phosvitin molecules should exist in a lipovitellin molecule as a dimer made up by fusion of two molecules and thus offer the typical reflection pattern in the X-ray diffraction. Calculations based upon measurement of the biochemical data indicated that a dimer of two phosvitin molecules would have a diameter of 49.8 A when it formed a spherical particle, and one lipovitellin molecule would increase its diameter from 101 to 104.6 A when it contained a dimer of phosvitin in itself. These values for the diameter, however, were for molecules in the dry state. If these chemical components were in the fresh or wet state, the diameters would be somewhat larger. These values calculated from the biochemical data were nearly in accordance with the diameter of 50 A in the dense dots seen in the present electron micrographs and with an axial length of 110.5 A in the present X-ray analysis. The axial ratio of 2.17 in the crystalline lattice of the yolk platelet suggested that the chemical component unit might be a slightly elliptical particle of the shape of an oblate spheroid, whose short axis was parallel to the c axis of the crystalline lattice. This might still be more favorable for understanding the weak birefringence observed in the yolk platelet. A model for the chemical component unit sectioned along its short axis was thus proposed in Fig. 8. An oblique view of a model for the arrangement of lipovitellin molecules packed in a hexagonal crystalline lattice is shown in Fig. 9.

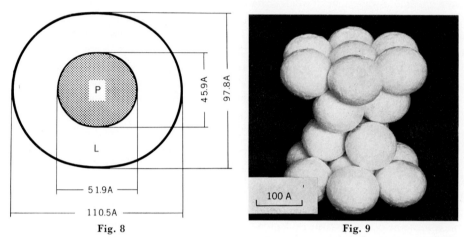

Fig. 8   Fig. 9

Fig. 8 A sectional model for the chemical component unit of the yolk proteins which is composed of one dimer of two phosvitin molecules and one lipovitellin molecule. The dimer of the phosvitin molecules (P) was enclosed in the lipovitellin molecule (L). The chemical component unit appeared as an oblate spheroid whose long and short axes were calculated to be 110.5 and 97.8A, respectively. The short axis was parallel to the c axis in the crystalline lattice. The dimer of the phosvitin molecules also appeared as an oblate spheroid and was calculated to have long and short axes of 51.9 and 45.9 A, respectively. (Modified from HONJIN R. & NAKAMURA T. (1969) J. Ultrastruct. Res., *20:* 400)

Fig. 9 An oblique view of a model for the arrangements of lipovitellin molecules packed in a hexagonal crystalline lattice, $a=110.5\pm1$A, $c=239.8\pm2$A, and $c/a=2.17$. Each lipovitellin molecule contained a dimer of the phosvitin molecules in itself.

Recently, based on both freeze-etch electron microscopy of isolated platelets and our X-ray diffraction data (HONJIN et al. 1965; HONJIN & NAKAMURA 1969), LEONARD et al. (1972) proposed a new closest-packing model of the platelet structure, using lipovitellin dimers as building blocks, with one molecule of the phosvitin associated with each monomer of the lipovitellin dimer. According to our study, however, the rectangular and semi-hexagonal patterns seen by them in fracture faces of the platelets corresponded to the cleavage faces along the (10·2) and (10·1) planes in our lipovitellin model shown in Fig. 9, respectively. It seemed that the findings in their freeze-etch electron micrographs could be fully explained by our model.

REFERENCES

[1]   ARMSTRONG P. B. (1972)   Unusal yolk platelets in embryos of *Xenopus laevis* (Amphibia). Z. Zellforsch., *129:* 320.

[2]   BARMAN T. E., BAI N. -K. and THOAI N. -V. (1964)   Studies on a herring-egg phosphoprotein. Biochem. J., *90:* 555.

[3]   BJURSTRÖM T. (1931)   Graphische Methoden zum Aufsuchen der quadratischen Form aus röntgenographischen Pulverphotogrammen. Z. Physik., *69:* 346.

[4]   BURLEY R. W. and COOK W. H. (1961)   Isolation and composition of avian egg yolk granules and their constituent α- and β-lipovitellins. Can. J. Biochem. Physiol., *39:* 1295.

[5] EAKIN R. M. and LEAMANN F. E. (1957) An electron microscopic study of developing amphibian ectoderm. Arch. Entwicklungsmech. Organ., *150:* 177.

[6] ESSNER E. S. (1954) The breakdown of isolated yolk granules by cations. Protoplasma, *43:* 79.

[7] GUINIER A. (1956) Théorie et Technique de la Radiocristallographie. Dunod, Paris.

[8] HOLTFRETER J. (1946a) Experiments on the formed inclusions of the amphibian egg. I. The effect of pH and electrolytes on yolk and lipochondria. J. Exptl. Zool., *101:* 355.

[9] HOLTFRETER J. (1946b) Experiments on the formed inclusions of the amphibian egg. II. Formative effects of hydration and dehydration on lipid bodies. J. Exptl. Zool., *102:* 51.

[10] HOLTFRETER J. (1946c) Experiments on the formed inclusions of the amphibian egg. III. Observations on microsomes, vacuoles, and on the process of yolk resorption. J. Exptl. Zool., *103:* 81.

[11] HONJIN R. (1961) Electron microscope and X-ray diffraction studies on the ultrastructure of yolk platelets. Proc. 17th Ann. Meet. Electr. Micr. Soc. Japan, *17:* 38.

[12] HONJIN R. (1966) Electron microscope and X-ray diffraction studies on the crystalline lattice structure of amphibian yolk platelets. Electron Microscopy, Proc. 6th Internat. Congr. Electr. Micr. 1966, Vol. II, Biology, pp. 133–134, Maruzen, Tokyo.

[13] HONJIN R. (1967) Polarization optic, electron microscopic and X-ray diffraction studies on the crystalline lattice structure of amphibian yolk platelets. J. Electr. Micr., *16:* 126.

[14] HONJIN R. (1968a) Higher-order crystalline structure of protein molecules in yolk platelets. J. Electr. Micr., *17:* 193.

[15] HONJIN R. (1968b) Crystalline lattice of yolk platelets. Technical Data, Hitachi, Tokyo, *6:* 1.

[16] HONJIN R. (1969) Again on the refinement of the values of the lattice parameters in the crystal structure of amphibian fresh yolk platelets by X-ray crystallography. Acta Anat. Nippon., *44:* 77.

[17] HONJIN R. and NAKAMURA T. (1961) X-ray diffraction and electron microscope studies on the crystaline lattice structure in the yolk granules. Report 1. Acta Anat. Nippon., *36:* 369.

[18] HONJIN R. and NAKAMURA T. (1969) A refinement of the values of the lattice parameters in crystal structure of amphibian fresh yolk platelets by X-ray crystallography. J. Ultrastruct. Res., *20:* 400.

[19] HONJIN R., NAKAMURA T. and SHIMASAKI S. (1965) X-ray diffraction and electron microscopic studies on the crystalline lattice structure of amphibian yolk platelets. J. Ultrastruct. Res., *12:* 404.

[20] HONJIN R. and SHIMASAKI S. (1962) X-ray diffraction and electron microscope studies on the crystalline lattice structure in the yolk granules. Report 2. On the influence of lipid solvents and protein denaturants on the molecular structure of the yolk granules. Acta Anat. Nippon., *37:* II-7.

[21] HONJIN R. and SHIMASAKI S. (1965) X-ray diffraction and electron microscope studies on the crystalline lattice structure in the yolk granules. Report 3. On the unit cell of the crystalline lattice. Acta Anat. Nippon., *40:* 52.

[22] HONJIN R. and TSUDA H. (1958) Electron microscopic studies on the amphibian larvae. Report I. On the ultrastructure of the cell components in the ectodermal area of *Rhacophorus schlegelii* and *Triturus pyrrhogaster*. Acta Anat. Nippon., *33:* 238.

[23] HULL A. W. and DAVEY W. P. (1921) Graphical determination of hexagonal and tetragonal crystal structures from X-ray data. Phys. Rev., *17:* 549.

[24] HUXLEY H. E. and ZUBAY G. (1961) Preferential staining of nucleic acid-containing structures for electron microscopy. J. Biophys. Biochem. Cytol., *11:* 273.

[25] ICHIKAWA M. (1931) On the development of the green frog. Mem. Coll. Sci. Univ. Kyoto, Ser. *B4:* 18.

[26] ITO Y., FUJII T. and YOSHIOKA R. (1963) On a phosphoprotein isolated from trout egg. J. Biochem. (Tokyo), *53:* 242.

[27] KARASAKI S. (1959) Electron microscopic studies on cytoplasmic structures of ectoderm cells of the *Triturus* embryo during the early phase of differentiation. Embryologia (Nagoya), *4:* 247.

[28] KARASAKI S. (1962) Ultrastructure of yolk platelets in the amphibian egg. Electron Microscopy, Proc. 5th Internat. Congr. Electr. Micr. 1962, Vol. 2, p. T-7, Academic Press, New York.

[29] KARASAKI S. (1963a) Studies on amphibian yolk 1. The ultrastructure of the yolk platelet. J. Cell Biol., *18:* 135.

[30] KARASAKI S. (1963b) Studies on amphibian yolk 5. Electron microscopic observctions on the utilization of yolk platelets during embryogenesis. J. Ultrastruct. Res., *9:* 225.

[31] KARASAKI S. and KOMODA T. (1958) Electron micrographs of a crystalline lattice structure in yolk platelets of the amphibian embryo (*Triturus pyrrhogaster*). Nature, *181:* 407.

[32] KEMP N. E. (1956) Electron microscopy of growing oocytes of *Rana pipiens*. J. Biophys. Biochem. Cytol., *2:* 281.

[33] LANZAVECCHIA G. (1960) L'origine des mitochondries pendant le développement embryonnaire de *Ranna esculenta* L. Elecron Microscopy, Proc. 4th Internat. Congr. Electr. Micr. 1958, Vol. 2, p. 270, Springer-Verlag, Berlin.

[34] LANZAVECCHIA G. (1965) Structure and demolition of yolk in *Rana esculenta* L. J. Ultrastruct. Res., *12:* 147.

[35] LEONARD R., DEAMER D. W. and ARMSTRONG P. (1972) Amphibian yolk platelet ultrastructure visualized by freeze-etching. J. Ultrastruct. Res., *40:* 1.

[36] MASSOVER W. H. (1970) Formation and fate of intramitochondrial yolk-crystals and hexagonal crystalloids in bullfrog oocytes. Microscopie Électronique, Septième Congr. Internat. Micr. Électr. 1970, Vol. III, Biologie, pp. 139–140.

[37] MASSOVER W. H. (1971) Nascent yolk platelets of anuran amphibian oocytes. J. Ultrastruct. Res., *37:* 574.

[38] OKADA Y. K. and ICHIKAWA M. (1947) Revised plates of the development of *Triturus pyrrhogaster* (Boie). Ann. Rept. Exptl. Morphol. (Tokyo), *3:* 1.

[39] RIGLE D. A. and GROSS P. R. (1962) Organization and composition of the amphibian yolk platelet. II. Investigations on yolk proteins. Biol. Bull., *122:* 281.

[40] SHIMASAKI S. (1965) X-ray diffraction and electron microscopic studies on the crystal structure of the yolk platelets of amphibian anura. J. Jûzen Med. Soc., *71:* 63.

[41] STOCKENIUS W. (1961) Electron microscopy of DNA molecules "stained" with heavy metal salts. J. Biophys. Biochem. Cytol., *11:* 297.

[42] SUGANO H. and WATANABE I. (1961) Isolations and some properties of native lipoproteins from egg yolk. J. Biochem. (Tokyo), *50:* 473.

[43] TSUDA H. (1961) Electron microscopic studies on the amphibian embryo, the ultrastructure of the normally developing neural and epidermal cell components in both *Triturus pyrrhogaster* (Boie) and *Rhacopholus schlegelii* var. *arborea*. Acta Anat. Nippon., *36:* 106.

[44] TSUDA H. and TAKAHASHI A. (1958) Electron microscopic studies on the amphibian larvae. Report II. On the ultrastructure of the yolk granules and the lipochondria. Acta Anat. Nippon., *33:* VI-7.

[45] WALD R. T. (1962) The origin of protein and fatty yolk in Rana pipiens. II. Electron microscopical and cytochemical observations of young and mature oocytes. J. Cell Biol., *14:* 309.

[46] WALLACE R. A. (1963a) Studies on amphibian yolk III. A resolution of yolk platelet components. Biochim. Biophys. Acta, *74:* 495.

[47] WALLACE R. A. (1963b) Studies on amphibian yolk IV. An analysis of the main-body component of yolk platelets. Biochim. Biophys. Acta, *74:* 505.

[48] Wallace R. A. and KARASAKI S. (1963) Studies on amphibian yolk. 2. The isolaion of yolk platelets from the eggs of *Rana pipiens*. J. Cell Biol., *18:* 153.

[49] WARTENBERG H. (1962) Electronen mikroskopische und histochemische Studien über die Oogenese der Amphibieneizelle (*Rana temporaria*). Z. Zellforsch., *58:* 427.

[50] WISCHNITZER S. (1957) The ultrastructure of yolk platelets of amphibian oocytes. J. Biophys. Biochem. Cytol., *3:* 1040.

[51] WISCHNITZER S. (1964) Ultrastructural changes in the cytoplasm of developing amphibian oocytes. J. Ultrastruct. Res., *10:* 14.

[52] ZOBEL D. R. and BEER M. (1961) Electron stains. I. Chemical studies on the interaction of DNA with uranyl salts. J. Biophys. Biochem. Cytol., *10:* 335.

Crystalline Structures Observed in Both the Exocrine and Endocrine Pancreatic Cells

Nakazo WATARI

Various kinds of crystalline structures have been observed in the pancreatic cells including both exocrine and endocrine elements with electron microscopy. Some of these crystalline structures may be functional elements for the cells or organs and others are not as significant while some are only disposable products of the cell.

In this paper the author will discuss such crystalline structures along with their origin, significance and fate.

Materials and Methods

Materials

Animals used. Pancreas from various normal vertebrates including the human, monkey (*Macacus fuscutus*), dog, mole (*Talpa wogura wogura*), guinea pig, rabbit, mouse ($C_{57}BL$ and ddY-strain), rat, bat (*Myotis lucifugus lucifugus*), dove, snake (*Elaphe climacophora*), carp (*Cyprinus carpio*) and the yellow tail (*Seriola quinqueradiata*) were used for this comparative study.

Experimental hibernation. Group I: Bats (*Myotis lucifugus lucifugus*) were kept in a cold room of about 10 °C with a high humidity of over 90% for up to 4 weeks and then sacrificed. Group II: Control bats kept at room temperature of about 20 °C were also sacrificed.

Administration of chemicals.

1) Alloxan administration: Dogs weighing about 2.7–6.6 kg were given 200 mg/kg BW of alloxan intravenously and sacrificed between 1 and 24 hours after injection.

2) Mice of the $C_{57}BL$-strain were given 50 mg/g BW of actinomycin D by an intramuscular injection and sacrificed between 1 and 24 hours after injection.

3) Mice of the ddY-strain were injected with high doses of vitamin A (200–400 IU/week) for a long period of between 1 and 18 months and then sacrificed at designated periods.

4) Mice (ddY-strain) were injected with an excess of vitamin D_2 (200,000 IU) twice a week for 3 months and then sacrificed.

5) Dehydroascorbic acid (DHA) administration: Dogs were injected intravenously with 700 mg/kg BW of DHA which is a diabetogenic agent (Merulini & Caramia 1965), and sacrificed at 3 to 5 hours after treatment (Watari 1968b; Watari 1970).

6) A female cat weighing 1.1 kg which was suffering from cystic fibrosis was sacrificed.

Methods

Pancreatic tissues from the above-mentioned animals were fixed either in a fixative

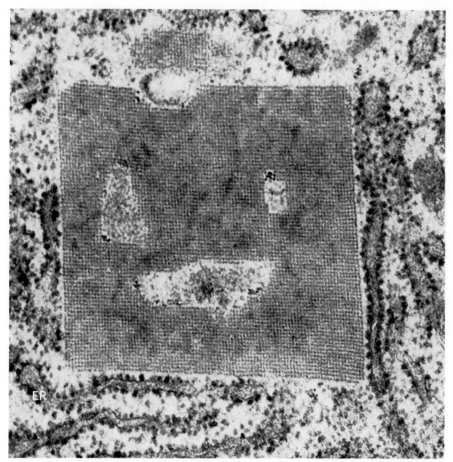

Fig. 1 A rectangular crystalline structure surrounded by rough-surfaced endoplasmic reticulum (*ER*) is observed in the pancreatic acinar cell of the mouse. The structure shows regularly arranged pattern of dotted texture. This structure has no limiting membrane. Fixed in a solution of glutaraldehyde and osmic acid. (×45,000)

containing 1% osmic acid buffered at pH 7.4 with veronal acetate buffer and made isotonic with 0.03 g (poikirothermal animals) or 0.045 g (homeothermal animals) sucrose/ml, or in a mixture of 2.5% glutaraldehyde and 2% osmic acid which was buffered at pH 7.4 and made isotonic as mentioned above (KUROSUMI 1970). Dehydration was performed through a series of ascending concentrations of alcohol. These samples were embedded in Epon 812 (LUFT 1961) after soaking in propylene oxide.

Thin sections for electron microscopy from the pancreatic blocks were cut with a Porter-Blum MT-2B ultramicrotome followed by adjacent sections cut at about 2 μm in thickness for correlative observations with light and electron microscopy (see WATARI et al. 1970).

Thin sections were doubly stained with uranyl acetate (WATSON 1958) and lead citrate (REYNOLDS 1957) and observed in HITACHI HU-11D and HU-12 electron microscopes.

Thick sections were stained with a further modification of a modified (MUNGER 1961) aldehyde thionine-hematoxylin-phloxin B staining method of GOMORI (1941). Light microscopic photographs taken from the stained thick sections were matched with the electron micrographs taken from an adjacent thin section. With this technique it was possible to identify each cell type, since differences in the ultrastructure of islet cells cor-

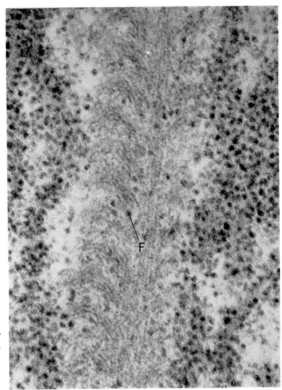

Fig. 2 Wavily arranged filamentous structures observed in a mouse pancreatic acinar cell following 3 months administration of vitamin A. The filaments *(F)* are probably elaborated by free ribosomes. Fixed in a solution of glutaraldehyde and osmic acid. (×78,000)

responded to those in the tinctorial characteristics of the light microscopic preparations (WATARI 1970; WATARI et al. 1970).

Some Epon-embedded thick sections of about 0.5–1.5 μm were doubly stained with uranyl acetate and lead citrate, and then observed with a Hitachi high voltage electron microscope (HU-1000) at 1,000 KV accelerating potential. Some electron micrographs were taken using tilting equipment in this instrument.

Observations and Discussion

Acinar cells

Cytoplasmic crystalline structures. The pancreatic acinar cells of the mouse commonly contain crystalline or crystalloid structures in the cytoplasm under normal conditions (BANNASCH 1966; SHIBATA 1967). This structure has no limiting membrane and the shape is usually rectangular (Fig. 1). High magnification electron microscopy reveals a line periodicity of about 110 A in these structures. In some cases, the lines are composed of small dots of about 50 A in diameter which are arranged about 60 A apart (Fig. 1). Under various experimental conditions, these structures are sometimes changed into fibrous structures, some of which are arranged in a whorled pattern containing a ribosome at the center of each whorl. These findings suggest that the filaments may be elaborated by the ribosomes (Fig. 2). With other experimental conditions such as an excess vitamin A administration in the mouse, some mitochondria (Fig. 3) are in close contact with the crystalline structure (WATARI et al. 1972). This may indicate that the

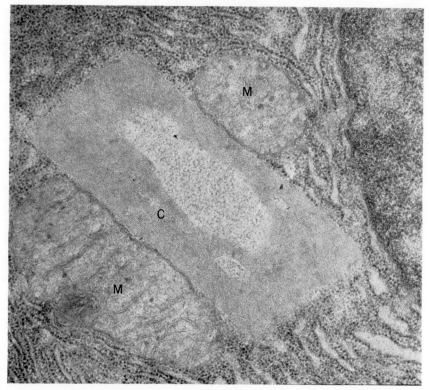

Fig. 3 A rectangular crystalline structure (C) in colse contact with mitochondria (M). Mouse exocrine pancreas 3 hours after the actinomycin D administration. Fixed only in osmic acid. (× 34,000)

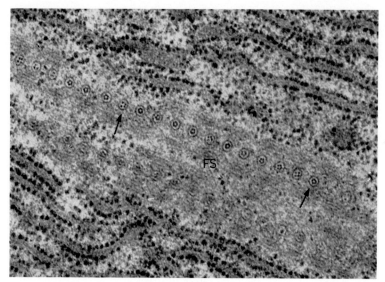

Fig. 4 Intracellular filamentous structure (FS) observed in an acinar cell of the mouse pancreas. Some round inclusions containing small particles or ringlets are regularly arranged in the structure (arrows). Twelve months administration of excessive amounts of vitamin A. Fixed in a solution of glutaraldehyde and osmic acid. (×50,000)

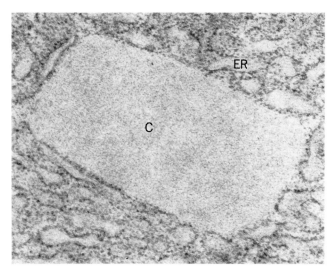

Fig. 5 A rectangular cytoplasmic area (*C*) filled with amorphous substance. Acinar cell of the mouse pancreas after an actinomycin D administration. *ER*; rough-surfaced endoplasmic reticulum. Fixed only in osmic acid. (×69,000)

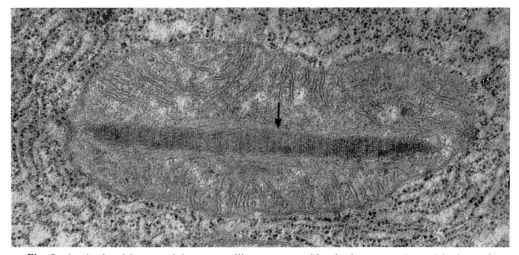

Fig. 6 A mitochondrion containing a crystalline structure with a lattice pattern (arrow) is observed in the pancreatic acinar cells of the dove. Fixed only in osmic acid. (×62,000)

mitochondria digest the contents of this structure to utilize them for energy source (FAWCETT 1966). Similar relationships between mitochondria and lipid droplets were also observed upon other experimental conditions with pancreatic acinar cells (JAMIESON & PALADE 1968; WATARI 1968a) and the hepatic cells (hibernating bat: WATARI, unpublished data). In a rare case of hypervitaminosis A, the filamentous structure rarely contained some small ring-shaped bodies (Fig. 4), which contained small particles or a ringlet in the center (WATARI et al. 1972).

In another study after actinomycin D administration, rectangular cytoplasmic areas filled with an amorphous substance were observed among the rough-surfaced endoplasmic reticulum. These areas could have been formerly filled with a crystalline substance, but were probably digested away before or during the specimen preparation, because their

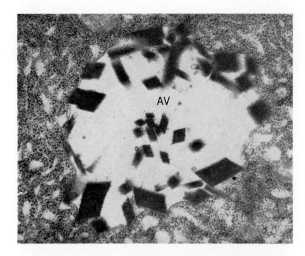

Fig. 7 A number of crystalline structures with rectangular or rhomboidal contours are seen in an autophagic vacuole (*AV*) observed in pancreatic acinar cells of a bat following 4 weeks hibernation. Fixed only in osmic acid. ($\times 49,000$)

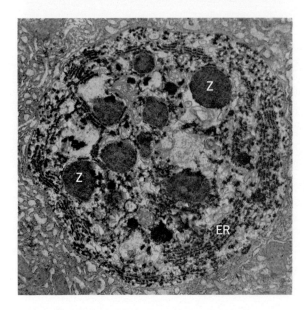

Fig. 8 A large autophagic vacuole containing degenerated cytoplasmic elements including rough-surfaced endoplasmic reticulum (*ER*), secretory granules (*Z*). Dog pancreatic acinar cell 24 hours after alloxan treatment. Fixed only in osmic acid. ($\times 14,000$)

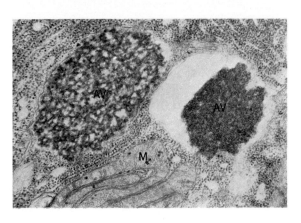

Fig. 9 Two autophagic vacuoles (*AV*) containing electron-dense amorphous, or reticular substance in a pancreatic acinar cell of a bat following 4 weeks of hibernation. *M*; distorted mitochondria. Fixed only in osmic acid. ($\times 12,000$)

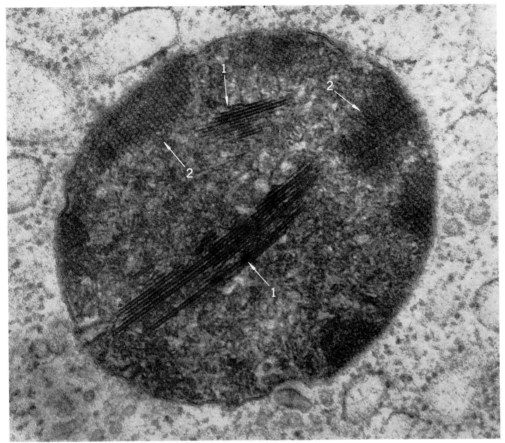

Fig. 10 A complex dense body observed in a pancreatic acinar cell of a dog 6 hours after alloxan treatment. The body contains tubular structures (arrow-1) and honeycomb structures (arrow-2). The latter may be a cross- or an oblique-cut view of the tubular structures. Fixed only in osmic acid. (×86,000)

outline was very similar to that of the crystalline structure observed previously (compare Fig. 5 with Fig. 3).

Mitochondria. In the pancreatic acinar cells of a dove some of the mitochondria contain crystalline inclusions in the matrix under normal condition (Fig. 6). The significance of this structure was not ascertained, although some investigators thought it to be an indication of a disturbed metabolism in the mitochondria (SPYCHER & RÜTTNER 1968).

Lysosomal crystalline inclusions. Other crystalline structures were observed in acinar cells of bats following 4 weeks of artificial hibernation. Some pancreatic acinar cells and rare centro-acinar cells contained electron-dense crystals having rectangular or rhomboidal shapes, which were usually surrounded by smooth limiting membranes (Fig. 7) (WATARI 1968a). This structure could have originated from digested substances in autophagic vacuoles, a number of which were observed in this condition. At the initial stage of this change, the autophagic vacuoles contained degenerated cytoplasmic elements including rough-surfaced endoplasmic reticulum, Golgi apparatus, mitochondria and the like, which were surrounded by a smooth limiting membrane (Fig. 8). During the second phase, these inclusions were digested by lysosomal hydrolases and transformed to amorphous or reticular substances (Fig. 9). They might be elementary substances such as proteins and lipids

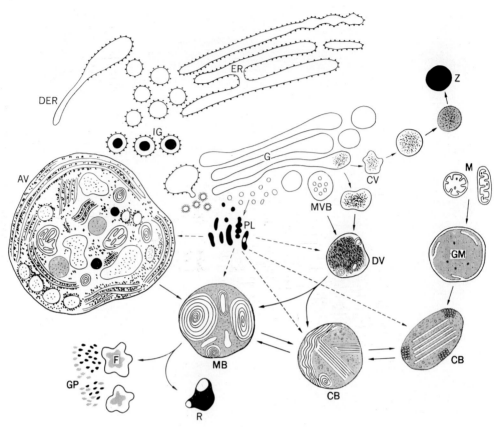

Fig. 11 The origin and fate of the complex dense bodies (*CB*) and myelin body (*MB*) are illustrated diagrammatically. The solid lines show the possible pathway of the bodies and the dotted lines show possible cooperations of the primary lysosomes (*PL*). *ER*; rough-surfaced endoplasmic reticulum, *DER*; degranulated ER, *IG*; intracisternal granules, *G*; Golgi apparatus, *MVB*; multivesicular body, *CV*; condensing vacuole, *DV*; dense vacuole, *AV*; autophagic vacuole, *Z*; zymogen granule, *M*; mitochondria, *GM*; huge degenerated mitochondrion, *RB*; residual body, *GP*; glycogen particles, *F*; fat droplets.

resulting from decomposition. Such substances could grow into crystalline structures (Fig. 7) or tubular structures (Fig. 10) and myelin figures. The majority of the substances comprising these structures might be utilized again by the cell itself. However, portions of them might be excreted to the outside of the cell by exocytosis as reported by WATARI (1973a). The origin, digesting sequence and the fate of the autophagic vacuoles are diagrammatically illustrated in Fig. 11.

Islet cells

Islet cell nucleus. Some islet cell nuclei contain filamentous structures under normal conditions (BOQUIST 1969). The filament bundle is composed of parallel-arranged fine subfilaments (Fig. 12). The significance of these structures has not been clarified.

Crystalline inclusions observed in the islet A-cell. A-cells of the bat pancreatic islets contain electron-dense crystalline structures under normal conditions (WATARI 1970). The shape of these structures is rectangular surrounded by a smooth limiting membrane (Fig. 13). This may suggest an extrastorage place for the secretory substances; however, in some cases, an electron-dense amorphous substance is contained inside of these structures,

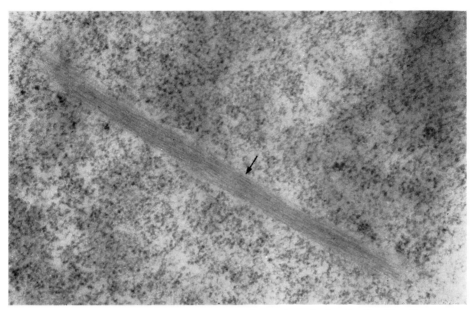

Fig. 12 Filamentous structure (arrow) observed in the nucleus of a bat B-cell. Fine filaments are arranged parallel to form a bundle. Fixed only in osmic acid. ($\times 38,000$)

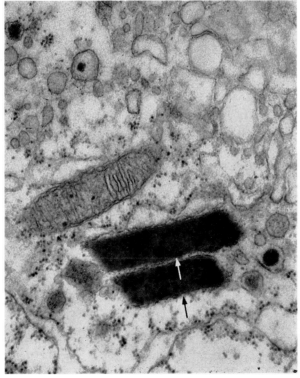

Fig. 13 Rectangular and rhomboidal crystalline structures (arrow) are seen in the cytoplasm of an A-cell in a bat pancreatic islet. The structures are surrounded by a smooth membrane. Fixed only in osmic acid. ($\times 36,000$)

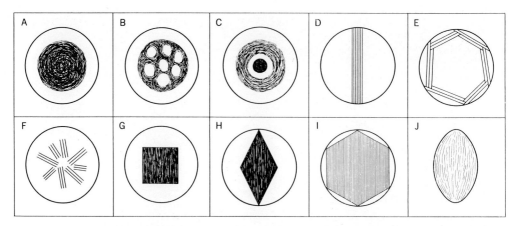

Fig. 14 Ten schematically illustrated typical types of B-cell granules. A: amorphous spherical type (rabbit, rat, mouse); B: reticular type (guinea pig); C: double-cored type (mouse following excess vitamin A and E administration); D: bar-shaped type (dog, bat, mole, mouse, monkey, yellow tail); E: peripherally arranged needle type (carp); F: centrally arranged needle type (chicken, dove, turkey); G: cuboidal type (human, yellow tail, snake); H: lozenge type (cat, mouse); I: hexagonal type (snake, yellow tail, mouse, rat, human); J: fibrillar-cored ellipsoidal type (one of teleosts).

which is similar to that of the core of the secretory granules. A similar crystalline structure was observed in the pancreatic A-cells of the tortoise (*Clemmys japonica*) (KANO 1961).

Crystalline secretory granules in the islet B-cell. As schematically illustrated by the author (1973b), the core of the B-cell granules are variable in shapes (Fig. 14). The majority of the cores reveal crystalline structures, especially when the tissues are fixed in a mixture of glutaraldehyde and osmic acid, which has been adjusted at pH 7.4 with veronal acetate buffer.

B-cell granules in the pancreas of the rodents are revealed to be less electron-dense and amorphous or reticular in shape, when the tissue is fixed singly with an osmic acid solution. After fixation in a mixture of glutaraldehyde and osmic acid, however, some of the granular cores are hexagonal or rod-shaped. B-cell granules of the dog (Fig. 15) (LACY 1957), and mole are mostly crystalline, but those of the monkey (Fig. 16), mouse (WATARI 1970) and yellow tail (Fig. 26a) are rarely bar- or rod-shaped (WATARI et al. 1970). Some granules contain one to several cores within a vacuole exhibiting an I-, V-, T- or N-like shape (Fig. 15). The cores of B-cell granules in the pancreas of birds including the chicken (HERMAN et al. 1964; SATO et al. 1966; MIKAMI & MUTOH 1971), turkey (SATO et al. 1966) and dove (Fig. 17) (WATARI 1970) show needle-like structures which are usually radially arranged forming a star shape (Fig. 17). In the carp, cores of B-cell granules also show needle-like structure, but the needles are peripherally arranged making a polygonal framework (Fig. 18) (WATARI 1971). A lozenge-formed crystalline core is frequently seen in each B-cell granule in the cat islets (BENCOSME & PEASE 1958; LEGG 1967) and rarely in the mouse islets (Fig. 19). With a cat suffering from cystic fibrosis (SUZUKI & WATARI, 1973), the majority of lozenge cores of the B-cell granules are less electron-dense than the peripheral portion of the granules, which contain electron-dense amorphous or small particulate substances (Fig. 20). The significance of this phenomenon is not clear, although with light microscopy the granules of such B-cells are less stained with aldehyde thionine, suggesting that the chemical nature of the granular content may have been altered.

Recently, BOQUIST and PATENT (1971) reported a very peculiar B-cell granule which was observed in the teleost, *Scorpaena scropha*. The majority of these granules revealed an ellipsoidal shape and contained fine filamentous elements (Fig. 14J).

CRYSTALLINE STRUCTURES IN THE PANCREAS

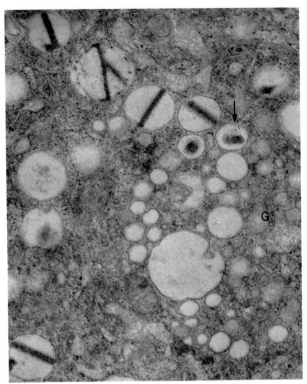

Fig. 15 Golgi area (G) of a B-cell in a dog pancreatic islet. Some secretory granules within the Golgi apparatus already contain bar-shaped core (arrow). Fixed only in osmic acid. (×25,000)

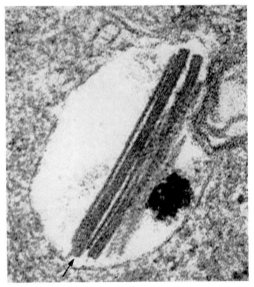

Fig. 16 Few of the secretory granules in B-cells of a monkey show the bar-shaped crystalline structure composed of parallel arranged needle-like substructures (arrow). Fixed in a solution of glutaraldehyde and osmic acid. (×124,000)

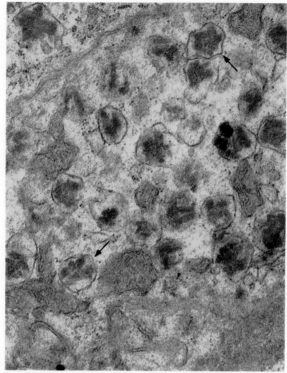

Fig. 17 B-cells of an islet of a dove containing secretory granules with needle-like core of star-like shape (arrows). Fixed only in osmic acid. (×27,000)

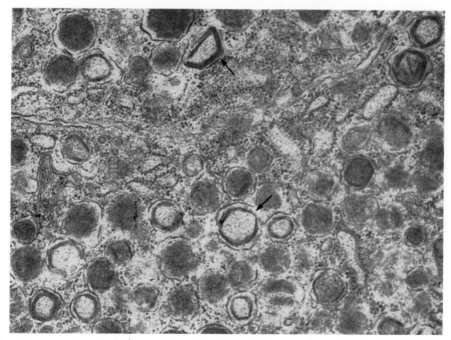

Fig. 18 B-cells of carp pancreatic islets. A number of the secretory granules contain bar-shaped cores which are arranged peripherally making a polygonal framework (arrows). Fixed only in osmic acid. (×36,000)

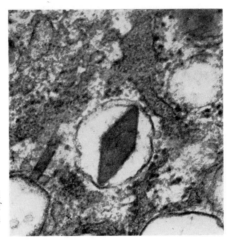

Fig. 19 B-cell granule with a lozenge core observed in a pancreatic islet of a mouse following 2 weeks of excess vitamin A administrations. Fixed in a solution of glutaraldehyde and osmic acid. ($\times 100{,}000$)

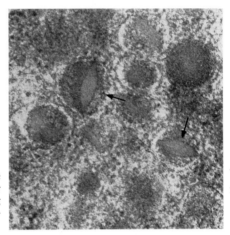

Fig. 20 B-cell granules with a less dense lozenge core (arrows) surrounded by electron-dense amorphous or particulate matrix. Pancreatic islet of a cat suffering from cystic fibrosis. Fixed in a solution of glutaraldehyde and osmic acid. ($\times 35{,}000$)

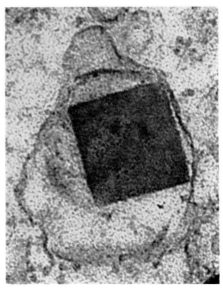

Fig. 21 A cuboidal secretory granule observed in an islet B_1-cell of a yellow tail pancreas. Fixed only in osmic acid. ($\times 165{,}000$)

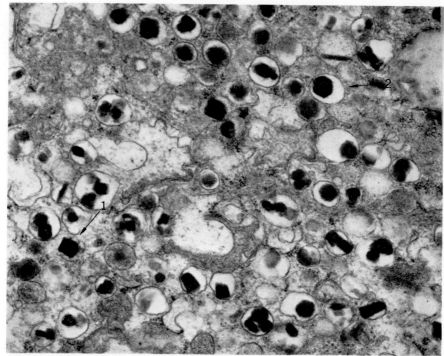

Fig. 22 B-cells of a human pancreatic islet containing multi-shaped crystalline granules. Some shapes are rectangular (arrow-1) or hexagonal (arrow-2), but others are irregular. 55-year-old man. Fixed in a solution of glutaraldehyde and osmic acid. (×25,000)

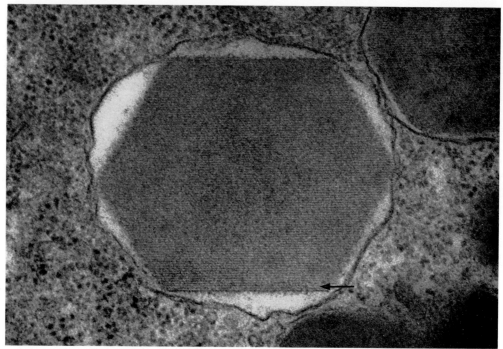

Fig. 23 A typical hexagonal crystalline granule observed in a B-cell of a snake pancreatic islet. The arrow indicates line periodicity. Fixed in a solution of glutaraldehyde and osmic acid. (×80,000)

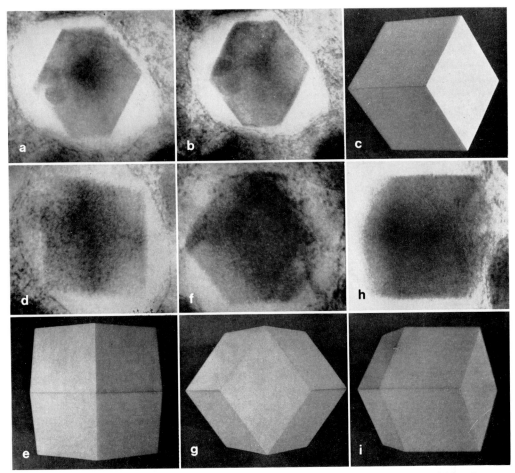

Fig. 24 a, b. Electron micrographs taken from a thick section of about 1.5 μm using a high voltage electron microscope with a 1,000 KV accelerating potential. The two pictures were taken of a section tilted at ±5°. Upon observing both pictures using a stereoscope, a three-dimensional shape can be observed. B-cell of a snake pancreatic islet. Fixed in a solution of glutaraldehyde and osmic acid. (×53,000) High voltage electron micrographs **d, f** and **h** were also taken from the same thick section of the snake pancreatic islet shown in Fig. 24a and b. Pictures **c, e, g** and **i** were taken of the cardboard model at four different angles. Comparison of these pictures to the electron micrographs indicated very similar structures respectively (c with a and b; e and d; g and f; i and h). (d; ×51,000, f; ×60,000, h; ×54,000)

In the yellow tail (Fig. 21) and snake, some B-cell granules contain rectangular cores. The majority of the B-cell granules of the lizard (*Iguana iguana*) (TITLBACH 1970) and human (SHIBASAKI & ITO 1969; GREIDER et al. 1970) pancreas contain rectangular core (Fig. 22). As reported by WATARI, TSUKAGOSHI and HONMA (1970), in the B-cells of the yellow tail (*Seriola quinqueradiata*), mouse and snake (*Elaphe climacophora*), some granules contain hexagonal crystalline cores (Figs. 23, 24a, b, d, f, h and 27). In the snake, after fixation in a mixture of glutaraldehyde and osmic acid, hexagonal granules reveal a substructure with a line pattern of about 90 Å periodicity (Fig. 23), and some of these exhibit dotted texture (WATARI, TSUKAGOSHI & HONMA 1970). In the yellow tail when fixed with osmic acid alone, at least two kinds of B-cells are distinguished, B_1- and B_2-cells (WATARI, TSUKAGOSHI & HONMA 1970). B_1-cells are characterized by an electron-dense crystalline core, and some of these are hexagonal in shape (Figs. 26b, 27). On the other hand, B_2-

cells are characterized by less electron-dense granules with a vague core, except for a few granules containing bar-shaped cores (Fig. 26a).

Recently GREIDER et al. (1969) demonstrated with electron microscopy a hexagonal crystalline core within the rat B-cell granules, when the granules were fractionated and negatively stained with phosphotungstic acid.

Analysis of the three-dimensional structure of crystalline B-cell granules. No one has reported thus far the three-dimensional shape of crystalline B-cell granules, although some investigators assumed that this was the reconstructed shape of B-cell granules from the variable cutting views which were observed by conventional electron microscopy. For example, TITLBACH (1970) supposed that the stereoscopic structure of the B-cell granules in the case of lizard might be an octahedron, while LANGE (1972, 1974) supposed it to be a rhombic dodecahedron in the case of the snake.

The author (WATARI 1974) observed sections of a snake pancreas more than one-micron thick using a high voltage electron microscope (Hitachi, HU-1000) with a 1,000 KV accelerating potential. In this method, some crystalline granules may be entirely embedded within the thick section, because the diameter of the granule is about 0.5–1.0 μm while the thickness of the section is more than 1 μm. Some electron micrographs were taken with a tilting apparatus set at + and −5°. When observing these paired pictures using stereoscopic glasses, a three-dimensional configuration of the crystalline granule is easily reconstructed (Fig. 24a, b).

On the other hand, some solid models such as pentagonal and rhombic dodecahedron and hexahedron were made with cardboard, and photographed from several different angles, after which they were compared with the electron micrographs of insulin granules, which were taken by high voltage electron microscope. Then, three-dimensional views of the rhombic dodecahedron model were matched with the four different shapes of the insulin granules (Fig. 24a–i). From these data, it is concluded that one of the stereoscopic shapes of the complete B-cell granule is a rhombic dodecahedron.

Table 1

	Mathematic ratio of cross-sectional shapes of rhombic dodecahedron (A)	Actual ratio of shapes observed in electron micrographs (B)
Hexagonal	30	30
Pentagonal	20	18
Rectangular	12	11
Triangular	5	0.5

Moreover, this result was also confirmed by the mathematic treatment as follows (WATARI et al. 1974): At first, a ratio of probable cross-sectional shapes of the rhombic dodecahedron was calculated mathematically; secondly, the ratio of the real numbers of the shapes of crystalline insulin granules was calculated as observed in the thin sections of the snake pancreatic islets. Both numbers (Table 1, A and B) are found to be almost similar, suggesting that one of the three-dimensional structures of the crystalline insulin granules might well be a rhombic dodecahedron as mentioned above. This rhombic dodecahedron shape was also demonstrated in the B-cells of some vertebrates including yellow tail, guinea pig, mouse, rat and human when observed with both the 1,000 KV and 100 KV electron microscopes applying the stereoscopic method (WATARI et al. 1974).

Another three-dimensional structure of the insulin granules might be a rhombic hexahedron, which was observed in the cases of human (Fig. 25), cat and mouse pancreatic islets under stereoscopic observations (WATARI et al. 1974).

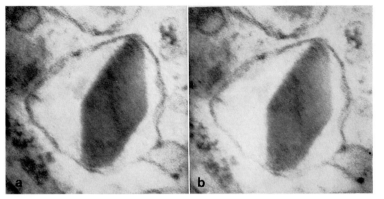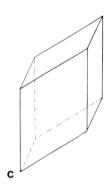

Fig. 25 **a, b.** The both paired electron micrographs of crystalline insulin granules were taken of the human pancreatic islet with a 125 KV conventional electron microscope at the tilted states of $\pm 10°$. Fixed with a solution of glutaraldehyde and osmic acid. (a and b; ×110,000) Under stereoscopic observations, they reveal a three-dimensional structure of the rhombic hexahedron. **c.** One side view of the rhombic hexahedron model. Compare this to a and b.

Formation, storage and extrusion of crystalline B-cell granules. It is known that the B-cell granule contains insulin. The insulin molecule is composed of 51 amino acids (polypeptide), and is elaborated in the cisternae of the rough-surfaced endoplasmic reticulum from the amino acids which are absorbed from the blood capillaries. Recently it has been shown, however, that ribosomes on the surface of the rough-surfaced endoplasmic reticulum initially elaborate proinsulin instead of insulin (STEINER et al. 1971). The place of the transformation of proinsulin into insulin has not been exactly ascertained, although it is thought to be in the Golgi apparatus or in the immature secretory granules (KANAZAWA 1973). The B-cell granules of the bat pancreatic islets following 4 weeks hibernation contain exclusively bar-shaped crystalloids. This suggests that the overproduced insulin may be stored as crystalline granules because the metabolic rate and the blood sugar level are very low and the insulin demand is quite limited during this period (WATARI 1968b).

In any event, the secretory substance (proinsulin) elaborated in the cisternae of the rough-surfaced endoplasmic reticulum is transferred to the Golgi apparatus via budding of the cisternae which are facing the paparatus (WATARI 1970). In the Golgi apparatus, the secretory substance is condensed and packed into secretory granules (Fig. 15). The Golgi apparatus works not only in the condensation of the secretory substance but also in the transformation from proinsulin to insulin.

The author assumes that the B_1-cells in the pancreatic islets of the yellow tail containing exclusively crystalline granules (Figs. 26b and 27) might be insulin-rich cells, and that B_2-cells primarily containing less dense amorphous granules (Fig. 26a) might be proinsulin-rich cells. However, further observations are needed.

At least three types of extrusion methods of B-cell granules have been reported. The first type is the eruptocrine secretion or emiocytosis (Type IV of KUROSUMI 1961) which has been suggested by many investigators (LACY 1962; BJÖRKMAN et al. 1963), but was first clearly demonstrated by the author and coworkers (WATARI et al. 1970) in a case of the yellow tail pancreas. The crystalline granules first move to the apical portion of the cell and make contact with the plasma membrane, and then a pore opens at the point of membrane contact, through which the secretory substance may be expelled. The granules sometimes are extruded sequentially to the outside of the cell in a bead-like formation (Fig. 27).

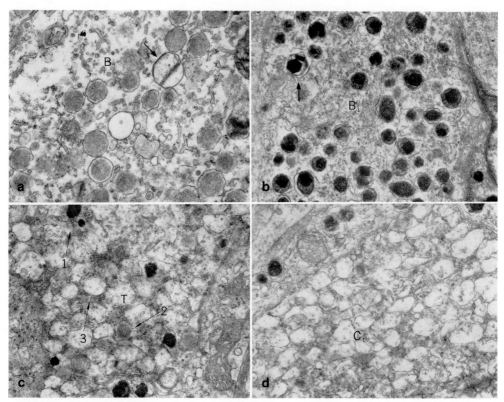

Fig. 26

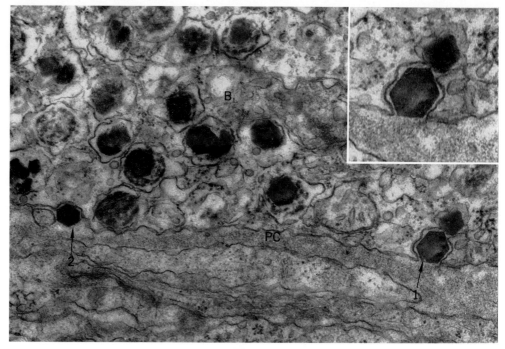

Fig. 27

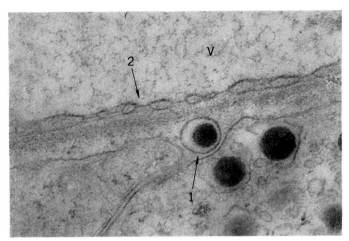

Fig. 28 Intact-secretion of B-cell granules in mouse pancreatic islet 12 hours after actinomycin D administration. Secretory granule with an intact limiting membrane is observed in the pericapillary space (arrow-1). An attenuated and fenestrated endothelial cell of a blood capillary (V) is seen (arrow-2). Fixed only in osmic acid. ($\times 46,000$)

The crystalline core of the granules sometimes is observed in an intact shape in the pericapillary space (Fig. 27). Shortly before the entrance of the insulin molecules into the capillary lumen, the core of the granules may be dissolved and thus separate the insulin molecules from zinc atoms.

The second secretion method is the diacrine secretion (Type V of KUROSUMI 1961). In dog pancreas after the administration of dehydroascorbic acid (DHA) (WATARI 1968b, 1970) which has a diabetogenic action similar to alloxan (MERULINI & CARAMIA 1965), most of the rod-like granule cores are dissolved and vacuolated, suggesting that the content might be excreted to the outside of the cell via the diacrine secretion. Another example of this is the B-cell of the yellow tail pancreas which was reported by WATARI et al. (1970) and WATARI (1973d). As touched upon above, the islets of the yellow tail pancreas contain two kinds of B-cells, B_1- and B_2-cells (Figs. 26a, b). However, they also have C_1-cells which contain only empty vacuoles with almost the same size as that of the B_1-granules (Fig. 26d). This may suggest that the B_1-cell might change into a C_1-cell after the cores of the B_1-granules are secreted via the diacrine secretion and converted into empty vacuoles. The yellow tail pancreas sometimes contains transitional types between B_1-, B_2- and C_1-cells, which contain simultaneously B_1- and B_2-granules and C_1-vacuoles (Fig. 26c). This also may indicate a possibility of the diacrine secretion. Some investigators also observed this diacrine mechanism under certain experimental conditions (rabbit: VOLK & LAZARUS 1963; rat: KOHAMA 1968; snake (*Natrix n. natrix*): CALGAREANU 1970; chicken: MIKAMI & MUTOH 1971).

Fig. 26 Pancreatic islet cells of the yellow tail. Fixed only in osmic acid. **a.** Secretory granules of the B_2-cell (B_2) characterized by less dense round cores, with a few exceptions that contain bar-shaped core (arrow). ($\times 18,000$) **b.** Typical B_1-cell (B_1) containing dense crystalline granules with variable shapes, some of which reveal hexagonal shapes (arrow). ($\times 14,000$) **c.** A transitional type cell (T) between B_1-, B_2- and C_1-cells containing B_1 (arrow-1), B_2 (arrow-2) granules and C_1 vacuoles (arrow-3). ($\times 16,000$) **d.** C_1-cell (C_1) containing vacuolated secretory granules. ($\times 16,000$)

Fig. 27 Hexagonal crystalline granules are being extruded to the pericapillary space (PC) via eruptocrine secretion (arrow-1 and 2). Inset is indicating the higher magnification of the part of arrow-1. ($\times 40,000$, inset; $\times 60,000$)

The third secretion method of B-cell granules is the intact secretion (WATARI 1973c) which was first demonstrated by HONJIN et al. (1965) in the case of insuloma. By this mechanism, the B-cell granules may be excreted to the outside of the cell without any deformation or loss of the granular core and its limiting membrane. Intact granules with a limiting membrane are sometimes observed in the capillary spaces and in the lumen of the blood capillaries. This secretion mode was also observed in the pancreatic islets of the mouse after an administration of actinomycin D (Fig. 28), or an excess of vitamin D_2. However, one must be very careful of the artefacts resulting from fixation, dehydration and embedding of the tissues.

Summary

The conventional and high voltage electron microscopes were used for the analyses of the crystalline structures which were observed in both exocrine and endocrine pancreatic cells under normal and experimental conditions.

Some crystalline structures such as those occurring within the acinar cells of the mouse as well as the B-cell granules are probably very important for cell function. However, some other crystalline inclusions, such as those in the acinar cells of the bat after artificial hibernation or chemical administration, may only be disposable substances. These are elaborated within the autophagic vacuoles due probably to the result of digestion by lysosomal enzymes.

Some crystalline B-cell granules in the islets of the snake, yellow tail, rat, mouse and human are hexagonal in shape when observed with the conventional electron microscope. Their three-dimensional structure might be that of a rhombic dodecahedron when observed stereoscopically with the high voltage electron microscope equipped with a tilting apparatus. Another three-dimensional structure of the B-cell granules might be a rhombic hexahedron as observed in the B-cells of the cat and human pancreatic islets.

The origin, significance and fate of these crystalline structures were discussed.

ACKNOWLEDGEMENTS

The author wishes to thank Professor Dr. RYOJI UYEDA and Associate Professor Dr. KAZUHIRO MIHAMA, Department of Applied Physics, Faculty of Engineering, Nagoya University and staff for the use of the 1,000 KV electron microscope.

REFERENCES

[1] BANNASCH P. (1966) Hülenlose Cytoplasmainclusionen und ihre Beziehung zur Sekretbildung im exokrinen Pankreas der Maus. J. Ultrastr. Res., *15:* 528–542.

[2] BENCOSME S. A. and PEASE D. C. (1958) Electron microscopy of the pancreatic islets. Endocrinol., *63:* 1–13.

[3] BJÖRKMAN N., HELLERSTRÖM C. S., HELLMAN B. and ROTHMAN U. (1963) Ultrastructure and enzyme histochemistry of the pancreatic islets in the horse. Z. Zellforsch., *59:* 535–554.

[4] BOQUIST L. (1969) Intranuclear rods in pancreatic islet β-cells. J. Cell Biol., *43:* 377–381.

[5] BOQUIST L. and PATENT G. (1971) The pancreatic islets of the teleost *Scorpaena scorpha*. An ultrastructural study with particular regard to fibrillar granules. Z. Zellforsch., *115:* 416–425.

[6] CALUGAREANU M. (1970) The behaviour of the endocrine cells in snake (*Natrix n. natrix*) pancreas, incubated "in vitro". The effect of glucose. Septimè Congr. Internat. de Micr. Électron., Grenoble, pp. 957–958.

[7] FAWCETT D.W. (1966) Mitochondria. *In:* An Atlas of Fine Structure. The Cell, Its Organelles and Inclusions. pp. 63–111, Saunders, Philadelphia & London.

[8] GOMORI G. (1941) Observations with differential stains on human islets of Langerhans. Amer. J. Path., *17:* 395–406.

[9] GREIDER M. H., BENCOSME S. A. and LECHAGO J. (1970) The human pancraetic islet cells and their tumors. I. Normal pancreatic islets. Lab. Invest., *22:* 344–354.

[10] GREIDER M. H., HOWELL S. L. and LACY P. E. (1969) Isolation and properties of secretory granules from rat islets of Langerhans. II. Ultrastructure of the beta granule. J. Cell Biol., *41:* 162–166.

[11] HERMAN L., SATAO T. and FITZGERALD P. J. (1964) The pancraes. *In:* Electron Microscopic Anatomy. KURTZ S. M. (ed.) Chapter 3, pp. 59–95, Academic Press, New York.

[12] HONJIN R., TAKAHASHI A, and MARUYAMA H. (1965) Electron microscopy of a case of insuloma. J. Electron Microsc., *14:* 183–188.

[13] JAMIESON J. D. and PALADE G. E. (1968) Intracellular transport of secretory proteins in the panceatic exocrine cell. IV. Metabolic requirments. J. Cell Biol., *39:* 589–603.

[14] KANAZAWA Y. (1973) Synthesis and secretion of insulin. Clin. Endocrinol., *21:* 701–707. (in Japanese)

[15] KANO K. (1961) Histologische, cytologische und elektronenmikroskopische Untersuchungen über die Langerhansschen Inseln der Schildkröte (*Clemmys japonica*). Arch. histol. jap., *22:* 123–180.

[16] KOBAYASHI K. (1966) Electron microscopic studies of the Langerhans islets in the toad pancreas. Arch. histol. jap., *26:* 439–482.

[17] KOHAMA M. (1968) Electron microscopic studies on the mechanism of insulin secretion from pancreatic β-cells. Med. J. Osaka Univ., *19:* 81–94.

[18] KUROSUMI K. (1961) Electron microscopic analysis of the secretion mechanism. Internat. Rev. Cytol., *11:* 1–124.

[19] KUROSUMI K. (1970) New fixation method using a mixture of glutaraldehyde and osmic acid. *In:* Textbook of Electron Microscopic Techniques, pp. 303–304, Seibundo-Shinko-Sha, Tokyo. (in Japanese)

[20] LACY P. E. (1957) Electron microscopic identification of different cell types in the islets of Langerhans of the guinea pig, rat, rabbit and dog. Anat. Rec., *128:* 255–267.

[21] LACY P. E. (1962) Electron microscopy of the islets of Langerhans. Diabetes, *11:* 509–513.

[22] LANGE R. H. (1974) Crystalline islet B-granules in the grass snake *Natrix natrix* (L.): Tilting experiments in the electron microscope. J. Ultrastr. Res., *46:* 301–307.

[23] LANGE R. H., BOSECK S. and ALI S. S. (1972) Kristallographische Interpretation der Feinstruktur der B-Granula in den Langerhansschen Inseln der Ringelnatter, *Natrix n. natrix* (L.). Z. Zellforsch., *131:* 559–570.

[24] LEGG P. G. (1967) The fine structure and innervation of the beta and delta cells in the islet of Langerhans of the cat. Z. Zellforsch., *80:* 307–321.

[25] LUFT J. H. (1961) Improvements in epoxy resin embedding methods. J. Biophys. Biochem. Cytol., *9:* 409–414.

[26] MERULINI D. and CARAMIA F. (1965) Effect of dehydroascorbic acid on the islets of Langerhans of the rat pancreas. J. Cell Biol., *26:* 245–261.

[27] MIKAMI S. and MUTOH K. (1971) Light- and electron-microscopic studies of the pancreatic islet cells in the chicken under normal and experimental conditions. Z. Zellforsch., *116:* 205–227.

[28] MUNGER B. L. (1961) Staining methods applicable to sections of osmium-fixed tissue for light microscopy. J. Biophys. Biochem. Cytol., *11:* 502–506.

[29] REYNOLDS E. S. (1957) The use of lead citrate at high pH as an electron opaque stain in electron microscopy. J. Cell Biol., *17:* 208–212.

[30] SATO T., HERMAN L. and FITZGERALD P. J. (1966) The comparative ultrastructure of the pancreatic islet of Langerhans. Gen. Comp. Endocrinol., *7:* 132–157.

[31] SHIBASAKI S. and ITO T. (1969) Electron microscopic study on the human pancreatic islets, Arch. histol. jap., *31:* 119–154.

[32] SHIBATA O. (1967) Crystalloids in the pancreatic acinar cells of the mouse. Exper. Cell Res., *47:* 655–657.

[33] SPYCHER M. A. and RÜTTNER J. R. (1968) Kristalloide Einschlüsse in menschlichen Lebermitochondrien. Virchows Arch. abt. B Zellpath., *1:* 211–221.

[34] STEINER D. F., CHO S., OYER P. E., TERRIS S., PETERSON J. D. and RUBENSTEIN A. H. (1971) Isolation and characterization of proinsulin C-peptide from bovine pancreas. J. Biol. Chem., *246:* 1365–1374.

[35] SUZUKI Y. and WATARI N. (1974) unpublished data.

[36] TITLBACH M. (1970) Licht- und elektronenmikroskopische Untersuchungen der Langerhansschen Inseln bei Grünlegeuan (*Iguana iguana*). Arch. histol. jap., *32:* 315–328.

[37] WATARI N. (1968a) Electron microscopy of the bat exocrine pancreas in hibernating and non-hibernating states, with specifil reference to the occurrence of intracisternal granules and crystalloids. Acta Anat. Nippon., *43:* 152–176.

[38] WATARI N. (1968b) Ultrastructure of the pancreatic islets under some experimental conditions, with special reference to the morphological changes of B-cells. Folia Endocrin. Jap., *44:* 721–727.

[39] WATARI N. (1970) Light and electron microscopy of the islets of Langerhans in the pancreas of some vertebrates, with special reference to the synthesis, storage and extrusion of the islet hormones. Gunma Symp. Endocrinol., *7:* 125–150.

[40] WATARI N. (1971) Ultrastructure and function of the islets of Langerhans. The Cell *3 (8):* 12–24. (in Japanese)

[41] WATARI N. (1973a) Ultrastructural alterations of the mouse liver after the prolonged administration of BHC. Japanese J. Clin. Electron Microsc., *5:* 1410–1420 & 1449–1456.

[42] WATARI N. (1973b) Islet cells as a member of the gastro-enteropancreatic endocrine system. Igaku-No-Ayumi, *85:* 421–433. (in Japanese)

[43] WATARI N. (1973c) Morphology related to the endocrine pancreas. Clin. Endocrinol., *21:* 663–677. (in Japanese)

[44] WATARI N. (1973d) The non-A and non-B cells of the endocrine pancreas. *In:* Gastro-Entero-Pancreatic Endocrine System. A Cell-Biological Aproach. FUJITA T. (ed.) pp. 71–89, Igaku-Shoin Ltd., Tokyo.

[45] WATARI N. (1974) Three-dimensional structure of crystalline insulin granules in B-cells of pancreatic islets. 8th Internat. Congr. on Electron Microsc., Canberra, 1974, Vol. II, pp. 434–435.

[46] WATARI N., HOTTA Y., SUZUKI Y., YURA J. and ESAKI R. (1974) Three-dimensional structures of crystalline insulin granules in situ under conventional electron microscopy. 6th Annual Meeting of Japanese Clin. Electron Microsc., 1974. *In:* J. Clin. Electron Microscopy, *7(3–4):* 271–272.

[47] WATARI N., TORIZAWA K., SAEKI S. and TAKADA K. (1972) Ultrastructural alterations of the mouse pancreas after the prolonged andover dose administration of vitamin A. J. Electron Microsc., *21:* 40–54.

[48] WATARI N., TSUKAGOSHI N. and HONMA Y. (1970) The correlative light and electron microscopy of the islets of Langerhans in some lower vertebrates. Arch. histol. jap., *31:* 371–392.

[49] WATSON M. L. (1958) Staining of tissue sections for electron microscopy with heavy metals. J. Biophys. Biochem. Cytol., *4:* 475–478.

II.

Electron Microscopic Histochemistry

1. Enzyme Histochemistry

Ultracytochemistry of Rat Hepatic Parenchymal Cells during the Mitotic Cycle

Takuma SAITO, Jeffrey P. CHANG and Kazuo OGAWA

Mitosis, one of the most fundamental functional phenomenon of the cell, provides for the growth and development of normal and pathological tissue cells. Although numerous reports on the investigations of the cell cycle at the light microscopic level have been made, only a few studies at the subcellular level utilizing electron microscopy have even been more rare (Chang & Saito 1969) despite the broad application of the ultracytochemical approach in the biological and medical sciences. The purpose of the investigation reported here was to examine the dynamics of peroxisomes, the Golgi apparatus, the endoplasmic reticulum and the nuclear envelope demonstrating catalase, thiamine pyrophosphatase, and glucose-6-phosphatase activities respectively during the cell cycle in a regenerating liver.

Materials and Methods

Male Sprague-Dawley rats weighing approximately 120 to 150 g were subjected to two thirds hepatectomies by the method of Higgins and Anderson (1931). All animals were maintained on a standard laboratory diet, Oriental NMF, and water *ad libitum* throughout the experiment. Hepatic tissue samples were excised and examined at 6, 12, 18, 24, 28, 30, and 32 hours after the operations. Mitosis appeared at from 28 to 30 hours after hepatectomy.

Fresh specimens were fixed in 2% glutaraldehyde in a 0.1 M cacodylate buffer at pH 7.2 (Sabatini et al. 1963) for 30 to 120 minutes at 0° to 4°C. After overnight storage in the same buffer, 40 μm frozen or nonfrozen sections were made. Specimens were stained for catalase (Novikoff & Goldfischer 1969), thiamine pyrophosphatase (Novikoff & Goldfischer 1961) or glucose-6-phosphatase (Wachstein & Meisel 1957) for the histochemical studies. Catalase was considered to be a marker for peroxisome, thiamine pyrophosphatase for Golgi apparatus, and glucose-6-phosphatase for endoplasmic reticulum and nuclear envelope. Distribution of these enzymes as followed throughout the various stages of the mitotic cycle.

After incubation, sections were postfixed in 1% osmium tetroxide (Caulfield 1957) for 60 minutes. Subsequently, the tissues were dehydrated and embedded in Epon (Luft 1961). Ultrathin sections were doubly stained with uranyl acetate (Watson 1958) and lead citrate (Reynolds 1963).

In order to select mitotic figures, 0.5 to 1.0 μm sections were stained with toluidine blue or methylene blue and Azur II and examined under the light microscope.

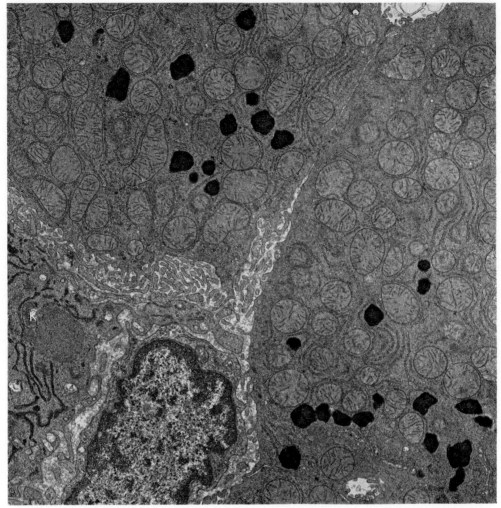

Fig. 1 Hepatic parenchymal cells of a normal rat. The catalase activity is seen in the peroxisome of two parenchymal cells. The endogenous peroxidase activity is seen in the cisternal space of the endoplasmic reticulum of the Kupffer cell (K) seen at the lower left. ($\times 11,000$)

Results and Discussion

Interphase

Observations made with light microscopy on the toluidine blue stained Epon sections indicated that the peroxisomes were scattered uniformly throughout the cytoplasm and were not concentrated in any particular zone of the hepatic lobule. Electron opaque reaction products due to the activity catalase were seen in the round and oval peroxisomes circumscribed by a single unit membrane (Fig. 1). In the normal hepatic parenchymal cells, the peroxisome was rather variable in size and shape, measuring 0.2 to 0.8 μm in diameter. In some instances, small DAB-stained granules of 0.1 to 0.2 μm in diameter were seen scattered among the elements of the endoplasmic reticulum (Fahimi 1969; Saito et al. 1973).

In the normal rat liver, the reaction products by the glucose-6-phosphatase activity were

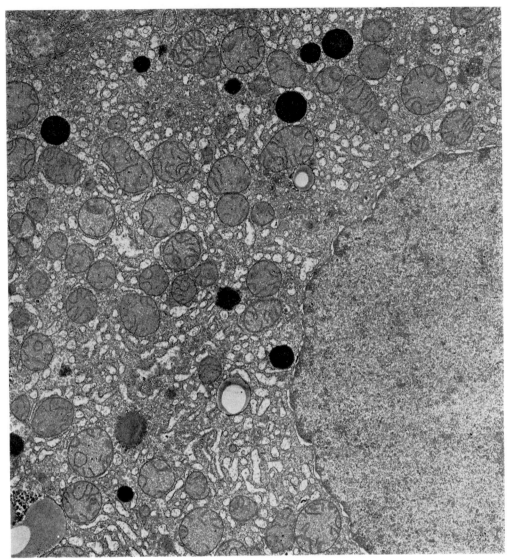

Fig. 2 The catalase activity in a hepatic parenchymal cell 24 hours after hepatectomy. The number of peroxisoms are markedly reduced with a pronounced disappearance of smaller ones. Abundant vesicular endoplasmic reticula are seen. (×13,000)

localized in the cisternal space of the endoplasmic reticulum and the nuclear envelopes. The glucose-6-phosphatase activity was positive only in the hepatic parenchymal cells. Endothelial cells, connective tissue cells, and epithelial cells of the bile duct were essentially devoid of reaction products (TICE & BARRNETT 1962; ORRENIUS & ERICSSON 1966; SAITO & OGAWA 1967).

The basic structural organization of the Golgi apparatus in normal liver cells consisted of 4 or 5 stacked cisternae, vacuoles in various sizes, and small vesicles which were 40 to 60 nm in diameter. Polarity in the organization of the Golgi apparatus has been frequently described. The localization of thiamine pyrophosphatase in 2 to 3 cisternae on the inner concave side demonstrated a gradient across the apparatus (NOVIKOFF et al. 1964).

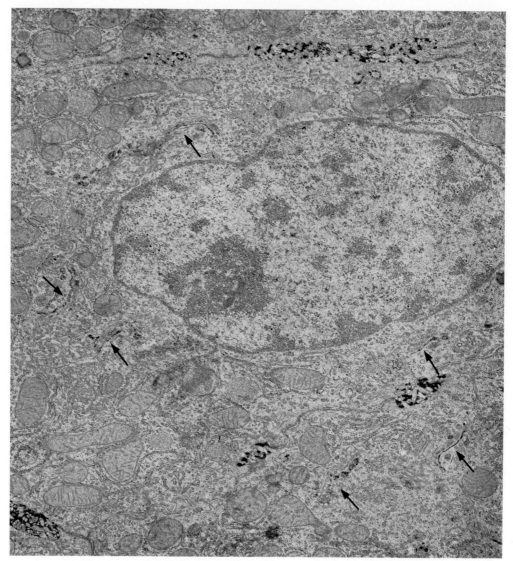

Fig. 3 The Golgi apparatus in early prophase of a hepatic parenchymal cell 30 hours after hepatectomy. Several, well-developed, Golgi apparatuses are seen. The thiamine pyrophosphatase activity is seen in the cisterna on the concave side of the Golgi apparatus (arrows) and on the plasma membrane of the bile canaliculi. ($\times 15,000$)

Soon after partial hepatectomy, nuclear and cytoplasmic changes occur in the liver cells. These changes are followed by the cell division (GRISHAM 1962). Numbers of peroxisomes at 6 hours after hepatectomy were slightly decreased; by 12 and 18 hours their numbers decreased moderately; and at 24 hours they were extremely low (Fig. 2). The smaller peroxisome, which has lesser electron opaque contents and shows weaker catalase activities than the larger peroxisome, were rarely observed in the regenerating hepatocyte, indicating that the numerical decreases in the peroxisome were mainly decreases in the smaller peroxisomes. The number of peroxisomes gradually began to recover at and from 28 hours, just before the regenerating liver cells entered into mitosis. Dilatation of the endoplasmic reticulum and the dissociation of ribosomes from the membrane of the rough endoplasmic reticulum were prominent features in a regenerating liver. Only a few scattered glycogen granules were observed in the endoplasmic reticulum.

In liver cells of a partially hepatectomized rat, the reaction products for the glucose-6-phosphatase activity in the cisterna of the endoplasmic reticulum and the space of the nuclear envelope were similar to those observed in the liver of a normal rat.

Golgi apparatuses in the regenerating liver cells were well developed in number and size, and it consisted of 4 to 5 flattened cisternae surrounded by small vesicles and dense granules. The thiamine pyrophosphatase activity was clearly demonstrated in 2 to 3 cisternae on the inner concave side of the Golgi apparatus.

Prophase

Prophase cells showed typical ultrastructural features as described by a number of investigators (ROBBINS & GONATAS 1964; CHANG & GIBLEY 1968; BRINKLEY & STUBBLEFIELD 1970; BAJER & MOLÉ-BAJER 1971; BRINKLEY & CARTWRIGHT 1971; NICKLAS 1971; BAJER & MOLÉ-BAJER 1972; CHANG et al. 1972). The appearance of spindle fibers, chromatin condensation at the nuclear membrane, nucleolar disintegration, break down of the nuclear membrane, and fragmentation of the nuclear envelope were observed.

The intracellular distribution of peroxisomes in the mitotic figures of a regenerating hepatic parenchymal cell was essentially similar to that observed in interphase cells at the same hours after hepatectomy.

In contrast to the interphase cell, the smooth endoplasmic reticulum was more prominently demonstrated by glucose-6-phosphatase activity in prophase cells.

The Golgi apparatuses maintained their normal organizational appearance during the early prophase (Fig. 3). Thiamine pyrophosphatase activity was clearly demonstrated in the cisternae on the concave side of the Golgi apparatuses. In late prophase, when the nuclear membrane disappeared and the chromosomes became more condensed, the Golgi apparatuses appeared smaller and had fewer cisternae. The parallel arrangement of the Golgi cisternae became disorganized. Thiamine pyrophosphatase activity was clearly evident in the cisternal spaces enable to positive identification of the Golgi apparatuses.

Metaphase

During this stage, chromosomes were aligned on the equatorial plane and mitotic spindles radiated from the poles. A fine structural localization of glucose-6-phosphatase activity in both the smooth and rough endoplasmic reticula was frequently seen in the mitotic figures (Fig. 4). Since descendants of the dissolved nuclear envelopes resembled endoplasmic reticulum and the glucose-6-phosphatase activity was positive in both organelles, it was difficult to determine their origin in metaphase cells.

The Golgi apparatus could be positively identified with reasonable certainty during metaphase. The size of the Golgi apparatus was much smaller than during interphase. It consisted of 1 to 2 flattened cisternae, about 0.5 to 0.2 μm long, associated with a few small vesicles (Fig. 5). Small cisternae without parallel arrangement were frequently observed. All of these small cisternae were thiamine pyrophosphatase positive providing a reasonable indication that they were descendant of the Golgi apparatuses.

Anaphase

During the early anaphase, the Golgi apparatus continuouslly decreased in size. The cisternae measured about 0.2 μm in length which was less than one tenth of that in the interphase sizes. Thiamine pyrophosphatase activity was consistently demonstrated in these small cisternae.

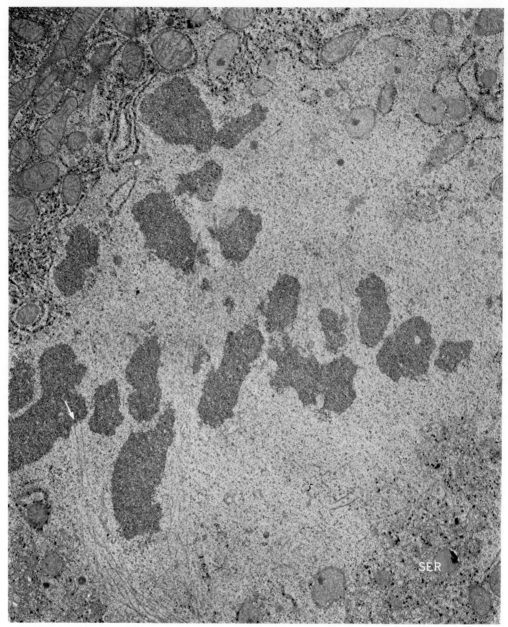

Fig. 4 A hepatic parenchymal cell in metaphase, showing the glucose-6-phosphatase activity in the cisternal space of the endoplasmic reticulum. Enzyme activity is also positive in the cisterna of the reticular smooth endoplasmic reticulum (SER). Spindle fiber microtubules are seen attached to a kinetochore (arrow). (×18,000)

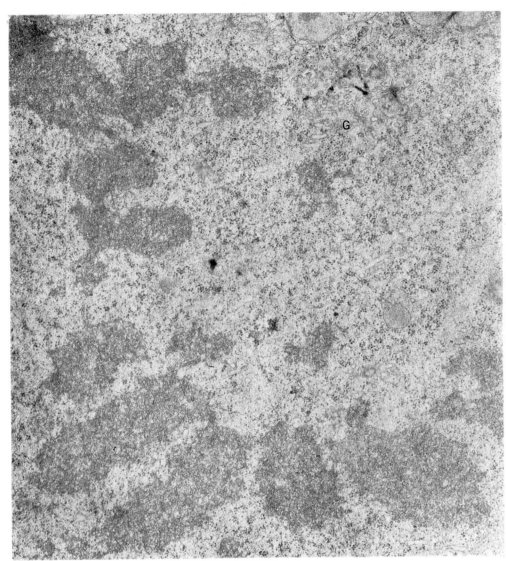

Fig. 5 The Golgi apparatus, in metaphase of a hepatic parenchymal cell 30 hours after partial hepatectomy, showing the chromosome arrangement on the equatorial plane. Thiamine pyrophosphatase activity is seen in the small (0.2 μm) cisternae of the Golgi apparatus (G). Small Golgi vesicles, 40 nm in diameter, are seen around the cisternae. (×30,000)

During the late anaphase, the chromosome was fused into a continuous mass; and in many areas, nuclear envelope reconstruction was about to occur through adhesion of the endoplasmic reticulum to the chromosomal surfaces (Fig. 6). The ribosomes were attached to the outer surfaces of the chromosome. Glucose-6-phosphatase activity was positive in the endoplasmic reticulum.

Telophase

The chromosomes condensed into highly electron opaque masses during the early telophase. Reconstruction of the nuclear envelopes by adhesion of pieces of endoplasmic

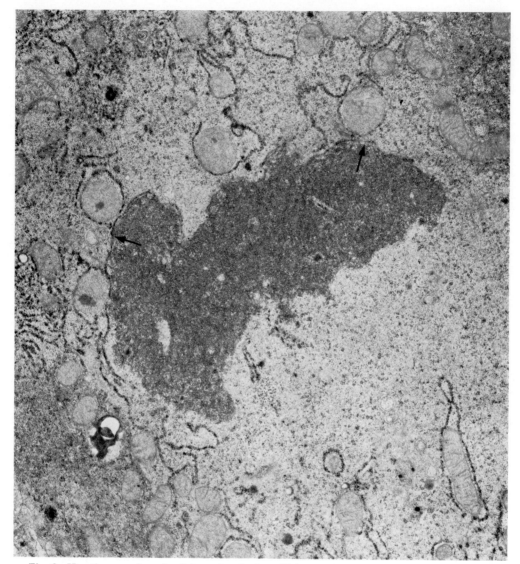

Fig. 6 Hepatic parenchymal cell in early telophase 30 hours after hepatectomy. The glucose-6-phosphatase activity is seen in the endoplasmic reticulum which is about to attach to the surface of the chromosomal mass, resulting in the formation of a nuclear envelope (arrows). ($\times 20,000$)

reticulum with positive glucose-6-phosphatase activity proceeded on the chromosomal masses (Fig. 7).

The Golgi apparatuses began to resume their normal complex forms as mitosis continued. Sizes and numbers of cisternae gradually became comparable to those seen during interphase (Fig. 8). Thiamine pyrophosphatase activity had become positive on the inner portion of the cisternae to indicate a polarity.

Since cell division is essential to normal development, and since the loss of its control is a characteristic of the neoplasm, the study of cell division in regenerating liver cells provides a good foundation for study of carcinogenesis.

Although ROUILLER and BERNHARD (1956) and others (DAVIS 1962; FRANKE & GOETZE

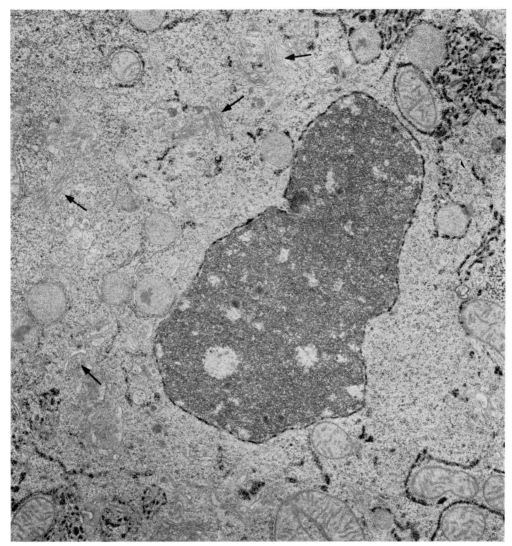

Fig. 7 Hepatic parenchymal cell in late telophase 30 hours after hepatectomy, showing the positive glucose-6-phosphatase activity in the cisterna of the endoplasmic reticulum and in the space of the nuclear envelope of the chromosome mass. Several of the Golgi apparatus (arrows) appear similar to those seen during interphase. ($\times 20{,}000$)

1966; STENGER & CONFER 1966; VIRÁGH & BARTÓK 1966) have reported numerical increases in peroxisomes in the regenerating liver cells, results of the present investigation clearly revealed their numerical decrease till approximately 24 hours after partial hepatectomy. This decrease is in accordance with the findings of TSUKADA et al. (1968) and DVOŘÁK and HORKY (1967). STEIN et al. (1951) reported a decrease in liver catalase activity during the initial period of hepatic regeneration. Sizes and complexities of microbodies have been reported to be proportional to the growth rate (DALTON 1964) or to the degree of differentiation (REUBER 1966). Ultrastructural studies of cells grown *in vitro* have revealed that microbodies are absent in Chang liver cells and in Novikoff hepatoma cells (HRUBAN et al. 1965). Similarly, very little or no catalase activity has

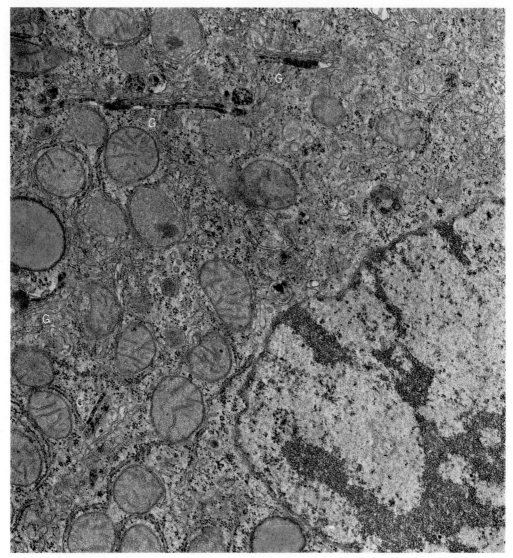

Fig. 8 Hepatic parenchymal cell in the late telophase. The Golgi apparatuses (G) have assumed a morphology similar to those seen during interphase. Thiamine pyrophosphatase activity can be seen as positive in 1 or 2 cisternae and in the vacuoles. ($\times 30,000$)

been observed in transplantable mouse hepatomas (Rechcigl et al. 1962; Ma & Webber 1966). Since the first peak in cell division appears after a reduction in peroxisomal number, this reduction is assumed to have some correlation with the early stage of the cell cycle.

Origin of the nuclear envelope has been discussed by Murray et al. (1963), Gondos and Bhiraleus (1970) and Szollosi et al. (1972), using stacked membranes as the morphological marker. By the time of the metaphase in rat hepatic cells, the paired nuclear envelope fragments separate into individual cisternae. Concequently, the nuclear envelope is indistinguishable from the other endoplasmic reticulum even with the aid of the glucose-6-phosphatase test. During the early telophase, formation of a new nuclear envelope seems to be brought about by a single endoplasmic reticulum with glucose-6-

phosphatase adhering to the surface of a chromosomal mass of the telophase nuclei, although one cannot rule out the possibility that this endoplasmic reticulum may have been derived from a previously paired endoplasmic reticulum.

One of the important findings in this investigation was the discovery of dynamic changes in the Golgi apparatus during a mitotic cycle. Evidently, there are diverse opinions among investigators on the fate of the Golgi apparatuses during mitosis. Diminution in their size and number during metaphase and anaphase has been demonstrated ultracytochemically with the aid of an enzyme marker. Similar results have been obtained from serial sections of melanoma cells showing progressive decreases in Golgi apparatuses up to the time of nuclear envelope formation during the early telophase (MAUL & BRINKLEY 1970). Recovery and organization of the normal structure occured in the late anaphase or early telophase.

The cellular site of synthesis, transport, storage, and secretion of pancreatic proteins have been investigated both by autoradiographic methods and cellular fractionation methods (SIEKEVITZ & PALADE 1960; WARSHAWSKY et al. 1963; JAMIESON & PALADE 1967, 1968). Newly synthesized proteins are transported from the endoplasmic reticulum to the Golgi apparatus, where they are accumulated, and thereafter, the labeled proteins are sent to the zymogen granules of the apical cytoplasm to be secreted.

Behavior of the fatty acids is also known to be localized and be transported through the Golgi apparatus as has been described in a series of papers by STEIN and STEIN (1967, 1968) utilizing electron microscopic autoradiography. The incorporation of labeled sugar into glycoprotein using tritiated sugars has also been shown (NEUTRA & LEBLOND 1966; WHUR et al. 1969) to be localized in the Golgi region and the nearby mucigen granules indicating that the Golgi apparatus is the site of complex carbohydrate synthesis.

Protein synthesis, which had been continued throughout the S and G_2 stages of the cell cycle at a high rate, had been reported to be abruptly decreased during mitosis by PRESCOTT and BENDER (1962). The abrupt interruption in RNA synthesis observed during mitosis (PRESCOTT & BENDER 1962; JOHNSTON & HOLLAND 1965; SALB & MARCUS 1965) was supposed to have resulted from the condensation of DNA in the chromosomes, making the DNA unavailable for transcription (JOHNSTON & HOLLAND 1965). Protein synthesis begins at the time of nuclear reconstruction from the late telophase to early G_1. RNA synthesis also commences about 30 minutes after nuclear reconstruction in a variety of cell types (SCHARFF & ROBBINS 1965; PFEIFFER & TOLMACH 1968).

The onset of regression in size and disorganization of the parallel arrangement of Golgi cisternae or lamellae from premetaphase to early telophase coincided with diminutions in protein and RNA syntheses occurring during the same phases. Furthermore, the resumption of macromolecular synthesis after nuclear envelope reconstitution is concomitant with the Golgi apparatus restoration taking place during the telophase. Therefore, the repression of Golgi apparatus may reasonably be assumed to be due to the passive result of the abrupt cessation of macromolecular synthesis which had passed through the Golgi apparatuses during interphase. In addition, FLICKINGER (1968, 1969) has reported a similar decline in the size of Golgi apparatuses of the *Amoeba proteus* after their enucleation and regeneration following nuclear transfer, a state comparable to the telophase in the mitotic cycle.

Summary

Changes in organelles during the cell cycle were observed in hepatic parenchymal cells after partial hepatectomy with the aid of ultracytochemistry. Three enzymes, catalase, thiamine pyrophosphatase, and glucose-6-phosphatase were used to mark (1) the peroxisomes, (2) the Golgi apparatuses, and (3) the endoplasmic reticulum and nuclear envelopes, respectively.

Just before the onset of the first cell division, the number of peroxisome markedly decreased. The endoplasmic reticulum serves to form nuclear envelopes for the daughter nuclei during early telophase. Great reductions in size and the distortions in shapes of the Golgi apparatuses coincide with the cessation of macromolecular synthesis known to take place through Golgi apparatuses.

REFERENCES

[1] BAJER A. and MOLÉ-BAJER J. (1971) Architecture and function of the mitotic spindle. *In:* Advances in Cell and Molecular Biology. DUPRAW E. J. (ed.) Vol. 1, pp. 213–266, Academic Press, New York.

[2] BAJER A. and MOLÉ-BAJER J. (1972) Spindle Dynamics and Chromosome Movements. Int. Rev. Cytol., Suppl. 3, BOURNE G. H. and DANIELLI S. F. (eds.), Academic Press, New York.

[3] BRINKLEY B. R. and CARTWRIGHT J., JR. (1971) Ultrastructural analysis of mitotic spindle elongation in mammalian cells *in vitro*. J. Cell Biol., *50:* 416.

[4] BRINKLEY B. R. and STUBBLEFIELD E. (1970) Ultrastructure and interaction of kinetochore and centriole in mitosis. *In:* Advances in Cell Biology. PRESCOTT D. M., GOLDSTEIN L. and MCCONKEY E. (eds.) Vol. 1, pp. 119–185, North-Holland, Amsterdam.

[5] CAULFIELD J. B. (1957) Effects of varying the vehicle for OsO_4 in tissue fixation. J. Biophys. Biochem. Cytol., *3:* 827.

[6] CHANG J. P. and GIBLEY C. W. (1968) Ultrastructure of tumor cells during mitosis. Cancer Res., *28:* 521.

[7] CHANG J. P. and SAITO T. (1969) Ultracytochemical study of mitosis in regenerating liver. J. Cell Biol., *43:* 19a.

[8] CHANG J. P., SAITO T., RAMSDAHL M. M. and RUSSELL W. O. (1972) Electron microscopy of human melanoma tissues and cells during mitosis. *In:* Pigmentation, Its Genesis and Biological Control. RILEY V. (ed.) pp. 51–70, Appleton Century Crofts, New York.

[9] DALTON A. J. (1964) An electron microscopical study of a series of chemically induced hepatomas. *In:* Cellular Control Mechanisms and Cancer. EMMELOT P. and MÜHLBOCK O. (eds.) pp. 211–225, Elsevier, Amsterdam.

[10] DAVIS J. M. G. (1962) Ultrastructure of rat liver cell cytoplasm during the process of regeneration after partial hepatectomy. Acta Radiol., *58:* 17.

[11] DVOŘÁK M. and HORKY D. (1967) Submikroscopische Struktur der Leberzelle nach Beeinflussung ihrer Sekretionstätigkeit. Z. Zellforsch., *76:* 486.

[12] FAHIMI H. D. (1969) Cytochemical localization of peroxidatic activity of catalase in rat hepatic microbodies (peroxisomes). J. Cell Biol., *43:* 275.

[13] FLICKINGER C. J. (1968) The effects of enucleation on the cytoplasmic membrane of *Amoeba proteus*. J. Cell Biol., *37:* 300.

[14] FLICKINGER C. J. (1969) The development of Golgi apparatus complexes and their dependence upon the nucleus in *Amebae*. J. Cell Biol., *43:* 250.

[15] FRANKE H. and GOETZE E. (1966) Electron microscopic studies on light and dark liver cells in regenerating rat liver. Acta Biol. Med. Ger., *17:* 99.

[16] GONDOS B. and BHIRALEUS P. (1970) Pronuclear relationship and association of maternal and paternal chromosomes in flushed rabbit ova. Z. Zellforsch., *111:* 149.

[17] GRISHAM J. W. (1962) A morphologic study of deoxyribonucleic acid synthesis and cell proliferation in regenerating rat liver. Autoradiography with thymidine-^{3}H. Cancer Res., *22:* 842.

[18] HIGGINS G. M. and ANDERSON R. M. (1931) Experimental pathology of the liver. I. Restoration of the liver of the white rat following partial surgical removal. Arch. Path., *12:* 186.

[19] Hruban Z., Swift H. and Rechcigl M. (1965)   Fine structure of transplantable hepatomas of the rat.   J. Nat. Cancer Inst., *35:* 459.

[20] Jamieson J. D. and Palade G. E. (1967)   Intracellular transport of secretory proteins in the pancreatic exocrine cell. II. Transport to condensing vacuoles and zymogen granules.   J. Cell Biol., *34:* 597.

[21] Jamieson J. D. and Palade G. E. (1968)   Intracellular transport of secretory proteins in the pancreatic exocrine cell. IV. Metabolic requirement.   J. Cell Biol., *39:* 589.

[22] Johnston J. C. and Holland J. J. (1965)   Ribonucleic acid and protein synthesis in mitotic HeLa cells.   J. Cell Biol., *27:* 565.

[23] Luft J. H. (1961)   Improvement in epoxy resin embedding methods.   Biophys. Biochem. Cytol., *9:* 409.

[24] Ma M. H. and Webber A. J. (1966)   Fine structure of liver tumors induced in the rat by 3′-methyl-4-dimethylaminoazobenzene.   Cancer Res., *26:* 935.

[25] Maul G. G. and Brinkley B. R. (1970)   The Golgi apparatus during mitosis in human melanoma cells *in vitro*.   Cancer Res., *30:* 2326.

[26] Murray R. B., Murray A. S. and Pizzo A. (1965)   The fine structure of mitosis in rat thymic lymphocytes.   J. Cell Biol., *26:* 601.

[27] Neutra M. and Leblond C. P. (1966)   Synthesis of the carbohydrate of mucous in the Golgi complex as shown by the electron microscopic autoradiography of goblet cells from rats injected with glucose-^{3}H.   J. Cell Biol., *30:* 137.

[28] Nicklas R. B. (1971)   Mitosis.   *In:* Advances in Cell Biology. Prescott D. M., Goldstein L. and McConkey E. (eds.) Vol. 2, pp. 225–297, Appleton Century Crofts, New York.

[29] Novikoff A. B. and Goldfischer S. (1961)   Nucleoside diphosphatase activity in the Golgi apparatus and its usefulness for cytological studies.   Proc. Nat. Acad. Sci., *47:* 802.

[30] Novikoff A. B., Essner E. and Quintana N. (1964)   Golgi apparatus and lysosomes.   Fed. Proc., *23:* 1010.

[31] Novikoff A. B. and Godfischer S. (1969)   Visualization of peroxisomes (microbodies) and mitochondria with diaminobenzidine.   J. Histochem. Cytochem., *17:* 675.

[32] Orrenius S. and Ericsson J. L. E. (1966)   On the relationship of liver glucose-6-phosphatase to the proliferation of endoplasmic reticulum in phenobarbital induction.   J. Cell Biol., *31:* 243.

[33] Pfeiffer S. E. and Tolmach L. J. (1968)   RNA synthesis in synchronously growing populations of HeLa S cells. I. Rate of total RNA synthesis and its relationship to RNA synthesis.   J. Cell Physiol., *71:* 77.

[34] Prescott D. M. and Bender M. A. (1962)   Synthesis of RNA and protein during mitosis in mammalian tissue culture cells.   Exp. Cell Res., *26:* 260.

[35] Rechcigl M. Jr., Price V. E. and Morris H. P. (1962)   Studies on the cachexia of tumor-bearing animals. II. Catalase activity in the tissues of hepatoma-bearing animals.   Cancer Res., *22:* 874.

[36] Reuber M. D. (1966)   Histopathology of transplantable hepatic carcinomas induced by chemical carcinogens in rats.   *In:* Biological and Biochemical Evaluation of Malignancy in Experimental Hepatomas. Yoshida T. (ed.) Gann Monograph 1, pp. 43–53, Japanese Cancer Association, Tokyo.

[37] Reynolds E. S. (1963)   The use of lead citrate at high pH as an electronopaque stain in electron microscopy.   J. Cell Biol., *17:* 208.

[38] Robbins E. and Gonatas N. K. (1964)   The ultrastructure of a mammalian cell during the mitotic cycle.   J. Cell Biol., *21:* 429.

[39] Rouiller C. and Bernhard W. (1956)   Microbodies and the problems of mitochondrial regeneration in liver cells.   J. Biophys. Biochem. Cytol., *2:* (Suppl.) 355.

[40] Sabatini D. D., Bensch K. and Barrnett R. J. (1963)   Cytochemistry and electron microscopy. The preservation of cellular ultrastructure and enzymatic activity by aldehyde fixation.   J. Cell Biol., *17:* 19.

[41] Saito T. and Ogawa K. (1967)   Ultracytochemical changes of the glucose-6-phosphatase (D-glucose-6-phosphate phosphohydrolase) activity in liver cells of the rat treated with phenobarbital.   Okajimas Folia Anat. Jap., *44:* 11.

[42] Saito T., Iwata K. and Ogawa K. (1973) Changes of peroxisomes in the regenerating hepatic parenchymal cells in the rat. Acta Histochem. Cytochem., *6:* 212.

[43] Slab J. and Marcus P. (1965) Translational inhibition in mitotic HeLa cells. Proc. Nat. Acad. Sci., *54:* 1353.

[44] Scharff M. W. and Robbins E. (1965) Synthesis of ribosomal RNA in synchronized HeLa cells. Nature, *208:* 464.

[45] Siekevitz P. and Palade G. E. (1960) A cytochemical study on the pancreas of the guinea pig. V. *In vivo* incorporation of leucine-C^{14} into the chymotrypsinogen of various cell fractions. J. Biophys. Biochem. Cytol., *7:* 619.

[46] Stein O. and Stein Y. (1967) Lipid synthesis, intercellular transport, storage and secretion. I. Electron microscopic radioautographic study of liver after injection of tritiated palmitate of glycerol infused and ethanol-treated rats. J. Cell Biol., *33:* 319.

[47] Stein O. and Stein Y. (1968) Lipid synthesis, intracellular transport and storage. III. Electron microscopic radiographic study of the rat heart perfused with tritiated oleic acid. J. Cell Biol., *36:* 63.

[48] Stein A. M., Skavinski E. R., Appleman D. and Shugarman P. M. (1951) Hepatectomy on liver catalase activity in normal and proteindepleted rats. Am. J. Physiol., *167:* 581.

[49] Stenger R. J. and Confer D. B. (1966) An electron microscopic study of the regeneration after subtotal hepatectomy. Exp. Mol. Path., *5:* 455.

[50] Szollosi D., Calarco P. G. and Donahue R. P. (1972) The nuclear envelope: Its breakdown and fate in mammalian oogonia and oocytes. Anat. Rec., *174:* 325.

[51] Tice L. W. and Barrnett R. J. (1962) The fine structural localization of glucose-6-phosphatase in rat liver. J. Histochem. Cytochem., *10:* 754.

[52] Virágh S. and Bartók I. (1966) An electron microscopic study of the regeneration of the following partial hepatectomy. Am. J. Path., *49:* 825.

[53] Wachstein M. and Meisel E. M. (1957) On the histochemical demonstration of glucose-6-phosphatase. J. Histochem. Cytochem., *4:* 592.

[54] Warshawsky H., Leblond C. P. and Droz B. (1968) Synthesis and migration of proteins in the cells of the exocrine pancreas as revealed by specific activity determinations from radioautographs. J. Cell Biol., *16:* 1.

[55] Watson M. L. (1958) Staining of tissue sections for electron microscopy with heavy metals. J. Biophys. Biochem. Cytol., *4:* 475.

[56] Whur P., Herscovies A. and Leblond P. C. (1969) Radioautographic visualization of the incorporation of galactose-^{3}H by rat thyroids *in vitro* in relation to the stages thyroglobulin synthesis. J. Cell Biol., *43:* 289.

2. AUTORADIOGRAPHY

Limits of Resolution in Electron Microscope Autoradiography

Vinci MIZUHIRA and Yutaka FUTAESAKU

Caro (1962), Caro and Schnös (1965), Pelc (1963), Granboulan (1963), Salpeter and Bachmann (1964), Bachmann and Salpeter (1965), Salpeter (1969), Salpeter and Bachmann (1969) published papers regarding resolution in electron microscope autoradiography (EM-AUT).

In their descriptions, there were two types of errors which were related to photographic process and geometric factors. These authors agreed that the minimum resolution of EM-AUT was estimated to be about 1,000 A for tritium (^{3}H) and 3,000 A for phosphorus (^{33}P). Recently, Salpeter (1969), Salpeter and Bachmann (1969) discussed resolution in EM-AUT using ^{3}H-polystyrene thin-films sandwiched between nonradioactive methacrylate and Epon-bases, covered with Kodak NTE and Ilford-L4 emulsions followed by various developing conditions.

They prepared histograms for both density (grains per unit area) and integrated distance (grains added consecutively) from the "hot" line (ultrathin layer of ^{3}H-polystyrene, 500 A in width). They calculated a value for the distance from the hot line within which 50% of the total grains fell and designated the distance as the "half distance" (HD). The value of the HD increased with factors which were expected to decrease the resolution, and they showed that the minimal HD was 800 A using 500 A thick sections covered with a monolayer of Kodak-NTE emulsion, developed with Dektol.

Recently, Salpeter and Salpeter (1971), Salpeter and Szabo (1972), Salpeter (1973), and Salpeter, Budd and Mattimoe (1974) have described an improved technique supported by theoretical background.

Salpeter (1973) found that the "self-absorption" to the β-ray exists in the osmium fixed and uranyl acetate block-stained sections which are coated with ^{3}H-polystyrene thin film.

The absorption increased by about 7% in contrast to that of non-stained Epon-sections. This resulted in approximately 10% increase electron microscope autoradiographic resolution.

Mizuhira and co-workers (Mizuhira & Kurotaki 1964; Mizuhira & Uchida 1966, 1969; Mizuhira, Uchida, Amakawa et al. 1968; Mizuhira, Uchida, Totsu et al. 1968; Mizuhira et al. 1970; Mizuhira 1971, 1972a, b, c, 1974; Mizuhira & Futaesaku 1971, 1972) on the other hand, investigated the theoretical problems of sensitization of silver haloid grains by β-rays using various kinds of nuclear sources and developmental conditions. A most remarkable result was obtained in experiments using a precipitate of labelled thiamine (B_1) with hydrogen platinum chloride (Pt) and developed under modified developing conditions with an Elon ascorbic acid (EAA) developer (17°C for 14 min), immediately after a brief treatment with 1% gold chloride solution (10 sec) for the Sakura NR-H2 or M2 emulsions.

Round, extremely finely developed silver grains ranging from 100 to 400 A in diameter were observed on the particles attached to the precipitate of ^{3}H-B_1 with Pt, and on ^{35}S-B_1, but they were never found without the precipitate (Figs. 1 and 2). In conventional EM-AUT, using, for example, ^{3}H-cholesterol (Fig. 5) or ^{3}H-pregnenolone (Fig. 4), we obtained the same high resolution as with B_1-Pt.

From these data, it was concluded that there were some unknown but important factors which improved the resolution of EM-AUT, which had not been discussed before, such as an absorption effect working against the β-rays in the sections. Therefore, we attempted to examine and compute the β-ray absorption ratio of the component elements of a section including the carbon film coating.

Experimental Methods

Model experiments for high resolution autoradiography

As we had previously described, a drop of collodion solution containing ^{3}H- or ^{35}S-labelled B_1-Pt precipitate was spread out on a water surface and a thin "hot" film was made. The radioactivity of ^{35}S-B_1·Pt-HCl was 65.32 mCi/mM, and of ^{3}H-B_1·Pt was 83.5 mCi/mM (MIZUHIRA & KUROTAKI 1964; MIZUHIRA, UCHIDA, AMAKAWA et al. 1968; MIZUHIRA, UCHIDA, TOTSU et al. 1968; SHIINA et al. 1969).

This thin "hot" film was picked up on a grid, thinly coated with carbon and then coated with a fine monolayer film of Sakura NR-H2 emulsion (with silver haloid grains of 700 to 900 A in diameter) for ^{3}H-B_1, and Sakura NR-M2 emulsion (grains 1500 A in diameter) for ^{35}S-B_1.

The grids were exposed in a dark-box and stored for four weeks in the cold.

Animal experiments for high resolution autoradiography

The absorption of ^{3}H-B_1 compound in the rat intestinal epithelium. We had previously reported the method in detail (MIZUHIRA, UCHIDA, AMAKAWA et al. 1968; MIZUHIRA, UCHIDA, TOTSU et al. 1968), thus it is being summarized here in a simplified diagram (Fig. 6).

The distribution of ^{3}H-pregnenolone and its derivatives (biosynthesis of testosterone in the interstitial cell). 10 μCi/g body weight of ^{3}H-pregnenolone was injected into the rat femoral vein, following a treatment with gonadotropic hormone (HCG) 70 IU/day/16 days, 2 weeks after X-ray irradiation of the testis at 1,000r.

Samples were excised at 15 and 30 min and treated as shown in Fig. 7. The samples were fixed with buffered osmium containing the saturated digitonine, a precipitant for the 3β-OH group (MIZUHIRA et al. 1970; MIZUHIRA 1972c).

The distribution of ^{3}H-cholesterol in the trophoblast cells of the rat placenta. ^{3}H-cholesterol was injected at the rate of 10 μCi/g body weight into the intraperitoneal cavity of a pregnant rat. Samples were extirpated within 1 to 2 hrs, fixed with 2% osmium tetroxide contained in a phosphate buffer solution (MIZUHIRA & UCHIDA 1969; MIZUHIRA et al. 1970; MIZUHIRA 1971, 1972a, b, c, 1974; MIZUHIRA & FUTAESAKU 1971, 1972; UCHIDA & MIZUHIRA 1970).

The demonstration of all latent images in a silver haloid crystal sensitized by a β-ray. The same sample was used as in the foregoing paragraph. A section was mounted with Sakura NR-H2 nuclear research emulsion, developed with EAA developer after a brief treatment with a gold chloride solution, at 17°C for 14 min. All procedures were done with care.

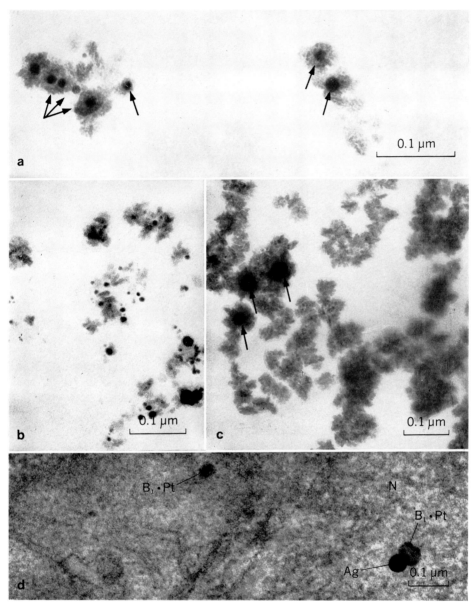

Fig. 1 A high resolution EM-AUT using ^{3}H-B$_1$·Pt and ^{35}S-B$_1$·Pt. **a.** Finely developed silver grains are located on the ^{3}H-B$_1$·Pt fine crystals in a thin collodion film (arrows). EAA, at 17°C for 14 min, Sakura NR-H2 emulsion. ($\times$260,000) **b, c.** The same kind of experient using ^{35}S-B$_1$·Pt. The results were the same as those obtained with ^{3}H-B$_1$·Pt. (b: $\times$184,000, c: $\times$144,000) **d.** The tritium-labeled thiamine was injected into the rat intestinal loop, and then samples were fixed with 2% osmium tetroxide containing saturated Pt. Thin sections were covered with diluted Sakura NR-H2. A few B$_1$·Pt crystals were scattered in the intestinal epithelial cell. In this picture, one of the crystals was associated with a finely developed silver grain in the nucleus. ($\times$132,000)

The treated section was observed under an electron microscope without removing the gelatin and staining with a lead solution (MIZUHIRA & UCHIDA 1966; MIZUHIRA 1972b; UCHIDA & MIZUHIRA 1970) (Figs. 3 and 21). These sections were stained with 2% uranyl acetate for 40 min before the emulsion-coating.

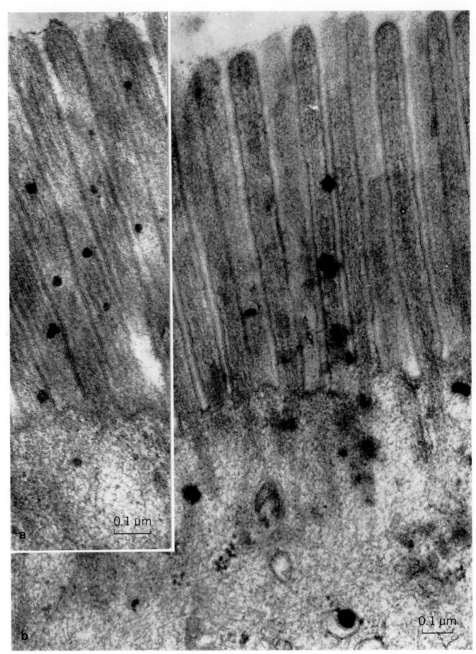

Fig. 2 a. A longitudinal section of the rat intestinal epithelium, with a fine precipitate of $B_1 \cdot Pt$ scattered in the microvilli and cytoplasm. This section was not coated with an emulsion. ($\times 124{,}000$)
b. The section was cut from the same block as Fig. 2a. After the emulsion was coated on a section, the grid was exposed for 4 weeks. After being developed with EAA (see the methods), many finely developed silver grains were scattered in the membranes and in the cytoplasm. They are located just on top of the labeled $B_1 \cdot Pt$ precipitate in the cell. This photograph shows one pathway of the thiamine passing through the intestinal epithelial cytoplasm from the microvilli to the capillaries. ($\times 100{,}000$)

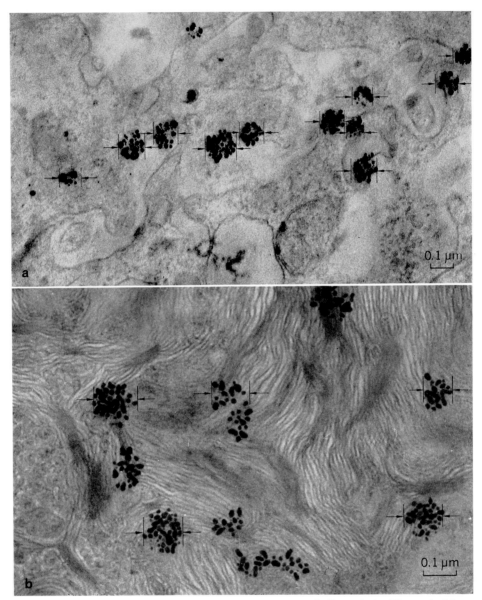

Fig. 3 a. This photograph shows a typical "cat's paw pattern" (foot-print). The method is explained in the text. Very tiny developed silver grains make a group on the fine structures. The average number of developed silver specks in a group was 10.5, and the average diameter of those group was 853 Å, which is the same as that of the original silver haloid crystals (Fig. 22). These finely developed silver specks originate from the section (see the text and Fig. 21). (×60,000) **b.** This figure also demonstrates the fine grain developed latent images by ^{3}H-uridine administrated to rice-insect tissue which contained a RNA-virus (Dr. Nasu). (×90,000)

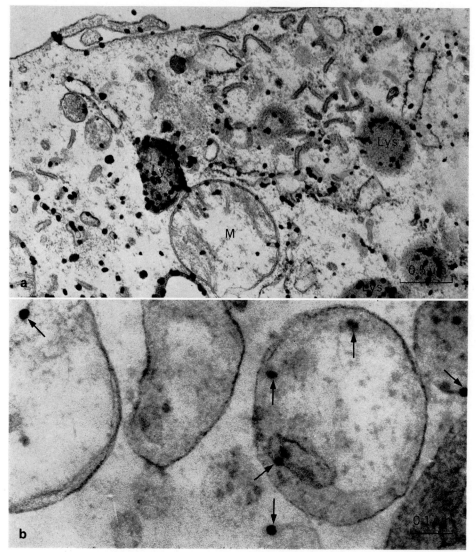

Fig. 4 a. This is an example of the extremely fine grain developed EM-AUT. ^{3}H-pregnenolone was injected into the rat peritoneal cavity. The testis was fixed with buffered 2% osmium tetroxide containing saturated digitonine. The 3β-OH group is fixed in the tissues and cells by the digitonine containing fixative (Mizuhira 1971, 1972a, b, c, 1974; Mizuhira et al. 1970; Mizuhira & Futaesaku 1972). A Sakura NR-H2 emulsion for 4 weeks at 0 °C, developed with EAA, at 17 °C for 14 min with the removal of gelatin and electron stained with Mizuhira-Kurotaki's solution (Table 2). Many finely developed silver grains occupy the membranous structures of the rat intestinal cell. Their location is very clear (see, Methods of ^{3}H-pregnenolone distribution; and Fig. 7). ($\times$36,000) **b.** This picture gives an example of a section of microsomal fractionation of the rat testis, which was administered ^{3}H-pregnenolone after being irradiated with 1,000 r of X-ray (see, Methods of ^{3}H-pregnenolone distribution; and Fig. 7). The sample was fixed in the same manner as mentioned above (Fig. 4a). The extremely finely developed silver grains (arrows) are located on the microsomal membranes.

($\times$140,000)

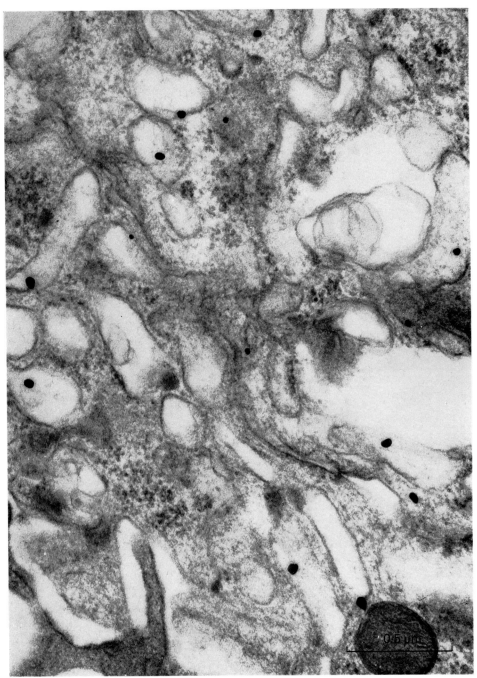

Fig. 5 This photograph shows a part of a syntitium cell of a rat placenta. The finely developed silver grains are located on the membraneous structures. 10 μCi/g body weight of ^{3}H-cholesterol was injected. After 2 hours the samples were excised and fixed with buffered 2% osmium tetroxide. They were developed with a modified EAA method. The scale indicates 0.5 μm. ($\times$65,000)

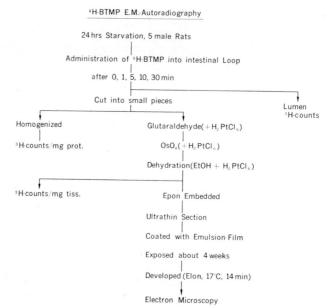

Fig. 6 Absorption of ^{3}H-B$_1$ compound in the rat intestinal epithelium.
BTMP: S-benzoylthiamine O-monophosphate, Sankyo Co. Ltd., Tokyo, ^{3}H-BTMP: 85.5 μCi/0.15 mg BTMP-^{3}H/0.3 ml, H$_2$PtCl$_1$: saturated "hydrogen platinum chloride", Emulsion: Sakura NR-H2, Elon: "Elon ascorbic acid" developer with gold-plating of the latent images before the developmental procedure (see the text, Methods of ^{3}H-B$_1$ compound absorption; and Developmental procedures for EAA).

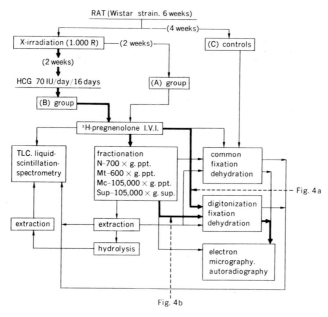

Fig. 7 This diagram shows the methods of an EM-autoradiographic study on the site of testosterone biosynthesis in the interstitial cell of a rat testis. Figs. 4a and 4b are shown as examples of the data (MIZUHIRA & UCHIDA 1969; MIZUHIRA 1972b).

Table 1 Extremely fine grain development with modified EAA.

---
1. Pretreatment with gold chloride* for about 10 seconds.
    ↓
2. Rinse for a few seconds with distilled water.
    ↓
3. Developed with EAA** at 17°C for 13 to 14 minutes.***
    ↓
4. Rinse for a few seconds with water.
    ↓
5. Fixed with a fixer (Kodak F-5) for 5 minutes.
    ↓
6. Rinse for 5 minutes with water.
    ↓
7. Removal of gelatin and electron staining with an alkaline lead solution (Mizuhira-Kurotaki's solution) (Table 2)
    ↓
8. Rinse with water.
    ↓
9. Carbon evaporation on a section. Observed under the electron microscpe.

* Pretreatment with gold chloride solution: Take 0.5 ml of 2% $AuCl_3 \cdot HCl \cdot 3H_2O$ solution, make a diluted 10 ml solution with distilled water. Add 0.125 g potassium thiocyanate and dissolve, then add 0.15 g potassium bromide, and dilute to make a total volume of 250 ml. This solution is very unstable, so the mixture must be made just before use.

** EAA (Elon ascorbic acid) developer: Elon (Metol) 0.045g, Ascorbic acid 0.3g, Borax 0.5g, Potassium bromide 0.1g. Make a 100 ml solution with distilled water. This EAA solution also unstable, so the solution (developer) must be used within 24 hrs after dilution.

*** This is a standard method for the Sakura NR-Hl, H2, M2 or M1 emulsions. The temperature and time for the development are adjustable according to the kind of emulsion or the lot number of the emulsion.

Table 2 Mizuhira-Kurotaki's solution (1964).

Dist. water	100 ml
NaOH	4 g
Potassium-sodium tartrate	1.6 g
Lead monoxide	4 g (approximately)

After the mixture boils, continue the boiling for 15 min, then cool with a Soda-lime container with a glass stopper. The mixture precipitates the remaining lead monoxide. The filtered mixture is kept in the refrigerator and can be used for more than half a year (as a stock solution). The stock solution is diluted about 30 to 50 times with distilled water, just before use. pH of the diluted solution is about 12. The removal of gelatin takes 20 to 30 min with a 50:1 dilution; and electron staining of conventional sections (not including EMAUT), takes 20 to 40 min with a 10:1 to 30:1 dilution.

Developmental procedures for EAA

After exposure, the grids were developed under the same conditions as the modified EAA method (Table 1). The grids were treated after development with a Mizuhira-Kurotaki's solution (MIZUHIRA & KUROTAKI 1964) (Table 2). This solution is a modified Karnovsky's lead solution and has two actions. One is the removal of gelatin from the section (acting as an alkali solution), and the other is an electron staining action (acting as a lead solution). After rinsing with water, the grids were coated with a thin carbon film and observed under the electron microscope. The biological sections were stained with 1–2% uranyl acetate for 20 to 40 min after being mounted on collodion stabilized copper grids. "Stabilized" indicates that the entire surface of a grid is coated with a thin collodion film without the grid holes. The sections adhere firmly to the grid and can resist the action of chemicals, including the developer and fixer. The method is very simple: clean grids are dipped in a 2% collodion solution, and then they are picked up on a filter paper with tweezers. They are then dried in an incubator for 10 to 20 min (Mizuhira & Kurotaki 1964; Mizuhira & Uchida 1966, 1969; Mizuhira, Uchida, Amakawa et al. 1968; Mizuhira, Uchida, Totsu et al. 1968; Mizuhira et al. 1970; Mizuhira 1971, 1972a, b, c, 1974; Mizuhira & Futaesaku 1971, 1972; Uchida & Mizuhira 1970).

The mounted grids with the section were coated with a thin carbon film less than 100 Å in thickness, and then rinsed well. They were mounted with a Sakura NR-H2 emulsion diluted about 14 times. The grids were processed for exposure and developed in the same manner as described above.

Results and Discussions

We attempted to examine and discuss the absorption of β-rays by the component elements of a section as the major problem of this experiment.

Interactions between β-rays and materials

Hypothesis 1-AH1. Once we assume that the energy of the β-ray is constant, the path length of the electron is shown as mg/cm^3 independent of the nature of each material. This is a well-known phenomenon.

Hypothesis 2-AH2. On the other hand, the absorbability of the tritium β-ray in water was measured by Robertson and Hughes (1955), as shown in Fig. 8 (Slack & Wag 1959; Tsuya et al. 1966).

Result 1-AR1. From the equation and the graph of AH2 (Fig. 8), we attempted to calculate the thickness of various materials needed to absorb 90% of the tritium β-ray, as shown in Fig. 9.

When we applied the vertical scale as the thickness (Å) and the horizontal scale as the averaged atomic mass of the material, each dot fell on a straight line. From this gradient, we could deduce an equation for calculating the thickness needed for a 90% absorption of the energy spectrum of the tritium β-ray.

$$\text{Thickness of material (Å)} = \frac{12,000^*}{\text{averaged atomic mass}} \quad (1)$$

Hypothesis 3-AH3. Fig. 10 shows the energy spectrum of the tritium β-ray (modified from Robertson & Hughes 1955; Slack & Wag 1959; Tsuya et al. 1966).

It is known that the maximum energy of the tritium β-ray is 18 KeV while the average value of its energy is 5.5 KeV. One can easily recognize the following facts from this Figure: 68% (σ^{**}) of the total β-ray radiation is included in 0 to 7 KeV of the energy spectrum, 90% is in the 0 to 11 KeV range, and 95% ($2\sigma^{**}$) is in the 12.5 KeV range.

Hypothesis 4-AH4. Morimoto (1962a) computed the value of the energy loss to the unit thickness of the silver bromide against the value of the electron energy from the approximate equation of Bethe-Block (Morimoto 1962a; Bethe 1933) (Fig. 11). This diagram shows that 10 KeV electrons will drop 6.5 KeV of their energy passing through

---

* 12,000 is a constant from Fig. 9.
** Sigma (σ) is a statistical standard deviation.

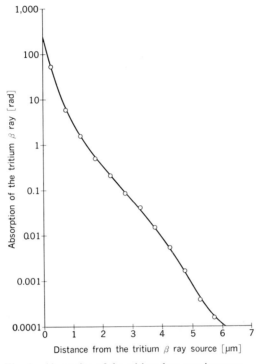

Fig. 8 Absorption of the tritium beta-ray in water.

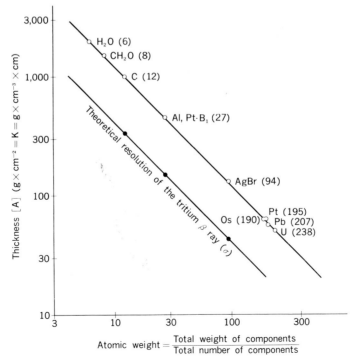

Fig. 9 90% absorption of the tritium beta-ray spectrum.

an unit micron of silver bromide, that is to say, the electron energy, which will lose 4 KeV passing through an unit micron of silver bromide, is 23 KeV.

Result 2-AR2. From these categories and equations (AR1, AH3 and AR2), one can deduce the absorption rate (%) of the tritium β-ray by the carbon film at various thicknesses (Fig. 12). For instance, from Fig. 10, one can easily see that 90% of the total tritium β-ray radiation is included between 0 and 11 KeV of the tritium β-ray energy spectrum. From Fig. 9, the thickness of the carbon film which absorbs 90% of the tritium β-ray is calculated as 1,000 A. On the other hand, the energy loss per unit thickness of silver bromide ((KeV/μm) e.l) of 11 KeV tritium β-ray is seen to be 6.2 from Fig. 11. Then we can postulate the following equation.

$$\frac{\text{KeV (^{3}H 90\%)}}{\text{thickness of carbon film }(\mu m)} \times C = \left(\frac{\text{KeV}}{\mu m}\right) \text{e.l}$$

Where, C is a proportional constant.

$$\therefore \quad C = \frac{6.2 \times 0.1}{11} = 0.05\dot{6}\dot{3}$$

From these results, we propose the following equation which will give the thickness of the carbon film.

$$\text{Thickness of the carbon film }(\mu m) = \frac{\text{KeV (^{3}H)}}{(\text{KeV}/\mu m)\text{ e.l}} \times 0.05\dot{6}\dot{3} \qquad (2)$$

Therefore, one obtains the thickness of the carbon film from the equation (2) and the reading (KeV/μm) e.l of the loss in energy value of the tritium from Fig. 11. From Fig. 10, the energy content of that value can be obtained for the total tritium.

Fig. 12, can be applied to other materials if we take the horizontal scale as C and we know the absorption rates (Ar) of the material from equation (3).

$$\text{Ar} = C(A) \times \frac{12^*}{\text{averaged value of atomic mass}} \qquad (3)$$

The converted constants in Fig. 12 show the atomic mass ratios of several elements.

Hypothesis 5-AH5. The 10 eV energy loss of a β-particle discharges one silver ion while that β-particle is passing through a silver haloid crystal (HIRATA 1967; GRANBOULAN 1963). It is also known that the formation of the latent image in a silver haloid crystal needs four reduced silver atoms as a minimum quantity.

Therefore, there must be a loss of 40 eV energy while the β-particle is passing through the silver haloid crystal as the minimum limit.

On the other hand, MIZUHIRA and his co-workers (MIZUHIRA & UCHIDA 1966, 1969; MIZUHIRA 1971, 1972a, b; MIZUHIRA & FUTAESAKU 1971; UCHIDA & MIZUHIRA 1970) reported previously that an average of 10.5 latent images were formed in a silver haloid crystal using an extremely fine grain development technique (Fig. 3a, b; 21, 22). From these experimental data, one may calculate that a β-particle has to lose 420 eV of energy to make the latent images in a silver haloid crystal. This is, however, the case only if we presume that all the lost energy of the β-particle is used effectively.

The exposure theories for silver haloid grains were developed in regards to the photons which passed through a silver haloid crystal during a 10^{-12} second interval. However, a β-particle can pass through in 10^{-13} second. Since a β-particle is ten times faster than a photon, one must consider the extinction of quite a few ion pairs which are formed by

---

* 12 is a proportional constant taken originally from the atomic mass of carbon.

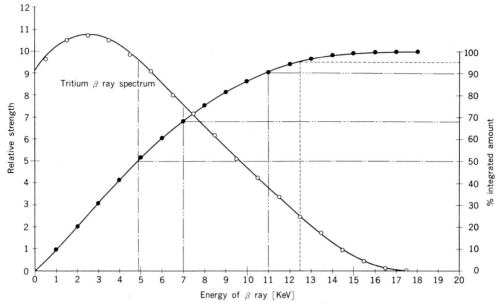

Fig. 10 Energy spectrum of the tritium beta-ray.

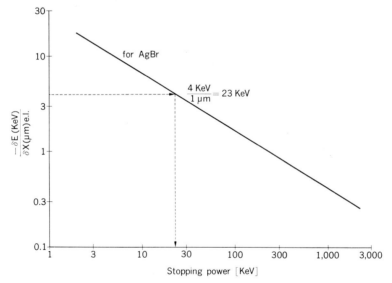

Fig. 11 Values of energy loss of electrons.

the β-particle when it hits a silver haloid crystal. Therefore, we can assume that the efficiency of latent image formation is less than 50%. Considering this factor, we assume that the energy loss is higher than 800 eV for a β-particle in order to develop the latensification of one silver haloid grain, and this value agrees with Morimoto's assumption (1,000 eV, 1962a).

Result 3-AR3. We can obtain Fig. 13 from Fig. 11 and the results of AH5 if we take the vertical scale as the path length of a silver haloid crystal and the horizontal scale as the energy loss value of the β-ray. Each line thus shows the calculations for 40, 200, 400,

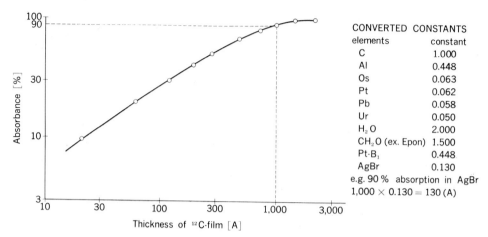

Fig. 12 Energy loss of the tritium beta-ray in carbon film.

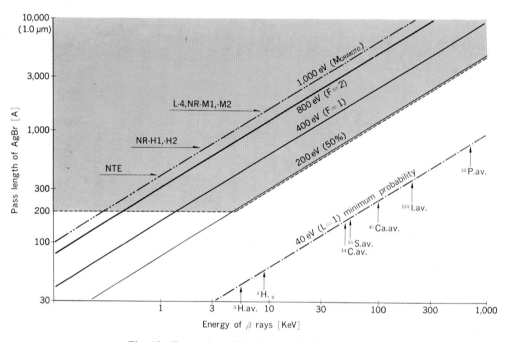

Fig. 13 Energy loss of the beta-ray in silver bromide.

800, and 1,000 eV respectively. This figure gives us the diameter of a silver haloid crystal which will be able to form an effective latent image at various β energies. For example, with electron microscopic autoradiography using the tritium β-ray theoretically, we can use 400 A silver haloid grains. But if we wish to achieve good efficiency, it is necessary to use 800 A grains. With ^{14}C or ^{35}S, we may use grain sizes of more than 1,500 or 3,000 A for the same reason as for tritium.

On the other hand, when ^{32}P is employed for EM-AUT, we cannot expect good efficiency.

The resolution and efficiency of the tritium β-ray in EM-AUT

Result 1-BR1. We now can discuss the geometrical relation between the β-ray source and a silver haloid grain as shown in Fig. 14. Here we take the horizontal axis X on the specimen surface and the vertical axis Y at right angles to X with the cross point of X and Y as 0. If we assume that the thickness of the specimen is 2 Bo and that the site of the β-ray source is P on the Y axis, then we obtain Bo=PO.

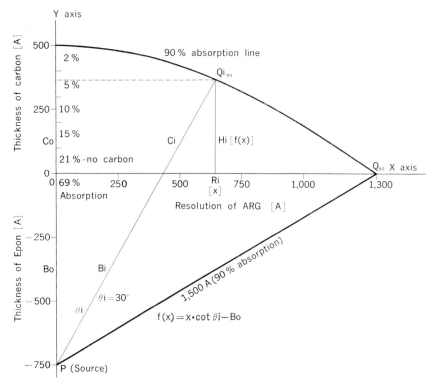

Fig. 14 Method of calculation for theoretical resolution.

Therefore, we assume that PQ_{90} is the distance at which 90% of the tritium β-ray is absorbed in the specimen. Furthermore, the β-ray is also absorbed by the carbon film layer which is coated on the specimen from P at the angles θi to the Y axis. While that β-ray particle passes through a distance Bi in the specimen and Ci in the coated carbon, it loses a part of its energy in Bi and Ci. If we decide the point Qi_{90} to be 90% of the sum of the energy loss of the tritium β-ray in Bi and Ci, then we can deduce equation (4).

$$\frac{Bi}{Ar} + Ci = K, \qquad Bi = \frac{Bo}{\cos \theta i}$$

$$\therefore \quad \frac{Bo}{Ar \times \cos \theta i} + Ci = K \tag{4}$$

Hence, K=1,000 A (which is the distance for absorbing 90% of the tritium β-ray in the carbon); Ar is the absorption rate from equation (3).
Therefore,

$$\therefore \quad Ci = 1,000 - \frac{Bo}{Ar \times \cos \theta i} \tag{5}$$

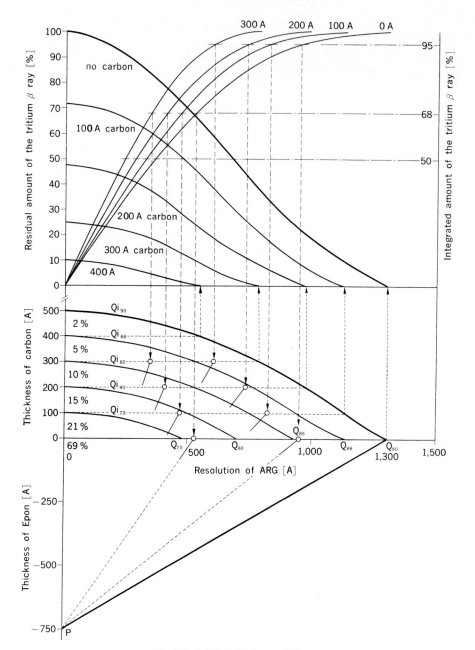

Fig. 15 1,500 A thickness of Epon.

If we express equation (5) on rectangular co-ordinates, Ri on the X axis is given by equation (6),

$$Ri = Bo \times \tan \theta i + Ci \times \sin \theta i \tag{6}$$

Hi on the Y axis is given by equation (7),

$$Hi = Ci \times \cos \theta i \tag{7}$$

The point Qi_{90} is expressed by Ri and Hi as Qi_{90} (Ri, Hi). The locus ($\overline{Q_{90} \, Qi_{90}}$) shows

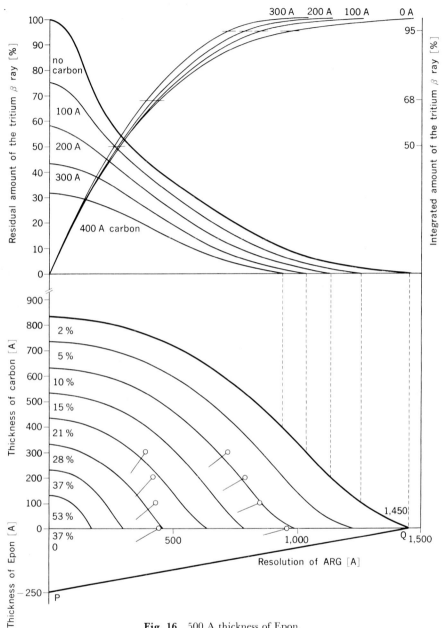

Fig. 16 500 A thickness of Epon.

a line of 90% tritium β-ray absorption. For convenience of the explanation on Fig. 14, it is shown as a 1,500 A Epon section.

When we next calculate equation (6) and (7) assuming K=900 A from equation (4), we obtain the locus $(\overline{Q_{88} \ Qi_{88}})$ of absorption during the tritium β-ray passage through a 900 A carbon layer and that absorption rate is deduced as 88% from Fig. 12.

If we continue to calculate equation (6) and (7), assuming K=800 A, 700 A, etc., these loci give curves at the right of the carbon thickness scale in Fig. 15.

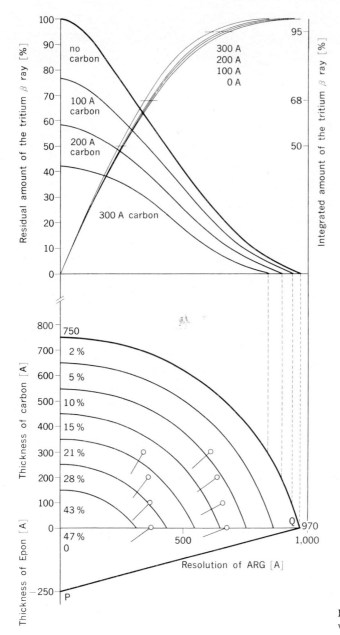

Fig. 17 500 A thickness of Epon with 1% Os and 1% U.

The curves on the left side of Fig. 15 are derived when we exchange the percentage to 100% of the residual amount of tritium β-ray which passes through the specimen and carbon layer of 0, 100, 200 A, etc., on the rectangular co-ordinates. Moreover we can express the integral values of the curves as percentages on the upper right side of Fig. 15.

From the upper left curve in Fig. 15, we can comprehend the distribution of the β-ray beyond the specimen surface or carbon surface, and the upper right curves give us the distribution of the tritium β-ray from the center 0 to each distance. For example, if we decide the resolution as a sigma (68%) without a carbon layer (C=0 A), it calculates to about 530 A, and in the case of 2 sigma (95%), it is about 950 A.

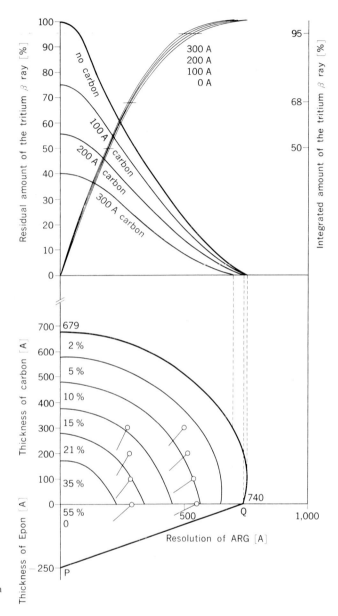

Fig. 18 500 A thickness of Epon with 2% Os and 2% U.

In the same way, we computed the data and made graphs for a 500 A thickness Epon, 1,000 and 500 A thicknesses of Epon with 1% osmium and 1% uranium, 778 and 500 A thicknesses of Epon with 2% osmium and 2% uranium and a 448 A thickness of labelled vitamin B_1 (thiamine) with hydrogen platinum chloride ($B_1 \cdot Pt$). Several examples of these diagrams and also the electron micrographs are shown in Figs. 1, 2, 3, 16–19, 21 and 22.

Result 2-BR2. Fig. 20 shows the relation between resolution and the nature of the specimens. This was obtained from Figs. 15, 16, 17, 18 and 19 and the same serial graphs which are shown in this report. These serial graphs played an important role in calculating the bases of Fig. 20 and Table 3. We used the horizontal scale as the thickness of the

Table 3 Theoretical resolution and efficiency.

Specimen	Thickness of carbon film			
	0A	100A	200A	300A
Epon-1,500A				
resolution σ (68%)	530	470	410	350A
efficiency	10.1	4.55	1.73	0.47%
resolution 2σ (95%)	950	830	740	590A
efficiency	21.8	10.4	4.41	1.13%
Epon-500A				
resolution σ (68%)	440	425	410	400A
efficiency	15.4	9.02	5.21	2.78%
resolution 2σ (95%)	950	880	800	730A
efficiency	22.4	15.2	9.84	5.79%
Epon-1,000A with 1% Os & 1% U				
resolution σ (68%)	350	330	310	270A
efficiency	6.71	3.61	1.22	0.385%
resolution 2σ (95%)	630	610	550	460A
efficiency	14.1	7.35	3.09	0.941%
Epon-500A with 1% Os & 1% U				
resolution σ (68%)	370	370	360	350A
efficiency	13.4	7.44	4.04	2.08%
resolution 2σ (95%)	770	760	750	730A
efficiency	19.8	12.7	7.84	4.47%
Epon-778A with 2% Os & 2% U				
resolution σ (68%)	260	250	240	230A
efficiency	4.95	2.22	0.804	0.293%
resolution 2σ (95%)	490	470	440	410A
efficiency	10.9	5.48	2.13	0.827%
Epon-500A with 2% Os & 2% U				
resolution σ (68%)	290	280	280	270A
efficiency	8.70	4.53	2.30	1.09%
resolution 2σ (95%)	550	540	530	500A
efficiency	14.7	9.32	5.53	2.81%
Pt-Bi-448A				
resolution σ (68%)	160	180	180	180A
efficiency	3.36	1.64	0.712	0.229%
resolution 2σ (95%)	300	330	340	310A
efficiency	6.72	4.00	1.97	0.597%

specimen absorbing 90% of the tritium β-ray and the vertical scale as the resolution from the calculation data.

Each point fell on a straight line through the 0 point. These results show that the resolution is inversely proportional to the "absorption section" of the specimen to the tritium β-ray irrespective of the thickness of the specimen and the carbon layer.

If we take the resolution as one sigma (σ, 68%), the gradient of this line is one third of the thickness of the specimen absorbing 90% of the tritium β-ray. If we take the resolution as two sigma (2σ, 95%), the line is two thirds. This is to say that we understand that the theoretical resolution of the tritium β-ray is one third of the vertical reading from Fig. 9 as one sigma and two thirds as two sigma. And we can express the resolution in the following equations, (8) and (9).

$$R_{\sigma} = \frac{12,000}{\text{averaged atomic mass}} \text{(equation-1)} \times \frac{1}{3} = \frac{4,000}{\text{averaged atomic mass}} \text{(A)} \qquad (8)$$

$$R_{2\sigma} = \frac{12,000}{\text{averaged atomic mass}} \times \frac{2}{3} = \frac{8,000}{\text{averaged atomic mass}} \text{(A)} \qquad (9)$$

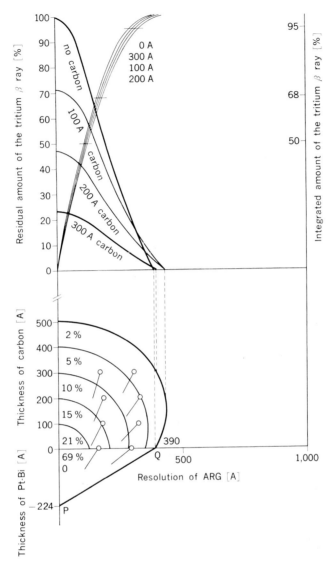

Fig. 19 448 A collodion film containing Pt-B₁ chrystal.

Result 3-BR3. The efficiency at that time is given by the equation (10).

$$\text{Efficiency (\%)} = \frac{1 - \sin \theta i}{2} \times \frac{\text{thickness of specimen (A)}}{500 \text{ (A)}}$$
$$\times \text{"the absorption rate of tritium } \beta\text{-ray in the specimen"}$$
$$\times \text{"residual rate of } \beta\text{-ray"} \tag{10}$$

Therefore, "Efficiency (Ef)" indicates the percentage of tritium β-ray, which is radiated from the point P and thus may make a latensification silver haloid grain within the theoretical resolution.

Thickness of the specimen (A)/500(A) indicates the factor of radiation value on the specimen surface which is moved by the thickness of the specimen, when we assume the concentration of the tritium β-ray to be distributed uniformly in the specimen.

$(1-\sin \theta i)/2$ indicates the rate of an angle including the resolution area toward a sphere. This (θi) was decided from Figs. 15–19 and their basic calculating data are not shown here. We can take an example from Fig. 15. The cross point, between the 95% horizontal line and the upper portion of the right curve which shows the integrated amount of the tritium β-ray, is moved to each surface of the specimen or carbon on the lower graph. Thus we can measure each angle toward P between O.

"The absorption rate of the tritium β-ray in the specimen" refers to the ratio of a tritium β-ray leaving the specimen surface to the total volume, because the absorption rate differs according to the nature of the specimen and according to the average depth of the tritium β-ray source P, while the β-ray passes through the specimen.

Finally, "the residual rate of the β-ray" refers to the absorption rate by the carbon layer of a tritium β-ray leaving the specimen surface.

The data from equation (10) are shown in Table 3.

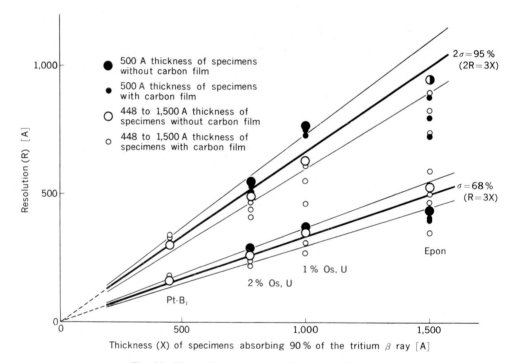

Fig. 20 Theoretical resolution of the tritium beta-ray.

A comparison between theoretical resolution and the experimental one

From our data, as shown in the serial figures (Figs. 9, 12, 15–20 and Table 3), it is very clear that "absorption efficiency" of a section of carbon film towards the β-ray is one of the most important factors in determining the resolution attained by EM-AUT, which never before has been discussed.

Result 1-CR1. Efficiency is generally decreased to less than one half when there is a 100 A thickness of the carbon film on a section, while the resolution is not necessarily twice as good without carbon. These considerations are shown in Fig. 20. It is certain that a thinner carbon layer on a section is necessary if one wishes to obtain higher efficiency and resolution with EM-AUT.

It has been said that higher resolution is obtained with a thinner section. This is true

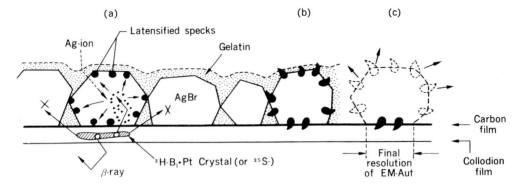

Fig. 21 High resolution EM-AUT with a collodion film containing ^{3}H- or ^{35}S-B$_1$·Pt. (See Figs. 1, 2, 3 and 22)
a. A silver bromide crystal is hit by a β-ray from just under the crystal. Ionized silver atoms transfer to the lattice holes which are locatedeat the surface of the silver bromide crystal, are stabilized there, making the latent images. According to our calculations, in this condition, the neighbouring silver bromide crystals would not be hit (Fig. 19). **b.** After the fine grain development with a modified EAA method (see the text, Methods of developmental procedures for EAA; and Fig. 7), the latent images are developed as extremely finely developed silver grains projecting their tips to the outside of the crystal surfaces. **c.** During fixation and gelatin removal, most of the finely developed silver grains are washed away with the exception of a few grains which are located in the basal face of the crystal. The remaining grains adhere to the carbon film on a section or to the section directly, as a high resolution developed silver grain on a section.

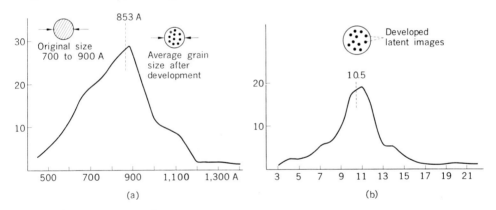

Fig. 22 The number of latent images in a hit silver bromide crystal. The average number of latent images (specks) in a silver haloid crystal was 10.5. The average diameter of the group of developed specks was 853 A, which coincides exactly with that of the original undeveloped haloid grain size. EAA development at 17 °C for 14 minutes, without removal of the gelatin after the developmental procedure. Sakura NR-H2 emulsion was applied. (See the text; and Figs. 3a, b, 21) **a.** The original and the average grain size after development. **b.** The average number of specks.

in the case of biological specimens embedded in pure Epon which has little absorbing (screening) effect; on the β-ray for example, if the section does not contain osmium and uranium and has no carbon film coating on it. But we can see that the resolution is not influenced by the thickness of the specimen when the screening effect is the same or more than that of carbon (Figs. 17–20 and Table 3).

When the thickness of the specimen was thicker than that which can absorb 90% of the tritium β-ray, the dose of radial rays of the tritium β-ray which left the specimen surface was constant and, had no relation with the thickness of the specimen. Therefore, the resolution obtained with EM-AUT was also fixed.

In this way, the sensitized and latensified silver haloid grain (latent images) can be developed by the developer after the β-ray hits the grain after passing through the specimen and the carbon film. As for 90% of the energy of the tritium β-ray, the ray enters only to a depth of 130 A in the silver bromide grain, if there has not been any absorption of the β-ray by the specimen and the carbon film (Fig. 12). For example, the average path length of the silver bromide crystals becomes 280 A, when we use grains 400 A in diameter (Figs. 13–20). It is assumed that the β-ray can not pass through this size grain, thus we need not worry about the possibility that the next silver bromide grain will be sensitized.

For this reason, the silver haloid grain is developed without fail, if the β-ray enters the silver bromide grain after passing through the specimen. Thus, in the case of a tritium β-ray, it is sufficient to consider the limited area where the β-ray enters the silver bromide grain and its size. That area is already calculated as the area of theoretical resolution (Fig. 9 and equation 8), and one may recognize the fact that the area depends upon the atomic composition of the specimen and that it is of a very limited area.

Result 2-CR2. The suitable size of silver bromide grains for the tritium β-ray is 400 to 800 A in diameter. Emulsions containing grain sizes of 400 to 1500 A are available on the commercial market.

We must next determine how the grain size of silver bromide influences the resolution of EM-AUT. According to the general sensitization theory, it is presumed that when approximately a dozen photons hit a silver bromide grain, numerous silver ions are excited around the struck photons, which gradually transfer to the lattice holes of the surface of a silver bromide crystal and are stabilized there as the nuclei of latent images. If the stabilization of the latent images by the tritium β-ray takes place in the same way as with photons, one β-particle enters a silver bromide crystal and stops about 100 A from the basal surface of the crystal and loses all its energy there. Since the form of a silver bromide crystal is a polyhedron, we may assume the size of the base plane of the crystal to be 200 A. Therefore it is certain that the first silver ionization occurs at a point where the β-particle enters the crystal on that surface or at a depth of 100 A from the base plane of the crystal.

Thereafter ionized silver atoms probably are transferred to the lattice holes near the vertex or side surfaces of the crystal and are stabilized as latent images (latensified specks).

But we wonder if the base plane of the silver haloid crystal which is latensified, protrudes beyond the theoretical resolution limit. The protruded area from our calculation, however, is 100 A at the maximum and 50 A on the average.

In the other instance, which is when the tritium β-particle enters the silver bromide grain at the limits of the resolution limit line, the β-particle will not penetrate deeply, because the energy of the β-particle will almost be completely absorbed by the specimen and the carbon film. Then the remaining energy of the particle will be lost at the point of entrance and the latent images will be stabilized there.

Result 3-CR3. By extremely fine grain developement (MIZUHIRA & KUROTAKI 1964; MIZUHIRA & UCHIDA 1966, 1969; MIZUHIRA, UCHIDA, AMAKAWA et al. 1968; MIZUHIRA, UCHIDA, TOTSU et al. 1968; MIZUHIRA et al. 1970; MIZUHIRA 1971, 1972a, b, c, 1974; MIZUHIRA & FUTAESAKU 1971, 1972; UCHIDA & MIZUHIRA 1970) (Fig. 7), dots similar to developed silver grains start growing from the latent images toward the outside from the highest point or surface of the silver bromide crystal. At the basal surface of the crystal which is contiguous with the section, such developing silver grains push or stick to the surface of the section. Thus we assume that the tip of such a developed grain remains on the surface of a specimen after treatment with the fixing solution and the removal of gelatin, but all the other finely developed silver grains are washed away during the developmental procedure (Fig. 21).

Such extremely finely developed silver grains ensure the higher resolution.

The efficiency with which the tritium β-particle enters a silver haloid grain from the plane surface of the polyhedron other than the base, is probably very low. Even if the β-particle enters a grain which is stabilized at the plane surfaces, the grain will be removed during the procedure of removal of the gelatin as mentioned above. A comparison between the average number (10.5) of finely developed silver grains appearing in a silver bromide crystal (Mizuhira & Uchida 1966, 1969; Mizuhira 1971, 1972b; Uchida & Mizuhira 1970) (Figs. 3a, b, 21, 22) without the removal of gelatin and a single well resolved extremely finely developed silver grain after the removal of gelatin (Figs. 1, 2, 3, 4, 21, 22) seems to support this "contact point theory" (Mizuhira & Kurotaki 1964; Mizuhira & Uchida 1966), as we have named it.

It seems very important that such an extremely finely developed silver is necessary for getting the highest resolution with EM-AUT. But if the developed silver grain size is less than 100 A, it would be very difficult to recognize such small grains in the electron micrographs. Thus we concluded that a suitable grain size would be 200 to 300 A for an extremely fine resolution autoradiography. This sort of fine grain may decrease the resolution about by 50 or 100 A on the average.

Result 4-CR4. There are many factors which can decrease the resolution, but the calculated theoretical resolution agrees well with the results of our experimental data. For example, the resolution that has been obtained in a measurement of the distance between the extremely finely developed silver grain and the cell membrane, in a biological specimen treated with tritiated cholesterol, pregnenolone or thiamine (Figs. 2–4), exhibited a value close to that of the Epon section with 1% osmium and 1% uranium (Fig. 20). And, with a precipitate of ^{3}H- or ^{35}S-thiamine on hydrogen platinum chloride ($B_1 \cdot Pt$), the extremely finely developed silver grain always appeared upon the precipitate of $B_1 \cdot Pt$ (Mizuhira, Uchida, Amakawa et al. 1968; Mizuhira & Uchida 1969; Mizuhira et al. 1970; Mizuhira 1971, 1972b, c; Mizuhira & Futaesaku 1971; Uchida & Mizuhira 1970) (Figs. 1, 2, 21 and 22).

Our theoretical data from the Epon base without any material to absorb the β-ray agrees well with the data of Salpeter (1969) and Salpeter et al. (1969) experiments. Furthermore, Salpeter and Bachmann pointed out that the developed silver grain which grew to coiled filaments made the resolution definitely worse, and this has also supported our "contact point theory".

Regarding these problems, Salpeter (1973) recently advanced her experiment and expressed also the same opinion as ours.

She discussed the "self-absorption" effect to the tritium β-ray which exists in the osmium fixed and uranyl acetate block-stained section coated on a hot thin-film on it. In this experiment, the resolution of electron microscope autoradiography was improved about 10% by the self-absorption effect. The experimental conditions are not the same as ours, but Salpeter's results are going to close to our theoretical data.

Result 5-CR5. In addition to these factors, one must consider the following points, if one wishes to get the highest resolution with EM-AUT.

1) As the β-ray source, one must choose tritiated materials as much as possible (Fig. 13).

2) It is essential to increase the absorbing effect of the specimen toward the β-ray by using several heavy metals, such as osmium, uranium and lead during the process of specimen preparation (Fig. 9).

3) One must choose the appropriate emulsion grain size which should be silver haloid grains 400 to 800 A in diameter contained in a nuclear research emulsion for tritium β-ray. A grain diameter of 800 to 1500 A is necessary for ^{14}C, ^{35}S, and ^{45}Ca.

4) It is essential to utilize a method of extremely fine-grain development, such as "Elon ascorbic acid (EAA)".

5) The coated carbon film on a section must be as thin as possible. The carbon evaporation must be kept to a minimum thickness for the purpose of antichemical fog. A thicker carbon film decreases the efficiency of EM-AUT and the ratio of meaningful silver grains which are latensified by the β-ray in the specimen to the background.

6) The carbon evaporation must be done in a clean vacuum. If the carbon film on a section is contaminated with oil, it would repel the emulsion.

7) As a conclusion from our experiments, the final resolution would be determined as to be the basal surface area of the silver bromide crystal, even though the developed silver grain size is extremely fine.

Future problems, especially freeze-dried sections

The occasion may soon present itself when we will have to face the problem of EM-AUT of diffusible substances by the method of freeze-dried sections, as reported by CHRISTENSEN (1969) and APPLETON (1972). The screening effect of a specimen of a freeze-dried section is calculated to be about one thirtieth of a carbon film, and this value is very near that of air (Fig. 9). We may assume the path length of a tritium β-ray in air and in freeze-dried sections to be about three micrometers.

With light microscope autoradiography, it is sufficient to be able to distinguish whether the developed silver grains are inside or outside the cells and nuclei. In fact, STUMPF and ROTH (1968, 1969) perfected a freeze-dried section method using a cryopump system for light microscope autoradiography.

There have been very few reports on freeze-dried sections for electron microscopy (CHRISTENSEN 1969; APPLETON 1972). However, since screening (or absorption) of the β-ray by a specimen of a freeze-dried section is very low, we have attempted to fix and stain with osmium vapors at the end of the freeze-dry process in order to improve the results. If the specimen should contain 1% osmium and 1% uranium, the path length of the tritium β-ray would be about 0.6 μm by our calculation.

Furthermore, there is a possibility that the resolution at a sigma would improve by 500 to 600 Å, if one should coat a carbon film, which is 200 to 300 Å in thickness, on an osmium-vapor treated freeze-dried section. The screening effect on the tritium β-ray of a carbon film is very effective for this purpose.

The resolution of such a freeze-dried section should be considerably improved if we alter the radiation angle (θ) from the β-ray source to the silver haloid grains. For example, if we assume the thickness of the freeze-dried section to be 1000 Å, the radiation angle (θ) from P is 43° when the area includes 68% of the β-ray (one sigma) leaving the specimen surface and the resolution at that time is calculated to be 470 Å.

If we place the silver haloid grain in a monolayer, we can be sure the resolution will be at least 500 Å under the conditions of osmium vapor fixation and with a carbon film of 200 to 300 Å even on such a freeze-dried section.

The tritium β-ray would not pass through a silver haloid grain under these conditions.

Finally, we can also calculate the use of ^{14}C, ^{35}S and others in the same way as for the tritium β-ray. But since the energy of ^{14}C and ^{35}S β-rays are about ten times stronger than that of tritium, the approximate resolution is estimated to be about 2.2 times less than tritium.

ACKNOWLEDGEMENT

This investigation has been supported in part by Scientific Research Grants from the

Ministry of Education of Japan, Sankyo Co., Ltd., and Hitachi Co., Ltd., granted to Dr. V. MIZUHIRA.

REFERENCES

[1] APPLETON T. C. (1972) "Dry" ultra-thin frozen sections for electron microscopy and X-ray microanalysis: The cryostat approach. Micron, *3:* 101–105.

[2] BACHMANN L. and SALPETER M. M. (1965) Autoradiography with the electron microscope: A quantitative evaluation. Lab. Invest., *14:* 1041–1053.

[3] BASERGA R. and MALAMUD D. (1969) Autoradiography: Techniques and Application. pp. 30–31, Hoeber, New York.

[4] BETHE H. A. (1933) Handbuch der Physik. *24:* 273, Springer, Berlin.

[5] CARO L. G. (1962) High resolution autoradiography: The problem of resolution. J. Cell Biol., *15:* 189–199.

[6] CARO L. G. and SCHNÖS M. (1965) Tritium and phosphorus-32 in high resolution autoradiography. Science, *149:* 60–62.

[7] CHRISTENSEN A. K. (1969) A way to prepare thin frozen sections of fresh tissue for electron microscopy. *In:* Autoradiography of Diffusible Substances. STUMPF W. E. and ROTH L. J. (eds.) pp. 349–362, Academic Press, New York.

[8] GRANBOULAN P. (1963) Resolving power and sensitivity of a new emulsion in electron microscopic autoradiography. J. Roy. Micr. Soc., *81:* 165–171.

[9] HIRATA A. (1967) The sensitization theory for autoradiography. Kagaku to Seibutsu, *5:* 109–113. (in Japanese)

[10] KONISHI K. (1969) Autoradiography of a biological specimen. Radioisotopes, *18:* 534. (in Japanese)

[11] MIZUHIRA V. and KUROTAKI M. (1964) High resolution electron microscopic autoradiography by means of a home-made nuclear research emulsion in Japan. Igaku no Ayumi, *49:* 725–733. (in Japanese)

[12] MIZUHIRA V. and UCHIDA K. (1966) High resolution electron microscopic autoradiography. J. Histochem. Cytochem., *14:* 765–766.

[12'] MIZUHIRA V. and UCHIDA K. (1967) High resolution electron microscopic autoradiography. Kagaku to Seibutsu, *5:* 178–182. (in Japanese)

[13] MIZUHIRA V., UCHIDA K., AMAKAWA T., SHINDO H., TOTSU J. and SUESADA I. (1968) High resolution electron microscopic autoradiography revealed by tritiated thiamine and its derivatives as a case of water-soluble tracer. Proc. 4th Europian Regional Confr. on Electron Microscopy, Rome, pp. 459–460.

[14] MIZUHIRA V., UCHIDA K., TOTSU J. and SHINDO H. (1968) Studies on the absorption of S-benzoylthiamine-^{3}H O-monophosphate. (IV) Electron microscopic autoradiography on the intestinal absorption of S-benzoylthiamine-^{3}H O-monophosphate in the rat. Vitamins, *38* (5): 334–346. (in Japanese with English abstract)

[15] MIZUHIRA V. and UCHIDA K. (1969) Electron microscopic autoradiography. (1) Radioisotopes, *18:* 338–343. (2) Radioisotopes, *18:* 402–421. (3) Radioisotopes, *18:* 473–484. (in Japanese)

[16] MIZUHIRA V., UCHIDA K., FUTAESAKU Y. and OKAZAKI K. (1970) The biosynthesis of testosterone in the testicular interstitial cells of rat. Proc. 7th, Internatl. Congr. on Electron Microscopy, Grenoble, II., pp. 521–522.

[17] MIZUHIRA V. (1971) Electron microscopic autoradiography, especially on the problems of diffusible substances and an attempt at quantitative analysis for EM-autoradiography. Igaku no Ayumi, *76:* 427–444. (in Japanese)

[18] MIZUHIRA V. and FUTAESAKU Y. (1971) On the new approach of tannic acid and digitonine to the biological fixatives. 29th Ann. Proc. Elect. Micro. Soc. Amer., Boston, pp. 494–495.

[19] MIZUHIRA V. (1972a) Chemical fixation of labeled radioisotopes (RI) in tissues to electron microscopic autoradiography (symposium lecture). Proc. 4th Internatl. Congr. Histochem. Cytochem., Kyoto, pp. 35–36.

[20] MIZUHIRA V. (1972b) Electron microscopic autoradiography and the problem of resolution. J. Soc. Photographic Sci. & Technol. Jap., *35* (5): 293–318. (in Japanese)

[21] Mizuhira V. (1972c)   Electron microscopic autoradiography and the basic problems in diffusible substances.   Rinsho Seiri, *2* (6): 528–553. (in Japanese)
[22] Mizuhira V. and Futaesaku Y. (1972)   An attempt of quantitative analysis for electron microscope autoradiography.   Acta Histochem. Cytochem., *5:* 195–200.
[23] Mizuhira V. (1974)   Electron Microscopy: For Biological Application.   Ishiyaku Shuppan, Tokyo. (in Japanese)
[24] Morimoto H. (1962a)   Effect of an electron beam on a photographic plate.   Appl. Physic, *31:* 137–142. (in Japanese)
[25] Morimoto H. (1962b)   Approximate formulas for electron penetration.   Appl. Physic, *31:* 306–310.
[26] Pelc S. R. (1963)   Theory of electron microscopic autoradiography.   J. Roy. Micr. Soc., *81:* 131–139.
[27] Robertson J. S. and Hughes W. L. (1955)   Personal letter to Dr. Tsuya A.
[28] Salpeter M. M. and Bachmann L. (1964)   Autoradiograhpy with the electron microscope: A procedure for improving resolution, sensitivity and contract.   J. Cell Biol., *22:* 469–477.
[29] Salpeter M. M. (1969)   Sensitivity and resolution in electron microscope autoradiography.   *In:* Autoradiography of Diffusible Substances. Roth L. J. and Stumpf W. E. (eds.) pp. 335–348, Academic Press, New York.
[30] Salpeter M. M., Bachmann L. and Salpeter E. E. (1969)   Resolution in electron microscope radioautography.   J. Cell Biol., *41:* 1–20.
[31] Salpeter M. M. and Salpeter E. E. (1971)   Resolution in electron microscope radioautography, II. carbon[14].   J. Cell Biol., *50:* 324–332.
[32] Salpeter M. M. and Szabo M. (1972)   Sensitivity in electron microscope autoradiography, I. The effect of radiation dose.   J. Histochem. Cytochem., *20 (6):* 425–434.
[33] Salpeter M. M. (1973)   Sensitivity in electron microscope autoradiography, II. Effect of heavy metal staining.   J. Histochem. Cytochem., *21 (7):* 623–627.
[34] Salpeter M. M., Budd G. C. and Mattimoe S. M. (1974)   Resolution in autoradiography using semithin sections.   J. Histochem. Cytochem., *22 (4):* 217–222.
[35] Shiina S., Mizuhira V., Uchida K. and Amakawa T. (1969)   Electron microscopic study on sodium ion distribution in cardiac ventricle cells.   Jap. Circulation J., *33:* 601–605.
[36] Slack L. and Wag K. (1959)   Radiations from radioactive atoms in frequent use.   U. S. Atomic Energy Comm., Washington.
[37] Stumpf W. E. and Roth L. J. (1968)   High-resolution autoradiography of ^{3}H-estradiol with unfixed, unembedded 1.0 μ freeze-dried frozen sections.   *In:* Advance in Tracer Methodology. Rothchild S. (ed.) Vol. 4, pp. 113–125, Plenum Press, New York.
[38] Stumpf W. E. and Roth L. J. (1969)   Autoradiography using dry-mounted freeze-dried sections.   *In:* Autoradiography of Diffusible Substances. Stumpf W. E. and Roth L. J. (eds.) pp. 69–80, Academic Press, New York.
[39] Tsuya A., Momose K. and Fukuyama K. (1966)   Textbook of Autoradiography.   pp. 573–574, Maruzen Press, Tokyo. (in Japanese)
[40] Uchida K. and Mizuhira V. (1970)   Electron microscope autoradiography with special reference to the problems of resolution.   Arch. histol. jap. *31* (No. 3/4), 291–320.

Application of Electron Microscopic Autoradiography of Radioactive Iodine for a Comparative Study of the Thyroid Gland

Hisao FUJITA

The thyroid consists of numerous ball-like structures named follicles and interfollicular connective tissue containing blood capillaries. Each follicle contains a homogeneous substance called a colloid filling the follicular lumen surrounded by the simple cuboidal epithelium which consists of numerous follicular epithelial cells, and a few parafollicular cells located at the basal part of the epithelium. The function of the thyroid gland has been understood to be the secretion of triiodothyronine, tetraiodothyronine (=thyroxine) and thyrocalcitonin. Triiodothyronine and thyroxine are hormones released from the follicular epithelial cells and the thyrocalcitonin is a hormone secreted from the parafollicular cells. Both thyroxine and triiodothyronine are amino acid derivatives consisting of two molecules of iodinated tyrosine bound through an ether linkage. However, the production process of these hormones is very complicated and the hormones are not synthesized directly from tyrosine and iodine. First, a high molecular glycoprotein designated thyroglobulin is synthesized in the follicular epithelial cell and is released into the follicular lumen. Some molecules of tyrosine in the thyroglobulin are iodinated and the coupling reaction occurs between two molecules of iodinated tyrosine. Monoiodotyrosine has one atom of iodide in a molecule of tyrosine, and diiodotyrosine has two atoms of iodide. Triiodothyronine (T_3) is made by the coupling of one molecule of monoiodotyrosine and one molecule of diiodotyrosine, while the tetraiodothyronine =thyroxine) (T_4) is made by the coupling of two molecules of diiodotyrosine. The site of iodination of the tyrosyl residue in thyroglobulin is one of the most important problems in recent thyroid research, and the site of the coupling reaction is also an interesting subject. Thyroglobulin stored in the follicular lumen is reabsorbed into the follicular epithelial cells and hydrolyzed to liberate triiodothyronine and thyroxine. Then these hormones are released from the basal portion of the cells into the pericapillary space. Autoradiography has been used for detecting the synthesis mechanism of thyroglobulin, the site of iodination of thyroglobulin and the reabsorption mechanism of the colloid in the thyroid gland. The present paper deals with the application of autoradiography of ^{125}I to clarify the site of iodination of thyroglobulin and the mechanism of the reabsorption of colloid.

Site of Iodination of Thyroglobulin

Higher vertebrates

When we inject inorganic radioactive iodine such as ^{125}I or ^{131}I into an animal, iodine is soon incorporated into the tyhroid gland. Inorganic iodide trapped in the follicular epithelial cell is combined with the tyrosyl residue in thyroglobulin in the thyroid gland. For detecting the site of iodination of (tyrosine residue in) thyroglobulin, autoradiography is the most useful method. For this purpose inorganic iodide must be completely washed away and only the iodide combined with thyroglobulin must be demonstrated. As the basis for solving this problem, the author performed the following experiments using paper chromatography.

Normal and methylmercaptoimidazole-treated mice were sacrificed 1, 4 and 24 hours after injections of 200 μCi of ^{131}I. Then the thyroid tissues were divided into two groups; (1) tissues without any fixation and dehydration, and (2) tissues fixed with 2% glutaraldehyde and 1% OsO_4 for 2 hours, and dehydrated with alcohol for 1 hour. The homogenates of both tissues were made and paper chromatography was carried out using a solvent system composed of butanol-acetic acid-water (40: 10: 20). X-ray film was used for the autoradiogram of this paper chromatography, and the γ-ray scintillation counter was used for the γ-ray count.

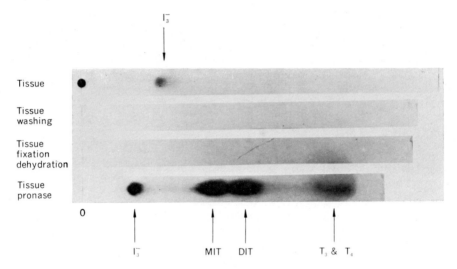

Fig. 1 Autoradiograms of ^{131}I-compounds separated by paperchromatography (solvent: butanol-acetic acid-water (40:10:20), in the rat thyroid 1 hour after injection of ^{131}I (200 μCi). Only inorganic iodide is detected in the homogenated fresh tissue, and is completely washed away by fixation and dehydration of the tissue. After the hydrolysis of the fresh tissue by pronase, MIT, DIT, T_3 and T_4 are separated.

As shown in Fig. 1, inorganic iodide trapped in the thyroid gland was completely washed away by the usual fixation and dehydration procedures for electron microscopy. This experiment had made it clear that only the organic iodide combined with thyroglobulin was detected by the usual autoradiographic method using fixed and dehydrated tissue. Since the energy conversion of ^{125}I was much lower as compared with that of ^{131}I, the former was more suitable for detecting the fine structural localization of radioactive iodine.

The author wishes to describe the results obtained in adult mice (FUJITA 1969). Animals were injected intraperitoneally with 200 μCi of Na^{125}I and sacrificed 3, 5, 7, 15 and 30 minutes and 1, 2, 4 and 24 hours following the injection. Some of the animals were treated with 1 unit of TSH (NIH-TSH B3 or Thytropar) 22 or 23 hours following the injection of Na^{125}I and were sacrificed one or two hours later. Sections cut on a Porter-Blum ultramicrotome, picked up on uncoated grids and coated with carbon, were dipped into the Ilford L-4 emulsion. Following exposure for 3 weeks at 4°C, the autoradiograms were

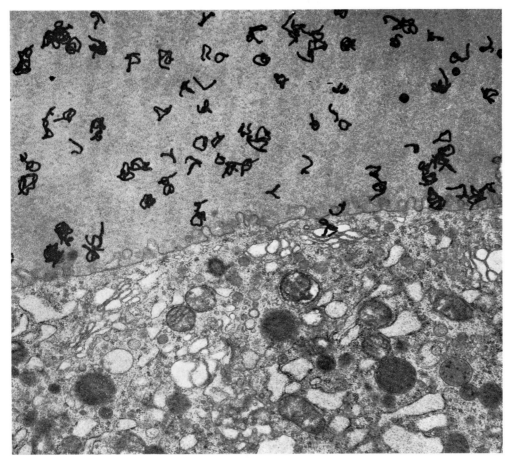

Fig. 2 A portion of the mouse thyroid 2 hours after an injection of 200 μCi of Na^{125}I. Numerous silver grains are localized over the follicular lumen, and one grain over the Golgi apparatus of the follicular epithelial cell. Emulsion: Ilford L4, Developer: Microdol X. (×17,000)

developed in Microdol X or phenidon for 5 minutes at 18°C and fixed with 20% sodium thiosulfate. The grids were stained with the Millonig's lead solution.

Silver grains of ^{125}I were already seen 3 minutes after an intraperitoneal injection over the follicular lumen of the mouse thyroid. The grains increased in number with the lapse of time, and the follicular lumen, especially its peripheral region was full of numerous grains 1, 2 and 4 hours after the injection, while the grains over the follicular epithelial cells were very few. This fact indicated that the iodination of thyroglobulin took place almost entirely in the follicular lumen especially in its peripheral region.

Similar data had already been published by STEIN and GROSS (1964), IBRAHIM and BUDD (1965), LUPULESCU and PETROVICI (1965), SIMON and DROZ (1965), EKHOLM (1966) and NADLER (1971). On the other hand TAKANO and HONJIN (1968) concluded that iodination occured only in the rough endoplasmic reticulum, though numerous silver grains were localized in the follicular lumen and very few in the cytoplasm in their photographs. They refused to consider the presence of numerous grains in the follicular lumen. TIXIER-VIDAL et al. (1969) studied the sheep thyroid cells isolated by trypsinization and incubated in the presence of ^{125}I exhibited a few grains in the rough endoplasmic reticulum, had the opinion that the site of iodination of thyroglobulin was intracellular, at the

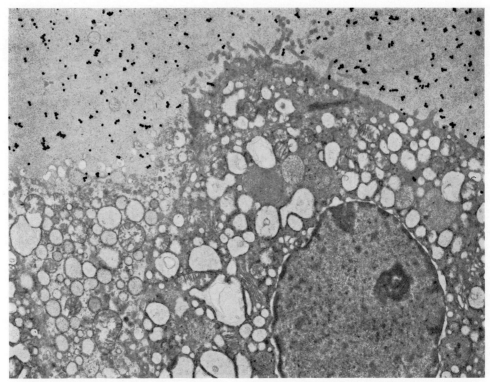

Fig. 3 A portion of the mouse thyroid fed with a low-iodine diet for 6 months, 15 minutes after an injection of 200 μCi of Na^{125}I. Silver grains are localized over the follicular lumen. Emulsion: Ilford L4, Developer: Phenidon. (×12,000)

level of the rough endoplasmic reticulum. TONG et al. (1962a, b) who found iodine incorporated in the sheep thyroid cells which had been dispersed and isolated by trypsinization, interpreted that the presence of colloid and the follicular structure were not essential for the iodine-concentrating function.

The present author (FUJITA 1969) did not deny the possibility that iodination of thyroglobulin could take place to a small degree in the cytoplasm. The thyroids of chick embryos 10 and 13 days old, whose follicular structure had not yet been completed, disclosed several silver grains over the cytoplasm, especially over the Golgi apparatus, the apical small vesicles, and rough endoplasmic reticulum as well as over the follicular lumen, 15, 30, 45 and 60 minutes after injection of 50 μCi of ^{125}I. These data suggested to us that iodination of thyroglobulin could take place not only in the follicular lumen but also in the cytoplasm though to a lesser extent in the latter. However, in the usual adult animals, thyroglobulin in the follicular lumen was in much greater in quantities than in the cytoplasm and the numerous molecules of thyroglobulin in the follicular lumen might not have been completely iodinated yet and therefore the injected iodine was considered to be combined preferentially with the luminal colloid. This seemed to be the reason why iodination took place almost entirely in the colloid lumen in adult animals having the usual large follicles.

The author and his coworker (unpublished data) also tried to make a dry mount light microscopic autoradiograph of inorganic iodide in the thyroid gland using freeze-dried sections. Mice were treated with a 2% methylmercaptoimidazole water solution substituted as the drinking water for 3 days. This chemical is used as an agent for blocking the iodination of thyroglobulin. The thyroid tissues of these animals were removed 1, 2 and 24 hours after the injection of 200 μCi of ^{131}I. Inorganic iodide trapped in the thyroid gland was localized mainly in the follicular lumen, though a few grains were found in the follicular epithelial cells. This fact indicated that inorganic iodide was also stored primarily in the follicular lumen. If so, it was more reasonable to assume that iodination of thyroglobulin occured primarily in the follicular lumen.

The enzyme necessary for the iodination of thyroglobulin was believed to be a peroxidase. The reaction had been considered as follows:

$$\text{NADH (or NADPH)} + \text{H}^+ + \text{O}_2 \xrightarrow{\text{peroxidase}} \text{NAD}^+ \text{ (or NADP}^+) + \text{H}_2\text{O}_2$$

$$\text{Oxidized iodide} + \text{tyrosine} \xrightarrow[\text{(or tyrosine iodinase)}]{\text{peroxidase}} \text{iodotyrosine}$$

The fine structural localization of an endogenous peroxidase had been demonstrated using the 3, 3′-diaminobenzidine tetrahydrochloride (DAB) method, by STRUM and KARNOVSKY (1970, 1971), NAKAI and FUJITA (1970), SHIN et al. (1970) and NANBA (1972). The reaction product for peroxidase was positive in the cisternae of the rough endoplasmic reticulum, dnese bodies, Golgi lamellae and vesicles, subapical vesicles, and the luminal colloid in the thyroids of both the rat and the frog. Peroxidase might also be synthesized in the rough endoplasmic reticulum, transported to the Golgi apparatus and secreted into the follicular lumen. However, it is a problem for the future whether the secretory granules containing thyroglobulin are identical or not with the subapical vesicles containing peroxidase. Nerverthless, it was easily understood that iodination of thyroglobulin might take place in many parts of the cell such as the cisternae of the rough endoplasmic reticulum, Golgi vesicles, subapical vesicles and the follicular lumen.

Once again the author wishes to repeat that the iodination of thyroglobulin takes place almost entirely in the follicular lumen, especially in its peripheral region, but iodination may occur to a small degree also in the cytoplasm such as in the rough endoplasmic reticulum, Golgi apparatus and the subapical vesicles.

Phylogenetic aspects

Fine structures of the thyroid glands are principally similar to one another throughout the animal kingdom from elasmobranchs to mammals. Autoradiography of ^{125}I had been performed for the amphibia, *Rana nigromaculata nigromaculata* (NAKAI et al. 1970) and the chimaeroid fish, *Hydrolagus colliei* (NAKAI & GORBMAN 1969). Thyroids of these animals exhibited quite similar patterns for the site of iodination of thyroglobulin to that seen in mammals. The principal site for iodination of thyroglobulin was the luminal colloid in these animals.

The cyclostome, the lowest order of the vertebrates, is the most interesting animal in the evolutionary development of the thyroid. One species of a cyclostome examined is the hagfish and another is the lamprey. Thyroids of both animals showed scattered follicles in the hypobranchial region. Follicles were diffusely distributed in a fairly large region and the interfollicular conncetive tissue was extremely abundant while the blood capillaries were very sparce in distribution as compared with those of higher vertebrates. The follicular epithelial cells of both animals had a relatively compact cytoplasm and a poorly developed cytomembrane system (FUJITA & HONMA 1966). The cisternae of the rough endoplasmic reticulum and Golgi apparatus were flattened both in the lamprey and the hagfish, though these elements in the thyroid of higher vertebrates were well dilated. These facts suggested that the thyroid glands in these animals were not as active when

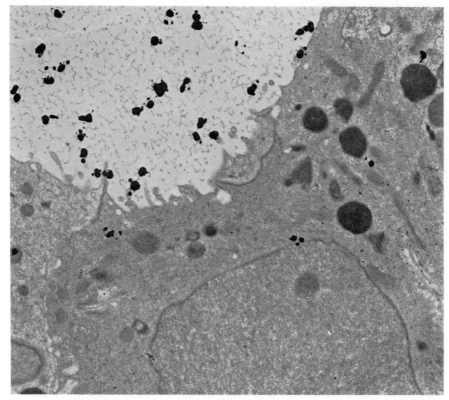

Fig. 4 A portion of the thyroid of a hagfish, *Eptatretus burgeri*, 2 hours after an injection of 1 mCi of Na^{125}I. Emulsion: Ilford L4, Developer: Microdol X. ($\times 12{,}000$)

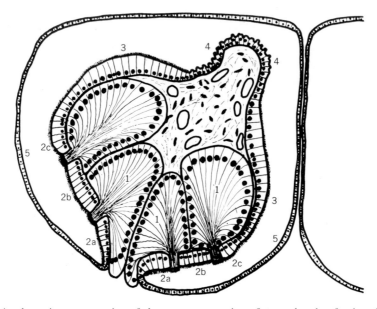

Fig. 5 A schematic representation of the transverse section of an endostyle of a larval lamprey, ammocoetes of *Lampetra japonica*. 1–5: type 1–5 cells. Type 2c and type 3 cells are homologous to the thyroid cell. (From FUJITA H.: Arch. histol. jap. *34*: 109–141, 1972).

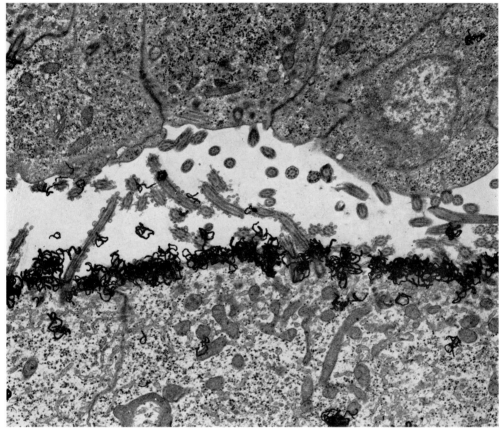

Fig. 6 A portion of an endostyle of a larval lamprey 2 hours after an injection of 200 μCi of Na^{125}I. Silver grains are localized over the apical portion of the type 3 cells. Upper: type 4 cells, lower: type 3 cells. Emulsion: Ilford L4, Developer: Microdol X. ($\times$14,500) (From Fujita H. & Honma Y.: Z. Zellforsch., *98*: 525–537, 1969).

compared with those of the higher vertebrates. Large lysosome-like dense bodies were also characteristic for this group. Electron microscopic autoradiography of the thyroid tissue of the hagfish 1 and 6 hours after an injection of 1 mCi of ^{125}I showed almost a similar pattern to that of higher vertebrates, though the silver grians in the follicular lumen were much less in number. These data indicated that the iodination of thyroglobulin in the cyclostome took place almost entirely in the follicular lumen as in the higher vertebrate.

The larval lamprey has a characteristic organ specified as the endostyle in thier subpharyngeal riegon. This organ, which does not show any follicular structure and communicates with the pharynx, has been considered to be homologous to the thyroid of higher vertebrates. Triiodothyronine and thyroxine had been demonstrated in extracts of the endostyle of the larval lamprey by Leloup and Berg (1951), Leloup (1955) and Roche et al. (1961). Epithelial cells of the endostyle of the larval lamprey had been classified into five or six types; type 1 to type 5 or 6 (Marine 1913; Leach 1939). Type 2 cells were divided into 3 subtypes; type 2a, type 2b, and type 2c cells. Among these, the type 2c cells and all the type 3 cells were considered to be homologous to the thryoid cell. At metamorphosis the type 3 cells were considered for the most part to become follicular epithelial cells of the thyroid (Sterba 1953; Clements-Merlini 1960; Honma 1960). Re-

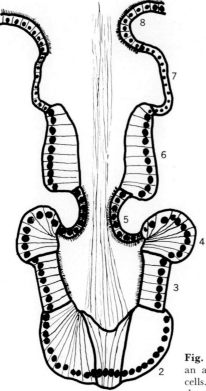

Fig. 7 Transverse section of an endostyle of an ascidian, *Ciona intestinalis*. 1–8: zone 1–8 cells. Zone 7 and 8 cells are homologous to the thyroid cell.

garding these cells of the larval lamprey, EGEBERG (1965), FUJITA and HONMA (1968) and HOHEISEL (1969) had reported on their fine structure. Type 2c and type 3 cells showed a great deal of similarity to the thyroid cell of the adult lamprey in their electron microscopic figures (FUJITA & HONMA 1969). Silver grains appeared within 30 minutes after an intraperitoneal injection of 200 μCi of ^{125}I in electron microscopic autoradiographs over the apical cell membrane region of the type 2c and the type 3 cells. These grains increased in number for two hours. This fact suggested that the main site of iodination of thyroglobulin was the apical cell membrane region of these cells. Since the materials in the endostylar lumen were washed away during fixation and dehydration of the tissue, the possibility of iodination also taking place in the endostylar lumen should not be ruled out.

The amphioxus and ascidians which belong to protochordate also have endostyles homologous to the thyroid. COVELLI et al. (1960) and TONG et al. (1962a, b) reported the existence of triiodothyronine and thyroxine in the amphioxus endostyle. Triiodothyronine and thyroxine had also been demonstrated in the ascidian blood by KENNEDY (1966) and thyroxine had been found in the extract of the endostyle of the ascidian, *Ciona intestinalis*, using the radiochromatography techniques by BARRINGTON and THORPE (1956) and BARRINGTON (1968). The endostylar cells of the ascidian have been divided into 8 types; zone 1 to 8 cells. The present author and his coworker (FUJITA & NANBA 1971) found numerous silver grains in the apical cell membrane region of zone 7 and zone 8 cell, especially of zone 8 cells, 1, 4, 6, 16 and 24 hours after immersion in sea water containing 1 mCi/L of ^{125}I.

This fact indicated to us that iodination of proteins took place primarily at the apical plasma membrane region of zone 7 and zone 8 cells of the ascidian endostyle. In addition, the possibility that iodination could occur also in the endostylar lumen should be considered, because the materials in the lumen which was open to the pharynx were washed away during the fixation and dehydration procedures. These data regarding the site of iodination of protein in the ascidian endostyle were quite similar to those of the larval lamprey.

Conclusion

Using electron microscopic autoradiography of ^{125}I, the site of iodination of thyroglobulin has been examined in the thyroids of mammals, birds, amphibians, elasmobranchs, and cyclostomes, and in the endostyles (homologous to the thyroid) of larval lampreys and ascidians. The author wishes to emphasize that iodination of thyroglobulin takes place almost entirely in the follicular lumen and the apical plasma membrane region of the thyroid, and in the apical plasma membrane region and perhaps also in the endostylar lumen of the endostyle. However, the author does not deny the possibility that iodination could also occur to a limited degree in the cytoplasm such as in the rough endoplasmic reticulum, Golgi apparatus and the subapical vesicles. All these organelles as well as the luminal colloid possess the peroxidase activity necessary for the iodination of thyroglobulin. However, thyroglobulin in the follicular lumen is far greater in quantity than that in the cytoplasm and the numerous tyrosine residues in thyroglobulin in the follicular lumen might not have yet been completely iodinated, and therefore the injected iodine is regarded as being preferentially combined with the luminal colloid.

Reabsorption of Colloid

Higher vertebrates

Colloid (thyroglobulin) stored in the follicular lumen is reabsorbed into the follicular epithelial cell. There are several kinds of granules and colloid droplets in the cytoplasm, and whether the large colloid droplet in the follicular epithelial cell is the reabsorbed material or a secretory one has also been an important question. Electron microscopic autoradiography has played an important role in solving this question (SHELDON et al. 1964; EKHOLM & SMED 1966; FUJITA 1969; SELJELID et al. 1970). Following are the data that the present author had already reported (FUJITA 1969). Twenty-four hours after the injection of ^{125}I into a rat or mouse, the radioactive substances were almost enclosed in the follicular lumen if thyroxine was treated. The thyroid stimulating hormone (TSH) had been known to stimulate the reabsorption of colloid into the cytoplasm. Five minutes or 1 hour after the injection of TSH into an animal already treated with ^{125}I 23 hours previously, numerous silver grains of ^{125}I combined with thyroglobulin were recognized in the large colloid droplet as well as in the follicular lumen. The population of the grains in the droplet was almost the same as that in the luminal colloid. This fact suggested to us that the large colloid droplet was not a secretory material but a reabsorbed one from the follicular lumen into the cell. Since the features for the engulfment of the follicular colloid were sometimes recognized after an injection of TSH, the colloid was considered to be reabsorbed into the cell by the phagocytic procedure. However, as SELJELID (1967a) and SELJELID et al. (1970) suggested, it was questionable whether or not the colloids were reabsorbed only by the phagocytic procedure. The present author observed silver grains not only in the large colloid droplets but also in some of the small vesicles located in the apical cytoplasm. The possibility that the colloid could also be

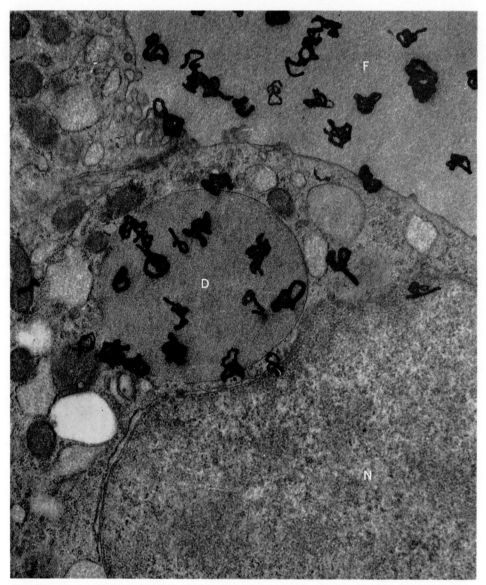

Fig. 8 A portion of the thyroid from the gland of a mouse which received 200 μCi of Na^{125}I 24 hours prior to fixation and 1 unit of TSH 1 hour before fixation. Silver grains are localized over the follicular colloid (F) and intracellular colloid droplet (D). N. nucleus. Emulsion: Ilford L4, Developer: Microdol X. (×22,000) (From FUJITA H.: Virchows Arch. B Zellpathol., 2: 265–279, 1969).

reabsorbed by a micropinocytotic mechanism and fused with one another to become a large droplet in the cytoplasm thus should not be denied. The reabsorbed colloid must be hydrolyzed to liberate triiodothyronine and thyroxine. As shown by EKHOLM and SMEDS (1966), SELJELID (1967b), WETZEL et al. (1965) and WOLLMAN et al. (1964), the primary lysosomes were fused with the reabsorbed colloid droplet and the hydrolysis of thyroglobulin occurred to liberate triiodothyronine and thyroxine. These hormone being amino acid derivatives of small molecules did not show any granular form. Hence, it was assumed that they might be transported through the cytoplasmic matrix and released from the

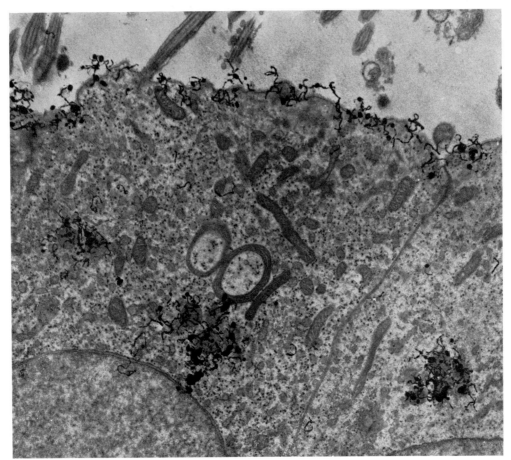

Fig. 9 Portions of type 3 cells of an endostyle of a larval lamprey (ammocoetes of *Lampetra japonica*) 6 hours after an injection of 200 μCi of Na^{125}I. Silver grains are located over the apical surface as well as the dense bodies in the cytoplasm. Emulsion: Ilford L4, Developer: microdol X. (×13,000) (From Fujita H. & Honma Y.: Z. Zellforsch., *98*: 525–532, 1969).

basal portion of the cell into the pericapillary space. As the amino acid derivatives were easily understood to be moved and washed away by fixation and dehydration of the tissue, it was difficult to detect the mechanism of release of triiodothyronine and thyroxine by the usual methods of electron microscopic autoradiography.

Phylogenetic aspects

The pattern of colloid reabsorption seen in the mouse and rat was common to that of all the vertebrate thyroids. Similar results had been published for amphibians (Nakai et al. 1970; Nanba 1972), teleosts (Fujita & Machino 1965; Fujita et al. 1966; Suemasa et al. 1968), and elasmobranchs (Nakai & Gorbman 1969). In the hagfish (*Eptatretus burgeri*), cyclostome, silver grains were mostly localized by electron microscopic autoradiography over the follicular lumen, apical plasma membrane region and over a few small vacuoles and a few large dense bodies in the cytoplasm. 24 hours after injection of ^{125}I (unpublished data). Therefore, these vacuoles and dense bodies might contain reabsorbed materials from the follicular lumen. If so, the hagfish thyroid also had a similar function to that of the higher vertebrates regarding the mechanism of reabsorption of colloid.

The data in reference to the endostyle of the larval lamprey were also very interesting (Fujita & Honma 1969). The silver grains are localized by electron microscopic autoradiography mostly in the apical plasma membrane region of type 2c and type 3 cells 30 minutes, 1 and 2 hours after an intraperitoneal injection of 200 μCi of ^{125}I. However, 6 and 24 hours after the injection, silver grains were localized not only over the apical plasma membrane region but also over the multivesicular bodies and lysosome-like dense bodies in the type 2c and type 3 cells. In addition, large or small vacuoles, dense or light in some type 5 cells, were also labeled 6 to 24 hours after an injection. This fact suggested that the large dense bodies and multivesicular bodies were involved in the reabsorption of iodinated thyroglobulin. Clements and Gorbman (1955), Gorbman (1959) and Thomas (1962) had considered that the thyroglobulin-like protein was transported to the alimentary canal to be hydrolyzed and absorbed through the digestive epithelium. The present author wishes to emphasize that iodinated protein is reabsorbed into type 2c, type 3 and type 5 cells of the endostyle, though the possibility that some portions of the iodinated protein is transported to the alimentary canal is not denied. Though the phagocytic figures, which were usually seen in TSH-treated higher vertebrates, were difficult to find in the endostylar cells of the larval lamprey as well as the thyroid cells of the hagfish and adult lamprey, the reabsorption of colloid was considered to be an important function for them. Since these animals were not as active in thyroid function, the colloid might be reasorbed not by phagocytosis but by micropinocytosis.

The author would like to show the result obtained from the endostyle of the ascidian, *Ciona intestinalis* (Fujita & Nanba 1971). As mentioned above the site of iodination of proteins was considered to be the apical plasma membrane region of zone 7 and zone 8 cells and perhaps the endostylar lumen in this animal. By autoradiography, the silver grains appeared in the apical plasma membrane region of zone 7 and zone 8 cells 1, 4, 6, 16 and 24 hours after immersion in sea water containing ^{125}I, and in the multivesicular bodies and lysosomes 4, 6, 16 and 24 hours, especially 16 and 24 hours, after immersion. From this fact it seemed that the organic iodide might be reabsorbed into the cytoplasm of these cells, though the phagocytic features were not recognized.

Conclusion

Iodinated thyroglobulin in the follicular lumen or the thyroglobulin-like protein in the endostylar lumen were reabsorbed into the thyroid follicular cell or into the endostylar cell throughout the vertebrates and protochordates having the ability to produce triiodothyronine and thyroxine. Though in higher vertebrates typical phagocytic features of the luminal colloid were easily seen after an injection of TSH, the colloid could be reabsorbed not only by this means but also by other mechanisms such as micropinocytosis. The reabsorbed protein was involved in a secondary lysosome to be hydrolyzed in order to liberate triiodothyronine and thyroxine throughout all the vertebrates and protochordates.

REFERENCES

[1] Barrington E. J. W. (1968) Phylogenetic perspectives in vertebrate endocrinology. *In:* Prespective in Endocrinology. Barrington E. J. W. and Jorgensen C. B. (eds.) pp. 1–46, Academic Press, London & New York.

[2] Barrington E. J. W. and Thorpe A. (1965) The identification of monoiodotyrosine, diiodotyrosine and thyroxine in extracts of the nedostyle of the ascidian, *Ciona intestinalis* L. Proc. Roy. Soc. Lond. B., *163:* 136–149.

[3] Clements M. and Gorbman A. (1955) Protease in ammocoetes endostyle. Biol. Bull., *108:* 258–263.

[4] CLEMENTS-MERLINI M. (1960) The secretory cycle of iodoproteins in ammocoetes. II. A radioautographic study of the transforming larval thyroid gland. J. Morphol., *107:* 357–364.

[5] COVELLI I., SALVATORE G., SENA L. and ROCHE J. (1960) Sur la formation d'hormones thyroidiennes et de leurs précurseurs par *Branchiostoma lanccolatum* Pallas (Amphioxus). C. R. Soc. Biol. (Paris), *154:* 1165–1189.

[6] EGEBERG J. (1965) Iodine-concentrating cells in the endostyle of ammocoetes. Z. Zellforsch., *68:* 102–115.

[7] EKHOLM R. (1966) Autoradiographic studies on the iodine metabolism of the guinea pig thyroid. J. Ultrastr. Res., *14:* 419–420.

[8] EKHOLM R. and SMEDS S. (1966) On dense bodies and droplets in the follicular cells of the guinea pig thyroid. J. Ultrastr. Res., *16:* 71–82.

[9] FUJITA H. (1969) Studies on the iodine metabolism of the thyroid gland as revealed by electron microscopic autoradiography of ^{125}I. Virchows Arch. Abt. B. Zellpathol., *2:* 265–279.

[10] FUJITA H. and HONMA Y. (1966) Electron microscopic studies on the thyroid of a cyclostome, *Lampetra japonica*, during its upstream migration. Z. Zellforsch., *73:* 559–575.

[11] FUJITA H. and HONMA Y. (1968) Some observations on the fine structure of the endostyle of larval lampreys, ammocoetes of *Lampetra japonica*. Gen. comp. Endocrinol., *11:* 111–131.

[12] FUJITA H. and HONMA Y. (1969) Iodine metabolism of the endostyle of larval lampreys, ammocoetes of *Lampetra japonica*. Electron microscopic autoradiography of ^{125}I. Z. Zellforsch., *98:* 525–537.

[13] FUJITA H. and MACHINO M. (1965) Electron microscopic studies on the thyroid gland of a teleost, *Seriola quinqueradiata*. Anat. Rec., *152:* 81–98.

[14] FUJITA H. and NANBA H. (1971) Fine structure and its functional properties of the endostyle of ascidians, *Ciona intestinalis*. A part of phylogenetic studies of the thyroid gland. Z. Zellforsch., *121:* 455–469.

[15] FUJITA H., SUEMASA H. and HONMA Y. (1966) An electron microscopic study on the thyroid gland of the silver eel, *Anguilla japonica*. Arch. histol. jap., *27:* 153–163.

[16] GORBMAN A. (1959) Problems in the comparative morphology and physiology of the vertebrate thyroid gland. *In:* Comparative Endocrinology. GORBMAN A. (ed.) Wiley, New York.

[17] HOHEISEL G. (1969) Untersuchungen zur funktionellen Morphologie des Endostyls und der Thyreoidea vom Bachneumauge (*Lampetra planeri* Bloch). I. Untersuchungen am Endostyl. Morphol. Jahrb., *114:* 204–240.

[18] HONMA Y. (1960) Studies on the morphology and role of the important endocrine glands in some Japanese cyclostomes and fishes. Niigata Daigaku Rigakubu Seibutsugaku Kyoshitsu. (in Japanese)

[19] IBRAHIM M. S. and BUDD G. C. (1965) An electron microscopic study of the site of iodine binding in the rat thyroid gland. Exp. Cell Res., *38:* 50–56.

[20] KENNEDY G. R. (1966) The distribution and nature of iodine compounds in ascidians. Gen. comp. Endocrinol., *7:* 500–511.

[21] LEACH W. J. (1939) The endostyle and thyroid gland of the brook lamprey, Ichthymyzon fossor. J. Morphol., *65:* 549–593.

[22] LELOUP J. (1955) Metabolism de l'iode et functionnement endostylaire chez l'ammocoete de *Lampetra planeri*. J. Physiol. (Paris), *47:* 671–677.

[23] LELOUP J. and BERG O. (1951) Sur la presence d'acides aminés iodés (monoiodotyrosine, di-iodotyrosine, et thyroxine) dans l'ammocoetes. C. R. Acad. (Paris), *238:* 1069–1071.

[24] LUPULESCU A. and PETROVICI A. (1965) Ètude radioautographique en microscopie electronique de I^{131} dans la thyroide du rat. Arch. Anat. microsc. Morphol. exp., *54:* 895–902.

[25] MARINE D. (1913) The metamorphosis of the endostyle of Ammocoetes branchialis. J. exp. Med., *17:* 379–395.

[26] NADLER N. J. (1971) The application of radioautography to the localization of the binding of iodine to thyroglobulin in rat thyroid follicles. Anat. Rec., *169:* 384–385.

[27] NAKAI Y. and FUJITA H. (1970) Fine structural localization of peroxidase in the rat thyroid. Z. Zellforsch., *107:* 104–110.

[28] NAKAI Y. and GORBMAN A. (1969) Cytological and functional properties of the thyroid of a chimaeroid fish, *Hydrolagus colliei*. Gen. comp. Endocrinol., *13:* 285–302.

[29] NAKAI Y., NANBA H. and FUJITA H. (1970) Fine structure and functional properties of the thyroid of the anura, *Rana nigromaculata nigromaculata*. (A part of the phylogenetic studies of the thyroid gland). Arch. histol. jap., *31:* 421–432.

[30] NANBA H. (1972) Endogeneous peroxidase activity in the frog thyroid gland. Histochemie, *32:* 99–105.

[31] ROCHE J., SALVATORE G. and COVELLI I. (1961) Mètabolism de ^{131}I et fonction thyroiodienne chez la larve (ammocoete) d'un cyclostome, *Petromyzon planeri* BL. Comp. Biochem. Physiol., *2:* 90–99.

[32] SELJELID R. (1967a) Endocytosis in thyroid follicle cells. II. A microinjection study of the origin of colloid droplets. J. Ultrastr. Res., *17:* 401–420.

[33] SELJELID R. (1967b) Endocytosis in thyroid follicle cells. IV. On the acid phosphatase activity in thyroid follicle cells, with special reference to the quantitative aspects. J. Ultrastr. Res., *18:* 237–256.

[34] SELJELID R., REITH A. and NAKKEN K. F. (1970) The early phase of endocytosis in rat thryoid follicle cells. Lab. Invest., *23:* 595–605.

[35] SHELDON H., MCKENZIE J. and NIMWEGAN D. (1964) Electron microscopic autoradiography The localization of ^{125}I in suppressed and thyrotropin-stimulated mouse thyroid gland. J. Cell Biol., *23:* 200–205.

[36] SHIN W. -X., MA M., QUINTANA N. and NOVIKOFF A. B. (1970) Organelle interrelations within rat thyroid epithelial cells. Congrès International de Microscopie Electroniuqe. Vol. III, pp. 7980–7981, Grenoble.

[37] SIMON C. and DROZ B. (1965) Iodine distribution in the thyroid follicle as determined by isotopic equilibrium and electron microscopic radioautography. *In:* Current Topics in Thyroid Research. CASSANO C. and ANDREOLI M. (eds.) pp. 77–84, Academic Press, New York & London.

[38] STEIN O. and GROSS J. (1964) Metabolism of ^{125}I in the thyroid gland studied with electron microscopic autoradiography. Endocrinology, *75:* 787–798.

[39] STERBA G. (1953) Die Physiologie und Histogenese der Schilddrüse und des Thymus beim Bachneunauge (*Lampetra planeri* BL.). Wiss. Z. Friedrich-Schiller-Univ. Jena, *2:* 239–298.

[40] STRUM J. M. and KARNOVSKY M. J. (1970) Cytochemical localization of endogenous peroxidase in thyroid follicular cells. J. Cell Biol., *44:* 655–666.

[41] STRUM J. M. and KARNOVSKY M. J. (1971) Aminotriazole goiter. Fine structure and localization of thyroid peroxidase activity. Lab. Invest., *24:* 1–12.

[42] SUEMASA H., HONMA Y. and FUJITA H. (1969) Electron microscopic observations on the thyroid gland of two species of teleost, *Semicossyphus reticulatus* and *Sebastiscus marmoratus*. (A part of phylogenetic studies of the fine structure of the thyroid). Arch. histol. jap., *29:* 363–375.

[43] TAKANO I. and HONJIN R. (1968) Uptake mode of I^{125} in the mouse thyroid follicle studied by light and electron microscopic autoradiography. Okajimas Fol. anat. jap., *44:* 173–201.

[44] THOMAS I. M. (1962) Some aspects of the evolution of thyroid structure and function. *In:* The Evolution of Living Organisms. LEEPER G. W. (ed.) Melbourne Univ. Press, Melbourne.

[45] TIXIER-VIDAL A., PICART R., RAPPAPORT L. and NUNEZ J. (1969) Ultrastructure et autoradiographie de cellules thyroidiennes isolees en présence de ^{125}I. J. Ultrastr. Res., *28:* 78–101.

[46] TONG W., KERKOF P. and CHAIKOFF I. L. (1962a) Identification of labeled thyroxine and triiodothyronine in amphioxus treated with ^{131}I. Biochim. biophys. Acta, *56:* 326–331.

[47] TONG W., KERKOF P. and CHAIKOFF I. L. (1962b) Iodine metabolism of dispersed thyroid cells obtained by trypsinization of sheep thyroid glands. Biochim. biophys. Acta., *60:* 1–19.

[48] WETZEL B. K., SPICER S. S. and WOLLMAN S. H. (1965) Changes in fine structure and acid phosphatase localization in rat thyroid cells following thyrotropin administration. J. Cell Biol., *25:* 593–618.

[49] WOLLMAN S. H., SPICER S. S. and BURSTONE M. S. (1964) Localization of esterase and acid phosphatase in granules and colloid droplets in rat thryoid epithelium. J. Cell Biol., *21:* 191–201.

3. IMMUNOHISTOCHEMISTRY

Application of Enzyme-labeled Antibody Methods for the Ultrastructural Localization of Hormones: Review

Paul K. NAKANE*

Almost a decade has passed since the introduction of the idea that enyzmes might be used as markers in immunohistochemistry (Nakane & Pierce 1966a, b, 1967; Papkoff 1971) and immunodiffusion (Avrameas & Uriel 1966). The labeling of antibodies by histochemically-localizable enzymes has permitted the morphological identification of many different antigens.

Enzyme Markers

There are several reasons to account for the successful use of enzymes as markers at the ultrastructural level (Nakane & Pierce 1967). Enzyme-labeled antibodies diffuse through tissues and cells more readily than the traditional ferritin-labeled antibodies since the enzymes employed are smaller than ferritin. Also unlike ferritin conjugates, enzyme-labeled antibodies can be visualized, by conventional histocytochemical methods, at both the light and electron microscopic levels. Since many molecules of the enzyme reaction product are deposited at the antigenic sites, the method is extremely sensitive. The enzymes chosen either should not exist in the material being studied or should be in well-defined areas so that the antigenic sites will not be confused with endogenous enzymes.

The most frequently used enzyme marker is horseradish peroxidase, a very stable enzyme with a high turnover rate. It is a hemecontaining glycoprotein with a molecular weight of 40,000, an amino acid content of about 300 residues, and a carbohydrate content of about 18% (Shannon et al. 1966; Welinder & Smillie 1972). Peroxidase is coupled to immunoglobulin through bifunctional reagents such as difluorodinitrodiphenylsulfone (Nakane & Pierce 1966a) or glutaraldehyde (Avrameas 1969) or through its own carbohydrate moiety (Kawaoi & Nakane 1973).

The hormones so far localized include insulin, gastrin, calcitonin, human chorionic gonadotropin (HCG), human somatomammotropic hormone (HCS), vasopressin, melanocyte stimulating hormone (MSH), growth hormone, prolactin, adrenocorticotropic hormone (ACTH), thyrotropic hormone (TSH), luteinizing hormone (LH) and follicle stimulating hormone (FSH).

---

* An awardee of USPHS Career Development Award # GM-46228.

Fixatives

The fixation of antigens without destroying their antigenicity is a common problem in immunohistochemistry. A fixative suitable for one antigen is not necessarily suitable for another. And a fixative which preserves a particular antigen well will not necessarily give the best ultrastructural preservation. Hence, the optimal fixation conditions for each particular antigen must be determined by experimentation.

For the fixation of endocrine tissues, p-formaldehyde-picric acid, glutaraldehyde, p-formaldehyde, and mixtures of glutaraldehyde and p-formaldehyde have been employed. P-formaldehyde-picric acid has been used to fix human placenta for the localization of HCG (Dreskin et al. 1970; Genbacev et al. 1972), porcine thyroid for calcitonin (De Grandi et al. 1970), and rat anterior pituitray for growth hormone, prolactin, ACTH, TSH, LH and FSH (Kawarai & Nakane 1970; Nakane 1970, 1971; Muzurkiewicz & Nakane 1972; Moriarty & Halmi 1972; Sétáló & Nakane 1972; Tougard et al. 1973). Various concentrations of glutaraldehyde have been used to fix amphibian and rat anterior pituitary for the localization of various hormones (Moriarty & Halmi 1972; Doerr-Schott & Dubois 1973), and human gut for gastrin (Greider et al. 1972). Various concentrations of p-formaldehyde have been used to fix human placenta for the localization of HCG (Okudaira et al. 1971) and HCS (De Ikonicoff 1973), human gut for gastrin (Greider et al. 1972), dog thyroid for calcitonin (Kalina & Pearce 1971), pancreas of various animals for insulin (Misugi et al. 1970), and rat anterior pituitary for ACTH (Moriatry & Halmi 1972). Mixtures of glutaraldehyde and p-formaldehyde have been used to fix human placenta for the localization of HCG (Hamanaka et al. 1971) and HCS (De Ikonicoff 1973), dog thyroid for calcitonin (Kalina & Pearce 1971), and rat posterior pituitary for vasopressin (Rufener & Nakane 1973).

Processing of Fixed Tissues

Several different approaches (Nakane 1974a, b) have been applied to the processing of fixed tissues for the ultrastructural localization of hormones:

1) The fixed tissues can be sliced or chopped in small pieces, exposed to antibodies, stained for enzyme activity, and embedded in plastic.

2) The fixed tissues can be frozen and sectioned in a cryostat. The thick sections (e.g., 20–50 μm), unmounted, are then exposed to antibodies, stained, and embedded in plastic.

3) The fixed tissues can be frozen and sectioned in a cryostat. The sections (6 to 8 μm) are mounted on albuminized microscope slides, exposed to antibodies, stained, and embedded in plastic.

4) The fixed tissues can be dehydrated, embedded in polyethylene glycol, and sectioned with a light microscopic microtome. The sections (6 to 8 μm) are mounted on light microscopic slides, exposed to antibodies, stained, and embedded in plastic.

5) The fixed tissues can be dehydratde, embedded in plastic, and the hormones localized on ultrathin sections of the embedded tissues.

Growth hormone. Growth hormone has been localized in the anterior pituitary gland of the rat (Kawarai & Nakane 1970; Nakane 1970; Mazurkiewicz & Nakane 1972) and of amphibians (Doerr-Schott & Dubois 1973). It was found in somatotropic cells, as described by routine electron microscopy (Rinehart & Farquhar 1953; Hedinger

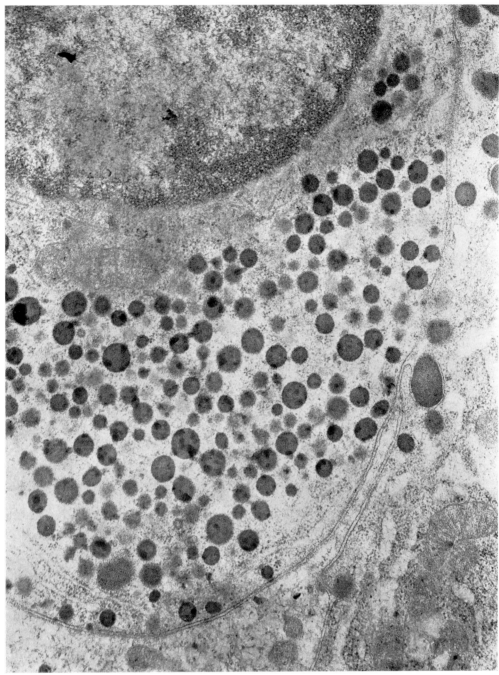

Fig. 1 Growth hormone cell in rat anterior pituitary (NAKANE 1971). The pituitary was fixed in p-formaldehyde-picric acid, postfixed in osmium tetroxide, and embedded in Epon. The hormone was localized on ultrathin sections using rabbit anti-rat growth hormone serum and peroxidase-labeled sheep anti-rabbit IgG immunoglobulin. 4-Cl-1-naphthol and hydrogen peroxide were used as substrates. The reaction product is visible over the secretion granules.

& FARQUHAR 1957), in secretion granules with an approximate diameter of 300–350 nm (Fig. 1).

Prolactin. Prolactin was also localized in the anterior pituitary gland of the rat (KAWARAI & NAKANE 1970; NAKANE 1970) and of amphibians (DOERR-SCHOTT & DUBOIS 1973). It was found in the secretion granules as well as in some endoplasmic reticulum of the cells which have been designated by routine electron microscopy as mammotropic cells (Fig. 2).

ACTH. Emphasis has been placed on the immunohistochemical localization of ACTH because of some controversy on identification of ACTH cells in anterior pituitary cells with routine electron microscopy (SIPERSTEIN & ALLISON 1965; KUROSUMI & KOBAYASHI 1966). Immunohistochemically, ACTH-containing cells appear to be generally stellate in shape and contain secretion granules with an approximate diameter of 200 nm (NAKANE 1970) or with an average diameter of 220 nm (MORIARTY & HALMI 1972). This subject has been reviewed in more detail by MORIARTY (1973). The immunoreactivity of ACTH in the anterior pituitary gland was preserved by 4% p-formaldehyde-picric acid (NAKANE 1970), by 2.5% glutaraldehyde, but not by osmium tetroxide (MORIARTY & HALMI 1972). It was preserved after embedding in methacrylate (NAKANE 1971) or araldite 6005 (MORIARTY & HALMI 1972). Because of the presence of a common N-terminal amino acid sequence in α-MSH, β-MSH and ACTH, usually an antiserum against the 17 through 39 amino acid sequence of ACTH has been employed for specific localization of ACTH. With this type of antiserum, ACTH was localized in the anterior pituitary gland of adult rats (MORIARTY 1973), fetal rats (SÉTÁLÓ & NAKANE 1972), humans (MORIARTY 1973), and frogs (DOERR-SCHOTT & DUBOIS 1973).

ACTH has been localized in the intermediate lobe of mice (NAIK 1973) and of rats (NAKANE 1970; MORIARTY & HALMI 1972; MORIARTY et al. 1973; NAIK 1973) at the ultrastructural level. ACTH cells with secretion granules were localized at the margin of the lobe. The secretion granules were dense and had an average diameter of 180 nm.

TSH. TSH was localized in the anterior pituitary gland of the rat (NAKANE 1970). The TSH cells were found to be angular or polyhedral with secretion granules of approximately 100–150 nm in diameter (Fig. 3).

Gonadotropins. LH and FSH have both been localized at the ultrastructural level in anterior pituitary (NAKANE 1970).

Specificities of the antisera used to localize these hormones have been questioned by some investigators (MORIARTY 1973) since it is now known that there are common amino acid sequences in the α subchain of TSH, and LH and possibly that of FSH (PAPKOFF 1971; CORNELL & PIERCE 1973). Antisera against these hormones may therefore cross-react. However, one must be reminded that animals immunized with these hormones will not always produce antibodies against the common subchain and do frequently produce antisera specific to the injected hormones. Particularly, when the antigens are glycoproteins, such as the case with these hormones, antisera may be produced against the carbohydrate moieties or the carbohydrate-amino acid complex specific to the given hormone. It has been well-established that an antiserum against 1–24 ACTH usually does not cross-react with α-MSH which is composed of the first 13 amino acids in an identical sequence to the first 13 amino acids of the ACTH and vice versa (MORIARTY & HALMI 1972). Therefore, if proper selection of antisera is made, proper controls are done, and the possibility of cross-reaction is eliminated, the antisera against whole hormones may be utilized (NAKANE 1973).

FSH and LH (Figs. 4, 5) were frequently found in morphologically-identical cell types and at the light microscopic level both hormones were found in the same cell (NAKANE 1970, 1973).

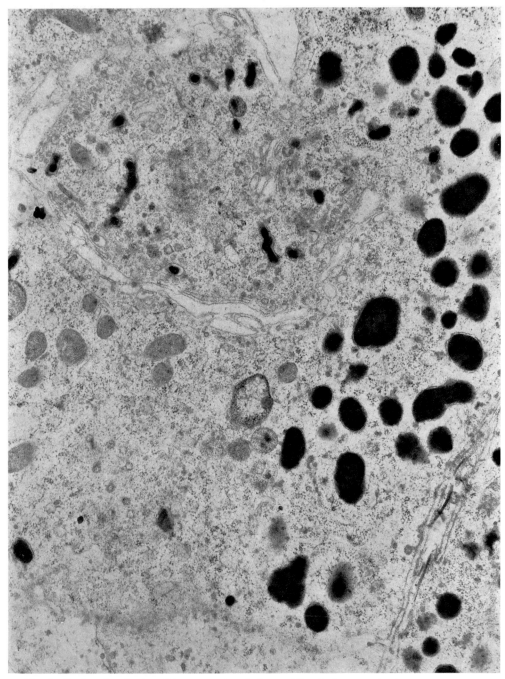

Fig. 2 Portion of prolactin cell in rat anterior pituitary. The pituitary was fixed in p-formaldehyde-picric acid, postfixed in osmium tetroxide, and embedded in Epon (NAKANE 1971). The hormone was localized on ultrathin sections using rabbit anti-rat prolactin serum and peroxidase-labeled sheep anti-rabbit IgG immunoglobulin. 4-Cl-1-naphthol and hydrogen peroxide were used as substrates. The reaction product is visible over the secretion granules in the cell periphery as well as over those in the matrix of the Golgi complex,

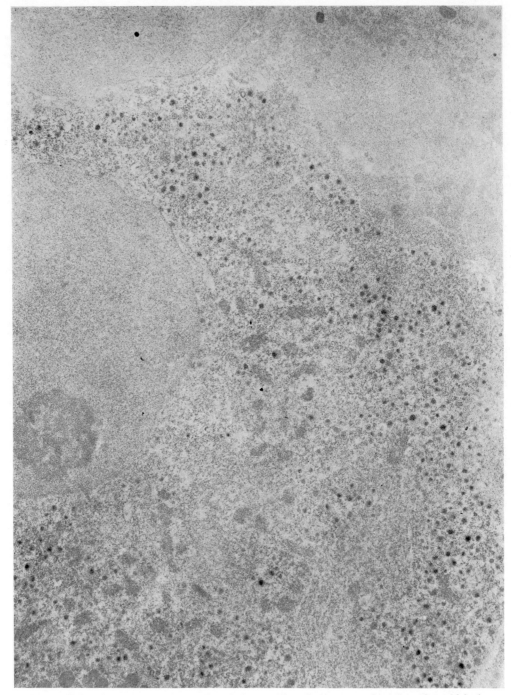

Fig. 3 Thyrotropic cell in rat anterior pituitary. The pituitary was fixed in p-formaldehyde-picric acid and embedded in polyethylene glycol. Six micron sections were mounted on slides (MAZURKIEWICZ & NAKANE 1972). TSH was localized using rabbit anti-human TSH serum and peroxidase-labeled sheep anti-rabbit IgG immunoglobulin. 3, 3'-diaminobenzidene and hydrogen peroxide were used as substrates. The hormone is present in the core of the secretion granules.

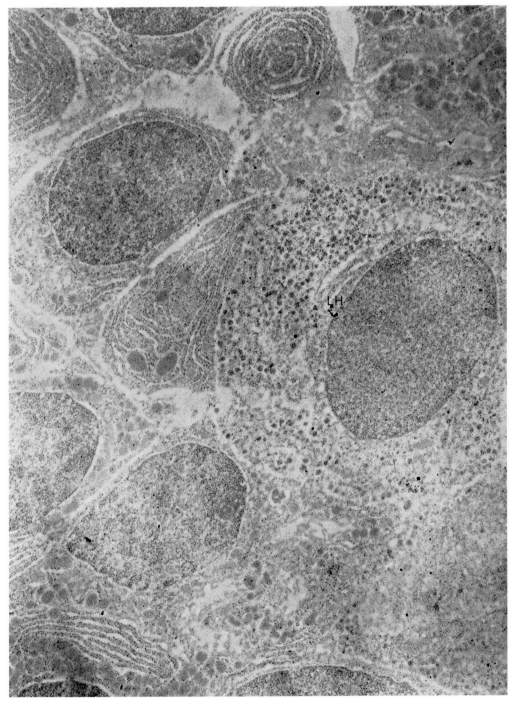

Fig. 4 LH cell in rat anterior pituitary. The gland was fixed in p-formaldehyde-picric acid and embedded in methacrylate (Kawarai & Nakane 1970). LH was localized on ultrathin sections using rabbit anti-HCG serum and peroxidase-labeled sheep anti-rabbit IgG immunoglobulin. 4-Cl-1-naphthol and hydrogen peroxide were used as substrates. The hormone is present in secretion granules of a gonadotropic cell (LH).

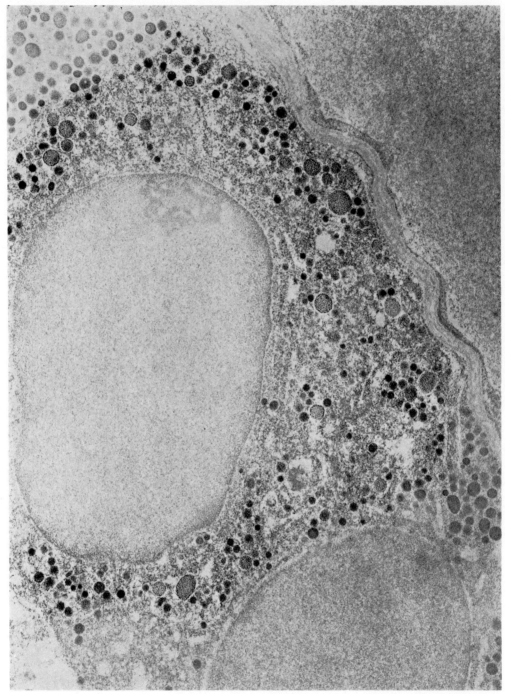

Fig. 5 LH cell in rat anterior pituitary (NAKANE 1974a). The gland was fixed in p-formaldehyde-picric acid and embedded in polyethylene glycol. Six micron sections were mounted on slides (MAZURKIEWICZ & NAKANE 1972). LH was localized using rabbit anti-HCG serum and peroxidase-labeled sheep anti-rabbit IgG immunoglobulin. 3, 3′-diaminobenzidine and hydrogen peroxide were used as substrates. The hormone is present in secretion granules of variable morphology.

β-MSH. β-MSH has been found in cells of the intermediate lobe which are different from ACTH-containing cells. (NAIK 1973). The MSH-positive secretion granules usually had a dense core and were larger than the secretion granules of ACTH cells (NAIK 1973). On the other hand, the same population of secretion granules reacted with anti β-MSH, anti 1–24 ACTH or anti 17–39 ACTH in the study conducted by MORIARTY (1973) and she suggested that these peptides were present in the same granules.

Vasopressin. Vasopressin has been localized in the secretion granules of Herring bodies in the posterior pituitary gland of the rat (RUFENER & NAKANE 1973).

HCG. HCG has been localized in human placenta by several laboratories (DRESKIN et al. 1970; HAMANAKA et al. 1971; GENBACEV et al. 1972). Irrespective of the method employed, the hormone was found on microvilli, on the membranes of the rough endoplasmic reticulum of syncytiotrophoblasts and on the basement membrane associated with these cells. In all cases the Golgi saccules were not stained.

HCS. HCS was found in identical sites to those of HCG in the syncytiotrophoblast of human placenta (DE IKONICOFF 1973).

Calcitonin. Calcitonin was localized with peroxidase-labeled rabbit anti-porcine calcitonin (synthetic) in dog thyroid (KALINA & PEARCE 1971) and with rabbit anti-porcine calcitonin (synthetic) followed by cytochrome c-labeled Fab' fraction of sheep anti-rabbit Fab' in porcine thyroid (DE GRANDI et al. 1971). The dog thyroid was fixed by perfusion either by 4% p-formaldehyde in 0.2 M cacodylate buffer or by a mixture of 0.5% glutaraldehyde and 4% p-formaldehyde in 0.2 M cacodylate buffer. The piglet thyroid was fixed by perfusion and the adult porcine thyroid by immersion in 2.7% formaldehyde with 0.2% picric acid in 5 mEq per liter $CaCl_2$ in 0.075 M cacodylate-HCl buffer, pH 7.4 adjusted to 1050 mos with sodium chloride. Both groups of investigators were successful in localizing calcitonin in C cells of thyroid, although the intracellular localization of the hormone varied considerably. In dog thyroid, the calcitonin was found within the secretion granules and the cisternae of the rough endoplasmic reticulum, but not in the Golgi saccules. On other hand, in porcine thyroid, it was found in the secretory granules and diffusely in cytoplasmic matrix, but not in the cisternae of rough endoplasmic reticulum, perinuclear space, or Golgi saccules. Several explanations may be considered for these discrepancies. For example, in the porcine thyroid, calcitonin was possibly not well-fixed and diffused through the cytoplasmic matrix or the cytochrome c-labeled antibody or rabbit anti-porcine calcitonin may have failed to penetrate into the cisternae of the endoplasmic reticulum as DE GRANDI et al. (1971) alluded.

Gastrin. Gastrin has been localized in human antral mucosa (GREIDER et al. 1972). The cells contained secretion granules with a diameter of 15–200 nm and were dispersed along the antral pyloric galnds, especially in the middle third of the glands.

Discussion

The immunocytochemical localization of hormones *in situ* relies heavily upon the effectiveness of the fixation method employed. The fixative must preserve ultrastructure without destroying antigenicity.

The rate of penetration of the fixative plays an important role in the precise localization of soluble antigens. If the fixative fails to reach and stabilize the hormones before they diffuse from their natural habitats, diffused hormone may subsequently be fixed at some other sites and result in artifactual localization. This is particularly significant when the hormones are small and highly soluble. The prevention of this diffusion artifact should always be in the minds of investigators. No reported investigation, including my own, has guaranteed that his has not occurred.

Diffuse immunocytochemical staining was observed in ACTH cells of anterior pituitary (MORIARTY 1973) and in C cells of thyroid (DE GRANDI et al. 1971). The former observation was interpreted to suggest either that ACTH is secreted by movement through secretion granule membranes and the plasma membrane or that diffusion of ACTH occurred during the fixation process (MORIARTY & HALMI 1972). Similarly, the latter observation was interpreted as being either the true physiologic situation or due to diffusion artifact (DE GRANDI et al. 1971). An improvement of the fixation procedure could eliminate the qualifying interpretations. In our laboratory, the possible use of cryostabilization at the time of fixation has been pursued for some time, but the results have not been encouraging.

Some correlations between choice of fixative and nature of the hormone antigen to be stabilized can be made. If the hormone is known to be a polypeptide, an aldehyde fixative should be used. If the hormone is known to be a glycoprotein, a fixative directed toward its carbohydrate moiety, such as that recently developed by McLEAN and NAKANE (1973), should be superior to an aldehyde fixative, since the antigenicity of the peptide portion should not be significantly altered. For the preservation of antigenicity, it is desirable that the fixative react with the antigen at a site distant from the antigenic determinant. For example, if an aldehyde fixative were employed to stabilize ACTH *in situ*, the anti-ACTH should be directed mainly against the C-terminal of the hormone rather than the mid-portion, since the aldehyde is expected to attach to the lysine residues in the 11, 15, 16, and 21st portions of the molecule and thus denature the antigenicity of this section.

The fixation and immunocytochemical localization of a hormone may be affected by the state of the hormone. For example, a hormone may be more easily denatured while it is being synthesized than during the process of secretion. Or, it may be complexed with other molecules such as enzymes or carrier proteins. The frequent lack of immunoreactivity in the saccules of the Golgi complex, and consistent lack of staining in mitochondria and nuclei should be reassessed from this point of view. Interpretation of results should therefore be based on positively-reacted areas unless the above possible complications are completely eliminated.

One other point to be stressed is the biologic and immunologic species cross-reactivity of hormones. Most frequently we employ an antiserum against a hormone from one species to localize the hormone in another. With this approach, the immunologic reaction is expected to be selective for the biologically-active portion of the hormone.

Because of the species cross-reactivity of antisera, antibodies are usually bound to endogenous hormone. The hormone-antihormone complexes can attach to hormone binding sites and thus result in false positive hormone staining (NAKANE, unpublished). It may often be desirable therefore to dissociate bound hormone from its antibody.

ACKNOWLEDGEMENTS

Supported in part by USPHS grants AI-09109 and AM-15663 and ACS grant DT-14. The technical assistance of Ms. B. WILSON, Drs. A. KAWAOI, Y. KAWARAI and J. MAZURKIEWICZ is greatly appreciated. The help for the preparation of this manuscript by Ms. B. WILSON and Ms. L. KING is also acknowledged.

REFERENCES

[1] AVRAMEAS S. (1968) Détection d'anticorps et d'antigenes à l'aide d'enzymes. Bull Soc. Chim. Biol., *50*: 1169.

[2] AVRAMEAS S. (1969) Coupling of enzymes to proteins with glutaraldehyde. Use of the conjugates for the detection of antigens and antibodies. Immunochemistry, 6: 43.

[3] AVRAMEAS S. and URIEL J. (1966) Mèthode de marquage d'antigèns et d'anticorps avec des enzymes et son application en immunodiffusion. C. R. Acad. Sci. Paris, Series D, 262: 2543.

[4] CORNELL J. S. and PIERCE J. G. (1973) The subunits of human pituitary thyroid-stimulating hormone. J. Biol. Chem., 248: 4327.

[5] DE GRANDI P. B., KRAEHENBUHL J. P. and CAMPICHE M. A. (1971) Ultrastructural localization of calcitonin in the parafollicular cells of pig thyroid gland with cytochrome c-labelled antibody fragments. J. Cell Biol., 50: 446.

[6] DE IKONICOFF L. K. (1973) Etude histochimique du placenta humain. Localisation ultrastructurale de l'hormone chorionique somatomammotrophique (HCS) par une technique immunocytochimique à la peroxidase. C. R. Acad. Sci. Paris, Series D, 276: 3355.

[7] DOERR-SCHOTT J. and DUBOIS M. P. (1973) Mise en évidence des hormones de l'hypophyse d'un amphibien par la cyto-immunoenzymologie au microscope électronique. C. R. Acad. Sci. Paris, Series D, 276: 2179.

[8] DRESKIN R. B., SPICER S. S. and GREENE W. B. (1970) Ultrastructural localization of chorionic gonadotropin in human term placenta. J. Histochem. Cytochem., 18: 862.

[9] GENBACEV O., ROBYN E. and PANTIC V. (1972) Localization of chorionic gonadotropin in human term placenta on ultrathin sections with peroxidase-labeled antibody. J. Microscopie, 15: 399.

[10] GREIDER M. H., STEINBERG V., McGUIGAN J. E. (1972) Electron microscopic identification of the gastrin cell of the human antral mucosa by means of immunocytochemistry. Gastroenterology, 63: 572.

[11] HAMANAKA N., TANIZAWA O., HASHIMOTO T., YOSHINARE S. and OKUDAIRA Y. (1971) Electron microscopic study on the localization of human chorionic gonadotropin (HCG) in the chorionic tissue by enzyme-labeled antibody technique. J. Electron Microscopy, 20: 128.

[12] HEDINGER C. E. and FARQUHAR M. G. (1957) Elektronenmikroskopische Untersuchungen von zwei Typen acidophiler Hypophysenvorderloppenzellen bei der Ratte. Schweiz Z. Pathol. Bakt., 20: 766.

[13] KALINA M. and PEARSE A. G. E. (1971) Ultrastructural localization of calcitonin in C-cells of dog thyroid: an immunocytochemical study. Histochemie, 26: 1.

[14] KAWAOI A. and NAKANE P. K. (1973) An improved method of conjugation of peroxidase with protein. Fed. Proc., 32: 840.

[15] KAWARAI Y. and NAKANE P. K. (1970) Localization of tissue antigens on the ultrathin sections with peroxidase-labeled antibody method. J. Histochem. Cytochem., 18: 161.

[16] KRAEHENBUHL J. P., GALARDY R. E. and JAMIESON J. D. (1974) Preparation and characterization of an immunoelectron microscope tracer consisting of a heme-octapeptide coupled to Fab. J. Exp. Med., 139: 208.

[17] KURISAKA M. (1972) Brain Tumor and human placental lactogen (HPL). The Nihon Univ. J. Med., 14: 187.

[18] KUROSUMI K. and KOBAYASHI Y. (1966) Corticotrophs in the anterior pituitary glands of normal and adrenalectomized rats as revealed by electron microscopy. Endocrinology, 78: 745.

[19] LEDUC E. H., SCOTT G. B. and AVRAMEAS S. (1969) Ultrastructural localization of intracellular immune globulins in plasma cells and lymphoblasts by enzyme-labeled antibodies. J. Histochem. Cytochem., 17: 211.

[20] MAZURKIEWICZ J. E. and NAKANE P. K. (1972) Light and electron microscopic localization of antigens in tissues embedded in polyethylene glycol with a peroxidase-labeled antibody method. J. Histochem. Cytochem., 20: 969.

[21] McLEAN I. W. and NAKANE P. K. (1973) Periodate-lysine-paraformaldehyde fixative: a new fixative for immunoelectron microscopy. J. Cell Biol., 59: 209.

[22] MISUGI K., HOWELL S. L., GREIDER M. H., LACY P. E. and SORENSON G. D. (1970) The pancreatic beta cell. Demonstration with peroxidase-labeled antibody technique. Arch. Path., 89: 97.

[23] MORIARTY G. C. (1973) Adenohypophysis: Ultrastructural cytochemistry. A review. J. Histochem. Cytochem., 21: 855.

[24] MORIARTY G. C. and HALMI N. S. (1972)   Electron microscopic study of the adrenocorticotropin-producing cell with the use of unlabeled antibody and the soluble peroxidase-antiperoxidase complex.   J. Histochem. Cytochem., *20:* 590.

[25] MORIARTY G. C., MORIARTY C. M. and STERNBERGER L. A. (1973)   Ultrastructural immunocytochemistry with unlabeled antibodies and the peroxidase-antiperoxidase complex. A technique more sensitive than radioimmunoassay.   J. Histochem. Cytochem., *21:* 825.

[26] NAIK D. V. (1973)   Electron microscopic-immunocytochemical localization of adrenocorticotropin and melanocyte stimulating hormone in the pars intermedia cells of rats and mice.   Z. Zellforsch., *142:* 305.

[27] NAKANE P. K. (1970)   Classifications of anterior pituitary cell types with immunoenzyme histochemistry.   J. Histochem. Cytochem., *18:* 9.

[28] NAKANE P. K. (1971)   Application of peroxidase-labeled antibodies to the intracellular localization of hormones.   Acta Endocrin. Suppl., *153:* 190.

[29] NAKANE P. K. (1973)   Distributions of gonadotropic cells in the anterior pituitary gland of the rat.   *In:* The Regulation of Mammalian Reproduction. SEGAL S. J., CROZIER R., CORFMAN P. A. and CONDLIFFE P. G. (eds.) pp. 79–88, C. C. Thomas, Springfield, Illinois.

[30] NAKANE P. K. (1974a)   Ultrastructural localization of tissue antigens with the peroxidase-labeled antibody method.   *In:* Proceedings 2nd International Symposium, Drienerlo, The Netherlands, June 25–29, 1973. WISSE E., DAEMS W. T., MOLENAAR I. and VAN DUIJM P. (eds.) pp. 129–143, North Holland Publishing Co., Amsterdam.

[31] NAKANE P. K. (1974b)   Localization of hormones with the peroxidase-labeled antibody method.   *In:* Methods in Enzymology. O'MALLEY and HARDMAN (eds.) *37:* Part B, in press.

[32] NAKANE P. K. and PIERCE G. B. (1966a)   Enzyme-labeled antibodies: Preparation and application for localization of antigens.   J. Histochem. Cytochem., *14:* 929.

[33] NAKANE P. K. and PIERCE G. B. (1966b)   Enzyme-labeled antibodies for the ultrastructural localization of antigens.   *In:* Proceedings of the Sixth International Congress of Electron Microscopy, *II:* 51.

[34] NAKANE P. K. and PIERCE G. B. (1967)   Enzyme-labeled antibodies for the light and electron microscopic localization of tissue antigens.   J. Cell Biol., *33:* 307.

[35] OKUDAIRA Y., HASHIMOTO T., HAMANAKA N. and YOSHINARE S. (1971)   Electron microscopic study on the trophoblastic cell column of human placenta.   J. Electron Microscopy, *20:* 93.

[36] PAPKOFF H. (1971)   The subunit nature of interstitial cell-stimulating hormone and follicle-stimulating hormone.   *In:* Structure-Activity Relationships of Protein and Polypeptide Hormones. MARGONLIES M. and GREENWOOD F. C. (eds.) Part I, pp. 73–79, Excerpta Medica, Amsterdam.

[37] RAM J. S., NAKANE P. K., RAWLINSON D. G. and PIERCE G. B. (1966)   Enzyme-labeled antibodies for ultrastructural studies.   Fed. Proc., *25:* 732.

[38] RINEHART J. F. and FARQUHAR M. G. (1953)   Electron microscopic studies of the anterior pituitary gland.   J. Histochem. Cytochem., *1:* 93.

[39] RUFENER C. and NAKANE P. K. (1973)   Light and electron microscopic localization of vasopressin in the supraoptic nucleus of hypothalamus and posterior pituitary gland of the rat. Anat. Rec., *175:* 432.

[40] SÉTÁLÓ G. and NAKANE P. K. (1972)   Studies on the functional differentiation of cells in fetal anterior pituitary glands of rats with peroxidase-labeled antibody method.   Anat. Rec., *172:* 403.

[41] SHANNON L. M., KAY E. and LIEW J. Y. (1966)   Peroxidase isozymes from horseradish roots. I. Isolation and physical properties.   J. Biol. Chem., *241:* 2166.

[42] SIPERSTEIN E. R. and ALLISON V. F. (1965)   Fine structure of the cells responsible for the secretion of adrenocorticotrophin in adrenalectomized rat.   Endocrinology, *76:* 70.

[43] TOUGARD C., KERDELHUE B., TIXIER-VIDAL A. and JUTISZ M. (1973)   Light and electron microscope localization of binding sites of antibodies against ovine luteinizing hormone and its two subunits in rat adenohypophysis using peroxidase-labeled antibody technique.   J. Cell Biol., *58:* 503.

[44] WELINDER K. G. and SMILLIE L. B. (1972)   Amino acid sequence studies of horseradish peroxidase. I. Thermolytic peptides.   Canad. J. Biochem., *50:* 63.

4. X-Ray Microanalysis

Microanalysis of Biological Sections in the TEM and SEM

ALAN O. SANDBORG and JOHN C. RUSS

One of the major goals of the cell biologist is tying together the morphology and chemical functions of subcellular structure. One method that can be used for this purpose has been the use of the TEM with thin sections stained with chemicals that deposit electron-dense precipitates. Staining methods unfortunately are beset with numerous artifacts and false positive reactions, poor specificity for exact chemical activity, and problems of reproducibility, and have very little capability to achieve quantitative results.

X-ray microanalysis does not have these drawbacks, but until recently has been little used in the biological sciences. The primary use of electron probe X-ray microanalysis has been in the various fields of materials science—metallurgy, geology and semiconductor technology. The problems in the past of applying the conventional electron probe microanalyzer to biological material can be summarized as follows:

1) Typical microprobe specimens are bulk samples, and the available images (backscattered or secondary electrons) represent surface topography. Since the surface images from bulk biological specimens give only a very crude indication of structure, it is difficult to identify features of interest. Furthermore, the depth of penetration of electrons into the sample produces most X-ray excitation below the surface features that can be observed.

2) The extent of electron penetration in biological material is great, and produces an excited volume of several μm^3. This is not a fine enough scale of spatial resolution for most applications to subcellular structures.

3) The existing quantitative models for electron probe microanalysis do not extrapolate well to biological matrices, and in addition the models require standards that are not generally available nor readily made.

4) The low X-ray fluxes produced by biological samples and low efficiency of conventional microprobe spectrometers require the use of high electron beam currents, which cause rapid damage to biological materials because of specimen heating.

5) The art of biological specimen preparation has not until recently developed methods suitable for preserving the natural distribution of elements in tissue.

These problems have been overcome in the past few years due to a combination of several factors: (1) The use of thin sections rather than bulk samples, imaged by transmitted electrons in the TEM or STEM gives an image familiar to the biologist and useful for recognizing internal structure; (2) Thin sections also produce better X-ray spatial resolution, generally better than 10^{-3} μm^3, and so features of interest can be selectively analyzed; (3) We have now developed a quantitative analysis method suitable for biological thin sections, requiring no standards; (4) The high efficiency of the energy dispersive X-ray analysis and the possibility of locating the detector close to the sample to collect a large fraction of the generated X-rays permits the use of low electron beam currents (and small diameter incident beams). The simultaneous analysis of all elements by the energy dispersive method further reduces the time of analysis and specimen damage; (5) Specimen preparation using cryo-sectioning immobilizes ions in tissue and permits their analysis. Also, more selective stains are being developed that selectively localize sites of chemical activity since an electron dense deposit is not required, but rather an analyzable precipitate of some element not naturally present.

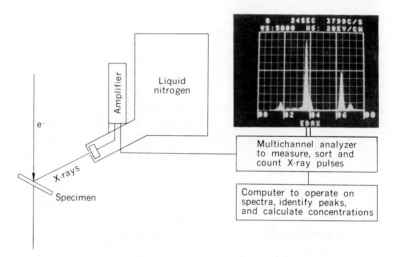

Fig. 1 Black diagram of energy-dispersive X-ray analysis system.

Energy Dispersive X-ray Analysis

Since it is still a relatively new technique, less well known than the older method of wavelength-dispersive (or diffractive) X-ray analysis which it has largely replaced, a description of the energy dispersive X-ray analyzer may be useful (Russ 1970). Fig. 1 shows a block diagram of the system. The X-ray detector is a highly purified single crystal of silicon, specially prepared to remove lattice imperfections and to fill any remaining vacancies and dislocations with small, mobile lithium atoms. As each X-ray photon enters the detector it loses its energy in a photoelectron shower that raises electrons to the conduction band in the intrinsic semiconductor. The number of conduction electrons is proportional to the X-ray photon energy, since 3.8 eV are required to raise each electron. An applied bias voltage of about 500 to 1,000 volts collects these electrons from the detector, which is then ready to receive the next X-ray. Since this process takes nanoseconds, all of the X-rays entering the detector are processed one at a time yet with apparent simultaneity to the user.

The electrons that are freed are collected by a field effect transistor connected to the detector and formed into a small voltage step, which is then amplified and differentiated to a pulse. The stream of pulses from the detector and amplifier represent the X-rays that entered the detector, and the height of each pulse is proportional to the energy of the corresponding X-ray. The detector and the first stages of the amplifier circuitry are cooled to liquid nitrogen temperature to reduce thermally generated electrical noise which would make the measurement of these extremely low currents difficult.

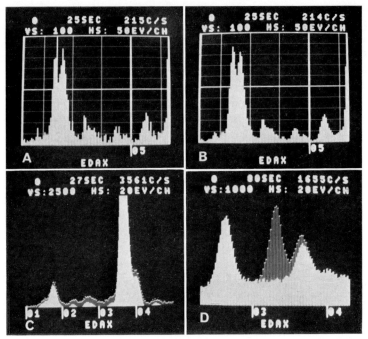

Fig. 2 Computer operations on spectra: A, B) before and after smoothing, which improves peak shape and definition; C) before (light) and after (dark) background subtraction, which gives true net peak areas; D) peak stripping to remove (light grey) potassium Kα and Kβ.

The pulses are measured and counted in the multichannel analyzer. The height of each pulse is converted to an address in the spectrum memory by an analog-to-digital converter, and a count stored at that address. Since many pulses are processed each second, from X-rays with different energies, this results in the continuous growth of several peaks in the spectrum. The position in the spectrum is a measure of the pulse height, and thus in turn of the X-ray energy, so that the horizontal scale is linear in X-ray energy and can be suitably calibrated and labelled. The size of each peak depends on the number of X-rays of that energy, which we can relate to the elemental concentration. The width of the peaks is the spectrometer resolution, and is the result of two factors causing the pulse height that is amplified and measured to have a small error instead of being exactly proportional to the original X-ray energy. One effect is the statistical nature of electron production in the detector itself, and the other is the electrical noise in the amplification process. The resulting peaks in the spectrum have a Gaussian shape and width that increases with the square root of energy. Since the X-ray energy from each electron shell (K, L, M) increases monotonically with atomic number, following Moseley's law, the position or energy of the peaks serves as a rapid and straightforward identification of the elements present.

The incorporation of a small computer in the system provides several important capabilities for data improvement. One is the stripping of overlapping peaks which can be present from different lines of various elements. Also, because of the statistical nature of X-ray production and the low X-ray intensities often encountered with biological samples, the shape of small peaks in the spectrum may be irregular and hence they may not be obvious to the user, without spectrum smoothing. A third, extremely important function is the separation of the characteristic X-ray peaks from the spectrum background, which is produced by the Bremsstrahlung X-rays generated by the incident electrons as they decelerate in the sample. These X-rays do not have characteristic energies related to the elements present in the sample, and make it difficult to detect small peaks from minor elements. Obtaining the net counts in each elemental peak is also the necessary first step in any quantitative analysis, and the integration of peak areas as well as each of the other steps of peak stripping and background subtraction can be carried out for one spectrum while the next one is being collected. Fig. 2 illustrates these operations for typical energy dispersive spectra.

Energy and Wavelength Dispersive Methods

In comparison to the wavelength-dispersive method of X-ray analysis, in which a diffractive crystal is used to separately select the X-ray photons from a given element, energy dispersive analysis of X-rays provides several important advantages, some of which are particularly significant for biological applications.

Since the detector processes X-rays of all energies, from all of the elements present, the spectrum that is accumulated gives information on all of the elements simultaneously. This not only represents a saving of time compared to one-at-a-time analysis of the elements (and hence reduces specimen damage), it also eliminates the possibility of missing unexpected elements. Electron-excited X-ray microanalysis with either energy or wavelength X-ray detection methods gives poor results for the light elements C, N and O, because of a low efficiency of X-ray production and the rapid absorption of these low energy photons in the sample. Fortunately their analysis in biological materials is rarely of interest, since they are present in abundance and it is low concentrations of heavier elements that can be used to investigate chemical processes.

The energy-dispersive X-ray analyzer does not have the focussing restrictions of a diffracting spectrometer, and therefore can be placed as close to the specimen as hardware restrictions permit to collect a large percentage of the generated X-rays, which are emitted in all directions. Furthermore, energy dispersive detection is 100% efficient over the range of energies of most elements and emission lines used, as compared to the 80–95% loss of intensity common to most diffracting crystals. These two factors together make energy dispersive X-ray analysis hundreds or thousands of times more efficient than the wavelength dispersive method, and hence make it possible to obtain useful data with small diameter (and therefore low current) incident beams and with thin sections of biological tissue, which produce low overall X-ray yields.

The simplicity and lack of moving parts, and the few controls of the energy dispersive X-ray analysis system provide reliability and ease of operation. Furthermore, since the system is absolutely stable and reproducible it is possible to extend the mathematics of quantitative analysis to eliminate the need for any standards.

Bulk Analysis with the SEM

The scanning electron microscope (SEM) is a descendant of the earlier electron probe microanalyzer, and is commonly used for surface imaging and microanalysis. The modes of operation for analytical purposes include mapping of the distribution of one element with respect to the surface features of the specimen, and complete analysis of the composition of a selected point. The first method can be performed in either an area map or line profile mode, and although it provides only qualitative information on concentration, it is pleasing and useful to the human eye. The low concentration of most elements in biological samples makes the mapping method of extremely limited usefulness in this field, however. The low intensities of X-rays that are generated produce low peak-to-background ratios, which are hard to discern in the traditional X-ray "dot" image and require very long scan times to collect useful statistics in the line profile mode. Additionally, variations in count rate due to topographic effects or to changes in Bremsstrahlung production due to variation in sample density can confuse the interpretation.

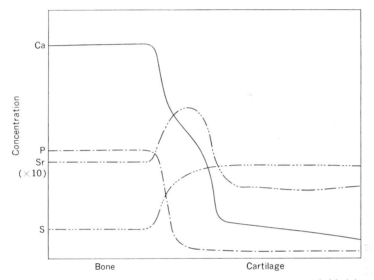

Fig. 3 Distribution of Ca, P, S and Sr across bone/cartilage interface in an arthritic joint. Note the variation of the Ca/P ratio in the interface and the accumulation of Strontium.

Point analysis has the advantage of giving complete and sensitive compositional information, either qualitative or quantitative, but requires the user to select locations for analysis based on his prior knowledge of the sample and application. It is possible to show concentration variations by a series of point analysis, as shown in Fig. 3; in this case correction for topographic and composition effects can be made and the information is presented in terms of concentration rather than just X-ray intensity.

The difficulty with this method as applied to most cases of biological tissue is, as mentioned before, the extent of electron penetration in the specimen. The range of 15 KV electrons, for example, in tissue is in excess of 6 μm, and most of the X-ray generation takes place in a roughly teardrop shaped volume that is several μm in diameter and more than 3 μm beneath the surface as shown in Fig. 4. This volume is larger than most of the structures that cell biologists want to analyze, and also lies beneath the surface morphological features which are visible in a microscope image.

Thin Sections and Spatial Resolution

The needs for imaging internal structure and improving analytical spatial resolution are both met by using thin sections as samples. Thin sections can be examined in either the

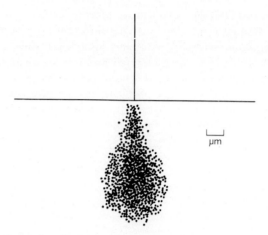

Fig. 4 Generation of calcium X-ray in bulk tissue specimen by 15 keV electrons, showing extent of penetration.

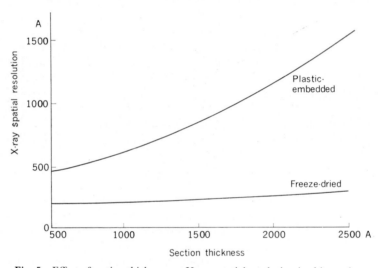

Fig. 5 Effect of section thickness on X-ray spatial resolution in thin sections.

conventional transmission electron microscope (TEM) or by using an SEM with a detector for transmitted electrons (STEM). The image in either case reveals internal morphological details in a way familiar to the biologist and with image resolution more than adequate to recognize and locate features to be analyzed.

The choice of the SEM with transmission capability or the TEM with attachments to form a fine incident beam (in which case it may too become an STEM) has implications beyond the scope of this paper. For analytical purposes the higher accelerating voltages and more versatile electron optics of the TEM allow better excitation of the sample, but the restricted space around the specimen requires great ingenuity in designing special cooling stages, which will be needed for some applications using cryo-sections of tissue.

With thin sections the analytical spatial resolution is still greater than the incident electron beam diameter because of transverse electron scattering, but this is much less than in bulk material. The curves in Fig. 5 show the results that we have predicted on the basis of

Monte-Carlo calculations and confirmed by experience for conventionally prepared Epon-embedded sections and freeze-dried sections (Russ 1972). The latter give less scattering because of the removal of the mass of the water, and so the spatial resolution is improved.

Since the excited volume in ultrathin sections can be less than 10^{-4} μm^3, and since electron excited analyses of most elements in an organic matrix (which is favorable by virtue of its low absorption and low production of Bremsstrahlung) can detect concentrations of a few hundred ppm or less, we can calculate a theoretical limit of mass detection of less than 10^{-22} g. This is only a few hundred atoms, and the problem in achieving such a detection lies in the extremely low count rate that can be obtained with samples of this type and the long times that would be required. More realistic and useful detection limits are 10^{-17} to 10^{-18} g, and to obtain higher count rates that give statistically meaningful results in practical analysis times of a few hundred seconds it is wise to use sections somewhat thicker than common for normal TEM viewing. Section thickness from 1,500 to 4,000 Å represent a useful compromise for most cases; they can be imaged by either TEM or STEM, although STEM is more satisfactory in providing adequate image contrast when the conventional staining chemicals are omitted.

Specimen Preparation in Different Applications*

Useful microanalysis can be performed with conventionally prepared—embedded and stained—sections for some purposes. If the elements of interest are well bound to internal structures, which is the case for many heavy metals of interest in pollution studies as well as some natural elements that are protein-bound, they may remain fixed in position during the extensive chemical treatments involved in normal preparation procedures. Even in this case it is wise to eliminate the heavy element stains, which produce X-ray peaks that can confuse or obscure the peaks of interest.

Another useful technique is the intentional introduction of chemicals that will selectively precipitate an analyzable deposit at the site of a specific chemical activity. Conventional staining chemicals are not highly selective; by removing the need for a massive deposit of a heavy electron-dense element a new class of marker stains is coming into being. Any analyzable element not already present in the tissue can be used, and the biochemical ingenuity of researchers will be given broad scope.

Many elements of analytical interest are highly mobile and special methods are required to preserve their distribution representative of the living condition. Most, if not all, of the satisfactory and generally useful methods developed to date involve the rapid freezing of the sample, as for example in melting nitrogen. Small bits of frozen tissue can then be sectioned in the cryo-ultramicrotome and the sections examined while still kept at or near liquid nitrogen temperature, or the sections can be further treated chemically and/or freeze-dried and examined at ambient temperature. Drying of course causes some translocation of the solute atoms but only to the nearest membrane or structural surface.

Obtaining Quantitative Results

Since most of the preparation methods involve either the removal of mass, primarily water, from the specimen, or the addition of mass in the form of stains or embedding media,

---

* Russ & Panessa 1972

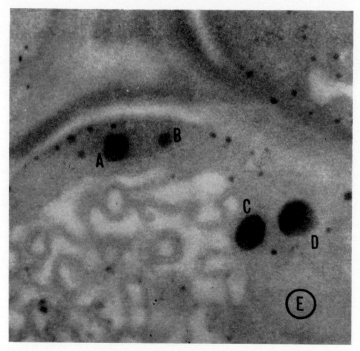

Fig. 6 Example of quantitative analysis of elemental ratios in structural granules of blue-green algae. Note the consistent levels of S and inverse relationship of K and Ca; point E is the control and Cl is present in the embedding plastic.

it is not practical to attempt to analyze elemental concentrations directly. Furthermore, an additional loss of organic mass occurs under the incident electron beam. It is most meaningful in the majority of cases to determine elemental ratios, perhaps using one element intentionally added as a reference (for instance the chlorine present in embedding plastics).

The Hall, or P/B model relates elemental concentration to the ratio of the net elemental intensity—a measure of the mass of the element present—to the background intensity—proportional to the total excited mass (HALL & WERBA 1968). This method requires the use of standards for each element and normalizes the effect of sample density and thickness. Elemental concentration ratios can be directly calculated without standards using a model we have recently developed (RUSS 1973). Standards are not needed since the energy dispersive system has a constant and calculable sensitivity for various elements. Variations in sample density and thickness cancel since the elements are all analyzed simultaneously in the same excited volume. The only requirement of the model is that the section thickness be small enough that X-ray absorption can be neglected in the sample. For organic material this condition is met in most cases by thicknesses up to a micrometer, and even for low energy X-rays from elements such as Na and Mg thicknesses to 1/2 μm can be used. Fig. 6 illustrates the application of the quantitative method.

Conclusion

Microanalysis of biological tissue can provide information linking together morphology and chemical function on a subcellular scale. The application of this technique in the biological sciences has resulted from the combination of several factors—the development of the TEM and STEM for thin section examination with fine incident beams; the high efficiency of the energy dispersive X-ray analyzer; and the development of new preparative methods, especially cryotechniques. Routine quantitative analysis of elemental concentration ratios is readily obtained.

REFERENCES

[1] HALL T. A. and WERBA P. (1968) The measurement of total mass per unit area and elemental weight-fractions along line scans in thin specimens, Proceedings of the 5th Int'l Congress on X-Ray Optics and Microanalysis, p. 93.

[2] Russ J. C. (ed.) (1970) Energy dispersion X-ray analysis; X-ray and electron probe analysis, ASTM Spec. Tech. Publ., *485*, Philadelphia, p. 217.

[3] Russ J. C. (1972) Resolution and sensitivity of X-ray microanalysis in biological sections by scanning and conventional transmission electron microscopy, Proceedings of the 5th Annual Scanning Electron Microscope Symposium, IITRI, Chicago, p. 73.

[4] Russ J. C. (1973) Microanalysis of thin sections, Coatings and rough surfaces, Proceedings of the 6th Annual Scanning Electron Microscope Symposium, IITRI, Chicago, p. 114.

[5] Russ J. C. and PANESSA B. J. (eds.) (1972) Proceedings of the Thin-Section Microanalysis Symposium, St. Louis, EDAX Laboratories.

III.

Freezing Technique for Electron Microscopy of Cells

Review of the Freezing Techniques and their Theories

Tokio NEI

The substance present in the greatest quantity as a component in most biological materials is water, which generally accounts for 70 to 80% of the total weight. Water plays an important role in maintaining the structures and functions characteristic of living matter. The removal of water exerts a great influence, causing some alterations in morphology and physiology of biological specimens. Standard procedures unfortunately require that specimens be dehydrated for observation in the electron microscope. Unlike light microscopy, it has long been believed impossible to observe hydrated specimens with the electron microscope, because of the need to place the specimens in a high vacuum.

When water associated with polymers such as proteins and lipids is removed, alterations in the molecular structure of the high polymers most likely occur and subsequently bring about a certain morphological change at the electron microscopic level. Since the essential aim of electron microscopy is to observe the fine structures of biological specimens, even a slight alteration in molecular structure should not be overlooked.

Various kinds of chemical and physical treatments are employed in the preparation of biological specimens; thin-sectioning, in particular, requires fixation, dehydration, embedding, sectioning and staining. In each of these procedures, there are possibilities that artifacts will arise to interfere with the morphological observation.

It has been a long-cherished desire of biologists, who have been engaged in electron microscopy, to observe hydrated materials in their native state without any artifacts caused by the various preparative treatments. A number of attempts have been made to examine living materials in controlled environmental devices. Dupouy et al. (1960) and Dupouy (1968) have done some pioneer work in this field. The recent developments in high voltage electron microscopy have promoted additional attempts to design new equipment, because of the higher transmission and lesser radiation damage to biological specimens. A differentially pumped hydration chamber was constructed for a high-voltage microscope and the observation of several types of cells in their wet state inside this chamber was made by Parsons and his colleagues (Parsons & Moretz 1970; Parsons et al. 1972). A closed type of environmental cell was also designed and several living organisms were examined by Nagata and Ishikawa (1972). Successful results, however, have not yet been obtained due to technical difficulties in lessening radiation damage and in increasing image contrast. Water molecules possess a very large cross-section for inelastic scattering. The brownian motion of tiny particles present in wet specimens hinders the attainment of a sharp image in electron microscopy. Even with the high voltage electron microscope, therefore, one can hardly expect to obtain high resolution in hydrated specimens (Glaeser 1971; Kobayashi 1971). As mentioned above, there still remain many problems to be solved in using environmental cells for observation of biological materials in their native state.

Cryotechniques such as freeze-drying and freeze-substitution methods have long been applied to electron microscopy. Owing largely to a recent appreciation of the great advantages found in the freezing and subsequent processing of biological materials, newly designed cryotechniques have been extensively developed in the past few years. Even soft specimens have been found to maintain their native state, if they are frozen under appropriate conditions. Such solidified specimens can easily be fractured or sectioned. Freeze-etching and cryosectioning techniques have thus been designed for this purpose and have brought about a most impressive list of progress in electron microscopy. Direct observations of frozen specimens have also been attempted and successful results obtained. Freezing has thus become an increasingly important method for stabilizing biological structures. In spite of such recent progress in cryotechniques, however, formation and suppression of ice crystals in frozen specimens is still a substantial problem to be investigated further.

Aims of the Freezing Techniques

Aims of the freezing techniques in electron microscopy are various. The first objective is to observe the specimen in its native state, as mentioned above. If the specimen is kept at sufficiently low temperatures under the microscope, there is no dehydration (sublimation) even under high vacuum. It should therefore be possible to observe the frozen specimen directly with a scanning electron microscope and, in special cases, with a transmission electron microscope, if the specimen is sufficiently thin. This objective also includes retaining of the three-dimensional structure of the specimen. Although most biological specimens are likely to shrink or be deformed when subjected to air-drying which is generally used in conventional methods, their original three-dimensional structures can almost be retained by freeze-drying or freeze-substitution. The second objective is to solidify the soft tissues by freezing. Solidification of the tissue makes sectioning or fracturing simple. The third objective is to lessen the release of chemical substances. Chemical treatment, especially dehydration with organic solvents employed in conventional procedures, particularly in thin-sectioning, may cause the release of some chemical constituents from the cells and tissues. Dehydration of specimens by sublimation or substitution from a frozen state prevents such losses. The fourth objective is to maintain the viability of the specimen. Activity or viability of biological specimens can be maintained more readily in the frozen state at low temperatures. This is an important problem viewed from a general cryobiological standpoint. With this approach, both morphological and physiological studies can be carried out concurrently.

Water and Ice

Water

Prior to understanding the formation of ice in frozen materials, it is necessary to have some knowledge of the structure and properties of water in biological systems (DORSEY 1940; WHIPPLE 1965; MERYMAN 1966). Water is the major component of biological specimens and plays an important role in maintaining structure and in promoting functional activities of living cells and tissues. The structure of water in biological systems has been discussed by many investigators (NEMETHY & SCHERAGA 1962a, b; FRANK 1965, 1970; KAVANAU 1965; KLOTZ 1965, 1970). There are two common terms used to define water in biological systems; solvent and nonsolvent water. The former is called free water or freezable water, which exists free in the bulk liquid; most of it can be frozen during the freezing process down to around $-20\,°C$. The latter is usually called bound water or unfreezable water, i.e., hydrated or immobilized water, which remains unfrozen below that same subzerotemperature. It has been generally assumed that removal of the solvent water is not harmful to the biological systems, but that removal of the nonsolvent water associated with high polymers affects the morphology and functions of the systems to some extent.

Crystallization of ice

Related to ice formation, there have been many research studies and numerous general reviews (EISENBERG & KAUZMAN 1969; RIEHL et al. 1969; FLETCHER 1970). The reader should refer to these reviews for further information on the physics of ice.

There are two kinds of freezing or solidification defined respectively as, (1) crystallization of ice which is the orderly arrangement of water molecules, (2) vitrification or amorphization which is a kind of supercooling in the sense that it appears to occur without an orderly arrangement of water. The crystallization process consists of a nucleation and crystal growth of ice. Nucleation is the combining of molecules into an ordered particle of a size sufficient to survive as a site for crystal growth. Crystal growth is simply the enlargement of the nucleus by the orderly addition of molecules (FLETCHER 1970; FENNEMA 1973). Nucleation which takes place in pure systems free from impurities is defined as homogeneous nucleation. The temperature limit to which water can be supercooled has been ascertained to be around $-40\,°C$. Heterogeneous nucleation occurs in water or solution when water molecules aggregate in a crystalline arrangement on nucleating agents such as suspended contaminating particles or on the walls of containers.

The nucleation temperature and the rate of ice crystal growth in the presence of various inorganic and organic solutions were examined by LUSENA. All solutes depressed the temperature and retarded the rate (LUSENA & COOK 1954; LUSENA 1955, 1960). Membranes also act as barriers to ice crystal growth under certain circumstances (LUSENA & COOK 1953; MAZUR 1970). A tremendous amount work has been conducted by LUYET and his colleagues dealing with the morphology of ice crystals (LUYET 1966). LUYET et al. (1958) and LUYET and RAPATZ (1958) studied crystalline formations encountered in various solutions and classified freezing patterns into four types; hexagonal forms, irregular dendrites, coarse spherulites and evanescent spherulites.

Vitrification and recrystallization

There have been no reports on the production of vitreous ice from bulk water, but only from the condensation of water vapor on a cold surface which was measured by X-ray diffraction or calorimetric methods by several investigators. Temperature ranges for the production of amorphous and crystalline ice were summarized by BLACKMAN and LISGARTEN (1958). According to their table, amorphous ice was produced at temperatures below approximately $-160\,°C$. Cubic ice was formed by warming vitreous ice or by condensing water vapor at temperatures between -140 and $-120\,°C$. Transformation of cubic ice to hexagonal ice was observed over a wide range of temperatures from about -130 to $-70\,°C$.

Amorphous ice produced at very low temperatures transforms into crystalline ice upon rewarming to higher temperatures. MERYMAN and KAFIG (1953, 1955) observed the grain growth of such ice over brief periods of time at temperatures between -100 and $-70\,°C$ by low temperature replication and electron microscopy. With X-ray analysis, DOWELL and RINFRET (1960) reported that conversion from amorphous to cubic ice occurred at -160 to $-135\,°C$ and from cubic to hexagonal at higher temperatures than $-130\,°C$.

Although pure amorphous ice is difficult to obtain from pure bulk water, as mentioned

Experimental Method	TEMPERATURE RANGES (°C) −180 −160 −140 −120 −100 −80 −60					Reference
X-ray diffraction	Amorphous			Semi-crystalline	Hexagonal		Burton and Oliver (1935)
Calorimetric	Amorphous		Crystalline				Staronka (1939)
X-ray diffraction	Small crystals	Intermediate range not investigated			Hexagonal		Vegard and Hillesund (1942)
Electron diffraction	Small crystals	Cubic			Hexagonal		König (1942)
Calorimetric	Vitreous		Crystalline				Pryde and Jones (1952)
Electron diffraction	Crystal growth poor	Cubic			Hexagonal		Honjo et al. (1956)
Calorimetric	Amorphous		Crystalline				Ghormley (1956)
Calorimetric	Vitreous			Crystalline			De Nordwall and Staveley (1956)
Electron diffraction	Amorphous or small crystals		Cubic	Hexagonal			Blackman and Lisgarten (1957)

Fig. 1 Summary of experimental results on deposits formed from water vapor on a base at low temperatures. Description of the deposits is that given by the respective authors. (From Blackman M. & Lisgarten N.D. (1958) Adv. Phys., 7: 189)

before it becomes easier when certain compounds, such as glycerol, sugars and protein which reduce the crystallization rates, are added to the water in the specimens. It also depends upon the cooling velocity and specimen size. MacKenzie and Luyet (1962) reported that crystallized patterns were found in very thin films of less than 0.1 μm of a 20% gelatin solution even at very high rates of cooling employed in their "collodion-sandwich-film" technique. Vitrified water was obtained in a gelatin solution of more than 50% (Dowell et al. 1962) or a glycerol solution of more than 70% (Luyet et al. 1958) at any cooling rate.

Freezing of Cells and Tissues

Freezing process and pattern (Rapatz et al. 1966)

Animal and plant cells contain proper structures and various kinds of constituents with which they conduct their own specific functions. The freezing process in cells or tissues is,

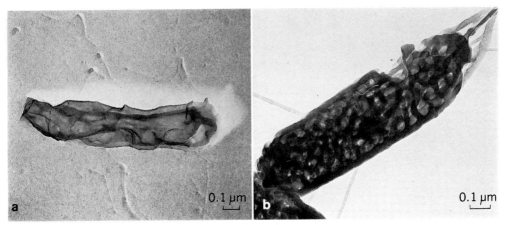

Fig. 2 Freeze-dried *Escherichia coli* cells. (NEI et al. 1969).
A droplet of the water suspension, mounted on a grid, was frozen slowly or rapidly and then freeze-dried. (×50,000) **a.** A shrunken cell, frozen slowly to –25 °C. **b.** Cells frozen rapidly by immersion into isopentane maintained at –150 °C with liquid nitrogen. Note numerous small cavities within the cells.

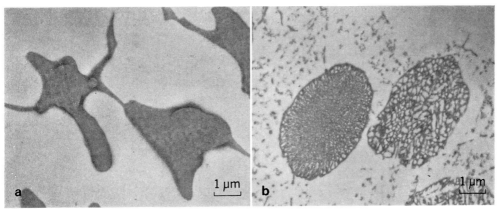

Fig. 3 Freeze-substituted and sectioned rabbit erythrocytes. (NEI et al. 1964).
0.02 ml of whole blood, thin-layered between a cover slip and an aluminum foil, was frozen slowly or rapidly, substituted with ethanol at –80 °C and then sectioned. (×8,000) **a.** Cells frozen slowly to –30 °C. Irregular shape by shrinkage, but no intracellular ice formation. **b.** Cells frozen very rapidly by immersion into isopentane maintained at –150 °C with liquid nitrogen. Numerous tiny intracellular cavities, probably left after sublimation of ice crystals.

consequently, quite different from that seen in solutions. Differences in nucleation sites, membrane permeability and other factors affect the crystallization of ice.

The freezing process had been revealed in cells by many investigators at the light microscopic level. When freezing was initiated in a cell suspension by seeding with ice below but near the critical freezing temperature and the system was cooled down gradually, ice grew in the medium. If the cells were surrounded by ice, intracellular water was withdrawn and froze extracellularly. Such withdrawal of cellular water caused dehydration and shrinkage of the cells. This type of freezing is called extracellular freezing; ice formation within the cells can not be demonstrated. If the cells were frozen more rapidly, ice was formed both outside and inside the cells, since there was insufficient time to withdraw all of the cellular water capable of being frozen. This is called intracellular freezing. These two types of freezing are the most typical ones found in cell suspensions.

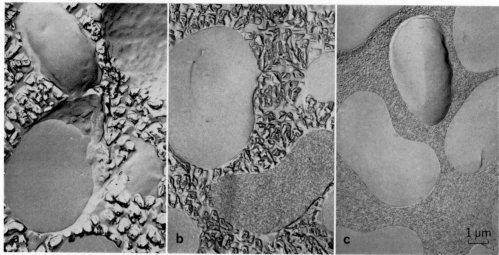

Fig. 4 Freeze-etched human erythrocytes. (NEI et al. 1971).
A droplet (0.01 ml) of cell suspension in 20% glycerol-saline was frozen by abrupt immersion in Freon 22 cooled to −150°C with liquid nitrogen and then the freeze-etch preparation was made. Observation was conducted on cells at three regions on a line from the center to the periphery of the frozen droplet. (×6,000) **a.** At the center, cells are shrunken, but not intracellularly frozen. **b.** At the intermediate region, some cells are filled with numerous tiny ice particles. **c.** At the periphery, cells are normally shaped and no intracellular ice particles are found.

Most of the intracellular water cannot avoid crystallizing even in the absence of external seeding, when the temperature is lowered beyond the cell's ability to supercool. About 10–15% of the total cell water consists of so-called bound water associated with macromolecules in the cells. Such water does not freeze even at low temperatures.

It has been reported possible to freeze very small droplets of yeast cells free of ice without prior modification of the cellular water (MOOR 1964), but it is difficult to avoid the formation of ice in large cells or tissues. Proponents of the likelihood that smaller cells do not freeze might argue that special interfacial effects (DERJAGUIN 1970) or capillary effects (KUNTZ et al. 1969) accounted for their findings. NEI et al. (1971) observed different freezing patterns in human red blood cells dependent upon the cooling rates in specimens prepared by the freeze-etching techniques. Three types of patterns were found in the cells scattered along a line from the center to the periphery of the frozen droplet; (1) shrunken cells without intracellular freezing, (2) cells filled with numerous tiny ice particles, (3) normally shaped, nonfrozen cells. These findings seemed to correspond to the different rates of cooling to be expected in different parts of the same specimen. FERNÁNDEZ-MORÁN (1960) obtained ice-free preparations by freezing frog sciatic nerve and guinea pig retina in helium II. MENZ and LUYET (1961) measured the ice crystal size in frozen muscle fibre preparations. No cavity was identified at the periphery of the fibre after sublimation of the ice, although the size of the cavities increased progressively towards the center of the specimen.

Freezing and viability

Since the cooling velocity affects the physico-chemical nature of cells during freezing, it follows that it will most probably also affect their viability. There have been many studies investigating changes in the viability of living cells due to freeze-thawing (MAZUR 1966, 1970). This is, in fact, one of the most important problems in cryobiology.

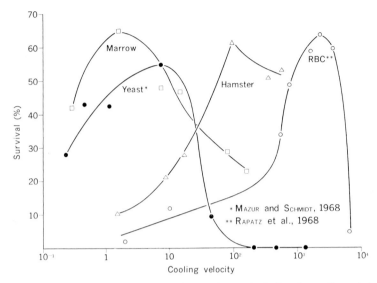

Fig. 5 Comparative effects of the cooling velocity on the survival of various cells cooled to −196 °C and thawed rapidly. (From Mazur et al. (1970) Interaction of cooling rate, warming rate and proctective additive on the survival of frozen mammalian cells. *In*: The Frozen Cell. Wolstenholme G.E.W. & O'Connor M. (eds.) J. & A. Churchill, London)

In comparative studies of cooling velocities, done by Mazur et al. (1970), affecting the survival of four diverse types of cells cooled to −196 °C and thawed rapidly, maximal survival was observed to occur at certain cooling rates in each specimen. As shown in Fig. 5, survival decreased in the range of cooling rates both above and below the optimal rate. This suggested that two factors dependent upon opposite cooling rates affected the cell survival. Sub-optimum rates caused damage by temperature-dependent removal of unbound water and its conversion into extracellular ice and by a concurrent increase in the concentration of intracellular electrolytes. Higher rates of cooling beyond the critical rate resulted in the production of intracellular ice which might cause a disruption of the cellular structure. Instracellular ice formation depended upon the rate of cooling and the permeability of the cell to water. As shown in the same figure, depending upon the different kinds of cells probably having different properties, the optimum rate varied over a wide range from 1.6 °C/min to 3,000 °C/min. It was assumed that the high survivals observed at those optimum rates just mentioned might be due to fortuitous circumstances in which the optimum rate was slow enough to prevent intracellular ice formation yet rapid enough to minimize the length of exposure time to concentrated solutions at critical temperatures.

Vitrification and recrystallization

For proper morphological observation, cells should be fixed in an unaltered state in which even the very fine intracellular ice formation resulting from very rapid freezing is avoided. The cooling rate which is assumed to be necessary to produce the vitrified state is 2 to 4 orders of magnitude higher than that at which most cells exhibit maximum survival. Until recently there were no reports except Moor's to examine the survival of such vitrified cells even though many had assumed that cells would be viable. Moor (1964) examined the relationship between the survival and morphology of yeast cells frozen at various rates of cooling. He observed that there were three different mechanisms operating during the freezing of cells, depending upon the cooling rates; (1) most yeast cells frozen at medium

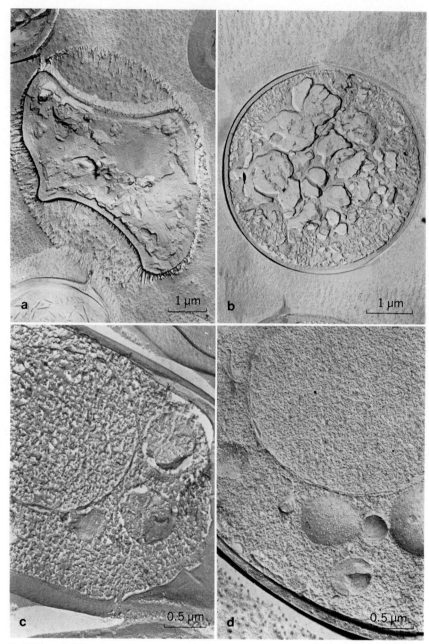

Fig. 6 Effects of different freezing rates on the fine structure of yeast cells. The final temperature reached is $-150\,°C$.
a. Freezing rate: $0.01\,°C/sec$. Extracellualr freezing. The cell is shrunken and surrounded by needle-shaped ice crystal which indicates the original cell contour. ($\times 12,000$) **b.** $1\,°C/sec$. Intracellular freezing. Ice crystals destroy the cytoplasmic structures. ($\times 14,000$) **c.** $10\,°C/sec$. Intracellular freezing. Cytoplasmic structures are recognizable, as intracellular ice crystals are smaller. ($\times 26,000$) **d.** $10,000\,°C/sec$. Vitreous ice. The absence of structural damage indicates the survival of such cells. ($\times 32,000$) (From Moor H. (1964) Zsch. Zellforsch., *62*: 546. These micrographs are by courtesy of Dr. H. Moor)

cooling rates of 1–10 °C/sec were killed by intracellular freezing, (2) extracellular freezing at lower cooling rates (slower than 10^{-2} °C/sec) protected cells from damage by dehydration, (3) by very rapid cooling (higher than 10^2 °C/sec), vitrification or extremely fine ice crystal formation took place within the cytoplasm and this inhibited cell damage (Fig. 6). He assumed that the minimum size of ice crystals which caused cell damage might be larger than 100 Å, even though it was not determined.

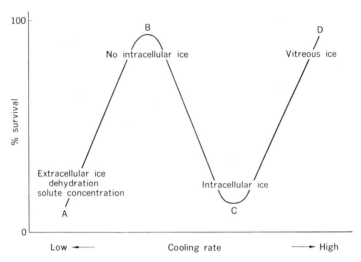

Fig. 7 Diagrammatic representation of the relationship between cooling rate and percentage survival.

From the results of studies thus far, on the morphology and survival of frozen cells, the relationship between freezing patterns and freeze-thaw survival is schematically illustrated as shown in Fig. 7, depending upon different cooling rates.

Moor (1964) also investigated the recrystallization of frozen yeast cells. Changes in freezing patterns and viability of vitrified cells were examined on rewarming to various temperatures. There was no change in survival when rewarmed to −40 °C, but beyond this the survival gradually decreased reaching a minimum at −25 °C. Ice crystal growth was observed in morphological investigations of vitrified cells starting at −35 °C and increasing with rising temperatures. This thus confirmed that there was a coincidence between intracellular ice crystal growth and cell damage (Fig. 8). Bank (1973) and Bank and Mazur (1973) also observed the freezing patterns of yeast cells suspended in distilled water frozen at various cooling rates and rewarmed to −45° and −20 °C. Cooling rates of 75,000 °C/min revealed frozen cells without intracellular ice formation, but at 1,500 °C/min the size of the intracellular ice particles was recognizable. Structural alterations such as ice growth occurred even though rewarmed to −45 °C. Sensitivity of the rapidly cooled cells to the rate of warming was believed to be due to the presence of small unstable intracellular ice crystals. These small crystals consolidated to form larger crystals if the cells were stored at a sufficiently high temperature (above −100 °C) or if they were rewarmed too slowly. MacKenzie and his colleagues (Albrecht et al. 1973) described a new procedure permitting a 100% ultra-rapid freeze thawing survival. Various yeasts and bacteria, suspended in distilled water and trapped by filtration on microporous filters, were frozen very rapidly by immersion into liquid or freezing nitrogen and then thawed. Electron microscopic observations of the specimens prepared by freeze-substitution showed that *Saccharomyces*, *Lactobacillus*, and *Leuconostoc* cells were observed to be filled with finely divided

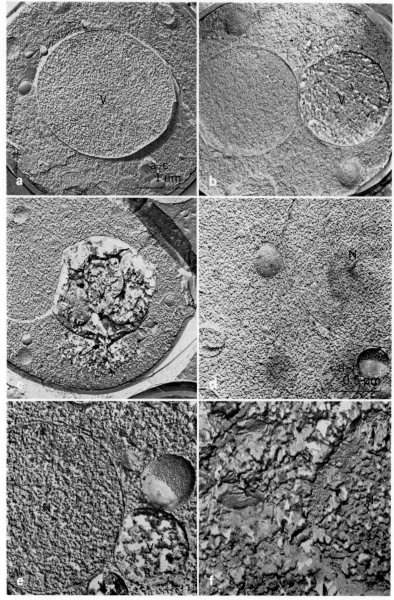

Fig. 8 Recrystallization in vitrified yeast cells.
a-c: Recrystallization in vacuole (V). (×15,000) d-f: Recrystallization in cytoplasm (N: nucleus). (×30,000) **a.** A vitrified cell, warmed up to −60°C, does not show any trace of recrystallization. **b.** A warming up to −50°C results in a coarser crystallization of the vacuole. **c.** At −35°C, large ice crystals are formed in the vacuole. At one point (arrow), ice crystals perforate vacuolar membrane and initiate recrystallization in cytoplasm. **d.** In a vitrified resting cell, containing no large vacuole, recrystallization does not take place by warming up to −40°C. **e.** Recrystallization is initiated at −35°C and results in coarser structures of cytoplasm and vacuolar content. **f.** Most cytoplasmic structures are completely destroyed at −30°C. (From Moor H. (1964) Zsch. Zellforsch., 62: 546. These micrographs are by courtesy of Dr. H. Moor)

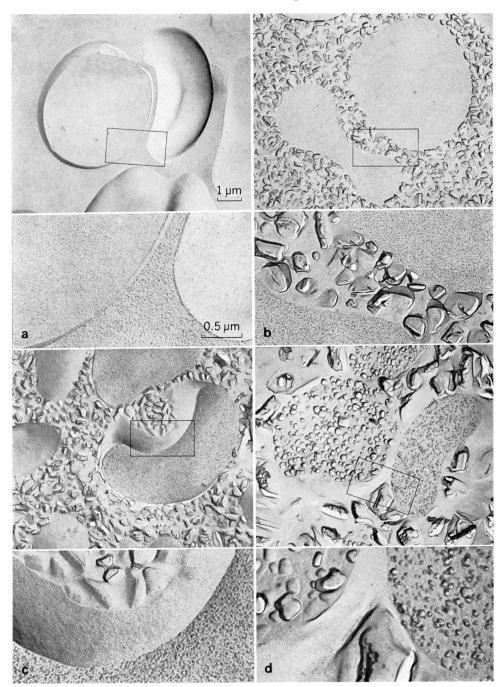

Fig. 9 Recrystallization of intracellular ice in human erythrocytes (From Nei T. (1973) J. Microscopy, *99*: 227). A droplet of cell suspension in 30% glycerol-saline was frozen by abrupt immersion into Freon 22 kept at –150°C and rewarmed to various temperatures between –80° and –60°C. After 30 minutes (60 minutes at –80°C) storage, freeze-etch preparations were made. ($\times$7,000 and $\times$25,000)

a. Control cells. No ice crystals are found inside and outside the cells. **b.** Cells warmed to –80°C. Ice particles have grown larger in the surrounding medium. Note intracellular tiny particles at the higher magnification. **c.** Cells warmed to –70°C. Intracellular particles are noticeable. **d.** Cells warmed to –60°C. Ice particles of various size are found within the cells.

ice while *Staphylococcus*, *Escherichia* and *Serratia marsescence* were found to be reduced in size and to have little or no intracellular ice. Higher cooling rates resulting from the use of thinner filters allowed intracellular freezing of the last three mentioned organisms. In all specimens, almost a 100% survival rate was obtained after freeze-thawing. NEI (1973) examined the formation of ice crystals which might have been possible artifacts because of cryotechniques used for electron microscopy during the process of rewarming rapidly frozen erythrocytes. Intracellular ice formation, which was usually found in cells suspended in saline by rapid freezing, was inhibited by the addition of 30% glycerol. When such glycerinated cells, having no ice crystals at liquid nitrogen temperature, were rewarmed to higher temperatures above $-80\,°C$, recrystallization of the ice occurred. Ice particles became visible within the cells even at $-80\,°C$ and grew larger with a temperature rise. From the results obtained in comparative morphological and physiological investigations, it also became evident that the recrystallization of ice appeared prior to an increase in hemolysis during the rewarming process. SHIMADA and ASAHINA (1972) observed very fine intracellular ice crystals in frozen rat ascites tumor cells which had been cooled very rapidly. The size of those ice crystals were approximately 0.03, 0.05 and 0.1 μm at cooling rates of 800, 500 and 300 $°C/sec$, respectively. Cells frozen at 800 and 500 $°C/sec$ revived after rapid rewarming. This suggested that the cells could survive intracellular freezing, if intracellular ice crystals were too small to be injurious to the cells during freezing and thawing. One should, therefore, recognize the possibility that ice crystals large enough to interfere with ultrastructural determinations are still not lethal and this possibility deserves very serious consideration.

Morphological alterations by cryoinjury

Structural alterations of tissues by freeze-thaw injury had thus far been described mostly at the light microscopic level. Working with the electron microscope, STOWELL et al. (1965) and TRUMP et al. (1965) made observations suggesting that the ultrastructural cryoinjury of the nuclei was accomplished by a decreased rate of amino acid incorporation into the nuclear protein. SHERMAN & KIM (1967) and SHERMAN (1972) compared *in situ* and *in vitro* cryoinjury and cryoprotection of mitochondria with emphasis on structural-function relationships. BANK and MAZUR's electron microscopic examination (1972) of Chinese hamster cells showed that freezing and thawing resulted in structural alterations. Cells exhibited different patterns depending upon the rate of cooling and warming. Freeze-thaw treatments producing similar percentages of cell survival did not necessarily cause similar structural alterations, nor was there a simple correlation between structural alteration and survival.

Freezing Technology: Factors Affecting the Morphology of Specimens

Freezing patterns of biological specimens depend largely upon the cooling conditions and the type and/or concentration of the antifreeze agents. The most serious artifact that might arise in a specimen prepared according to known cryotechniques is ice crystal formation. Since a high resolving power is required in electron microscopy, the main requirement is to produce much smaller ice crystals or no crystals in the specimen if possible. The formation of ice crystals larger than the resolution limit of the electron microscope used, 100 Å crystals at the largest, should be avoided; otherwise the images obtained would be incorrectly interpreted. The ideal cryotechnique is to attain a vitreous (amorphous or glassy) state by controlling the rate of cooling and the concentration of the antifreeze agents. The natural configuration of hydrated specimens could be observed even under the electron microscope, if this could be accomplished. The following factors should be considered to achieve this goal.

Rate of cooling

  Specimen size. Total mass of the specimen should be reduced to a size as small as possible in order to achieve a high rate of cooling. For this purpose, various kinds of methods have been attempted. WILLIAMS (1954) designed an equipment to be installed in a freeze-drying apparatus. Extremely small droplets of suspended material were made to impinge at a high velocity with a nebulizer upon a thin collodion film which was in intimate thermal contact with a massive heat reservoir maintained at a low temperature in this method. The droplets were almost instantaneously frozen and vacuum sublimation was then carried out. It was estimated that the entire droplet should change from the liquid to the ice phase in about 3×10^{-5} sec. He prepared his viral and bacterial specimens by this method. NEI (1966) reported on a specially devised cooling attachment located on the forward side of the specimen chamber of an electron microscope. Bacterial suspensions were deposited with a specially designed spray gun on a specimen grid precooled to various temperatures with liquid nitrogen in a vacuum. The specimen thus frozen rapidly was then transferred to a cold stage in the specimen chamber. Observation was carried out in the frozen state or after freeze-drying. BACHMANN and SCHMITT (1971, 1972) applied a spray method to the freeze-etching technique. Cell suspensions were sprayed into 1 ml liquid propane at $-190\,°C$. After the evaporation of propane at $-85\,°C$, the frozen droplets were mixed with a drop of butylbenzene ($-85\,°C$) and transferred onto the specimen holder followed by dipping in liquid nitrogen to solidify the butylbenzene. These specimens were then handled by the conventional freeze-etching procedure. This method of spray freeze-etching was successful without cryoprotectants. BUCHHEIM (1972) attempted a simple cryofixation of an aqueous solution for the freeze-etching procedure. The solution was dispersed in a viscous paraffine oil using a high-speed stirrer. The diameter of the droplets produced by the stirring procedure should not exceed 10 μm. Samples of about 1 mm³ were frozen by immersing into Frigen ($-150\,°C$) and subsequently treated in the freeze-etch apparatus in the usual manner. RIEHLE (1968) reported that by optimizing the heat transfer and freezing under a hydrostatic pressure load of $p=2.1$ kb, the so-called "vitreous state" with a particle size of 50 to 100 A was obtained for the first time in a 5% by weight glycerol water solution. Additional experiments suggested that all biological materials exhibited a critical cooling rate less than 2.10^4 K/s and might thus be vitrified in samples of a few hundred μm in thickness. The critical freezing rate was estimated to be 10^6 K/s in pure water and 10^4 K/s in living cells. He concluded that the normally encountered formation of ice crystals 1 to 10 μm in size destroyed the fine structure and viability of frozen cell materials. His apparatus is not commercially available because it is complicated and expensive.

  Animal and plant tissues should be thin-sectioned to minimize the size and to obtain a high rate of freezing. MONROE et al. (1968) proposed a ballistic cryofixation. Small samples were obtained *in vitro* from a pulsating dog heart by shooting a section of a small gauge hypodermic needle through it with a modified rifle and then frozen in liquid propane.

  Coolant. The ideal coolant is characterized by having a high boiling point with a low freezing point as well as a high thermal conductivity, high heat capacity and a low viscosity. Suitable coolants include propane, isopentane, Freon 12 and Freon 22 as shown in Table 1. Propane possesses a low melting point and a high cooling rate, but is inflammable. The most usable coolant in the cryotechnique method is Freon 22. Liquid helium II yields exceedingly high freezing rates because of its superfluid properties and high velocity of heat conduction (FERNÁNDEZ-MORÁN 1960), but the thermocouple measurements actually have not supported this suggestion (BULLIVANT 1965). Liquid helium II also has some drawbacks, i.e., a large expense and a chance of accidental "dry-freezing". Liquid nitrogen is not an ideal direct coolant because evaporated nitrogen gas insulates the specimen and reduces the rate of cooling.

Table 1 Properties of fluoro- and hydrocarbon quenching agents (From REBHUN L.I. (1972). Freeze-substitution and freeze-drying. In: Principles and Techniques of Electron Microscopy. HAYAT M.A.(ed.) published by Van Nostrand Reinhold Co., copyright 1972 by Litton Educational Publishing, Inc.)

Agent	Boiling point (°C)	Melting point (°C)	Molecular weight	Cooling rate (°C/sec measured at -79°C)
Propane	−42.12	−187.1	44.1	5,860
Isopentane	27.9	−159.9	—	2,415
Freon 12	−29.79	−158.0	120.93	2.940
Freon 22	−40.8	−160.0	86.5	3,976

Other factors. A method for rapid freezing was described by VAN HARREVELD and CROWELL (1964) in which use was made of the excellent heat conduction properties of silver. Freezing was accomplished by bringing the tissue into contact with a polished silver surface at the liquid nitrogen temperature. REBHUN (1972) examined insulator coatings and powders used as nucleating sites. The thickness and type of insulator coat affected the heat transfer rates to varying degrees and different powders exhibited different abilities to effect nucleation. No increase in cooling velocity was obtained with coats or nucleating sites when propane, Freon 12 or Freon 22 were used. COWLEY et al. (1961) and LUYET (1961) investigated the nature of coating substances, the physical state of the surface, the compactness of the coating materials and the heat conductivity of the coat. It was confirmed that coating increased the cooling rate by immersion in liquid nitrogen.

Inhibition of ice crystal formation

Partial dehydration. Partial dehydration of the specimen is sometimes effective in lowering the freezing point and reducing the formation of ice crystals (REBHUN & SANDER 1971). It is undesirable, however, when it causes cell shrinkage due to dehydration. Ice crystals were formed in human red blood cells, partially dehydrated by exposure to a hypertonic solution (NEI, unpublished).

Antifreeze agents. Cryoprotectants, such as glycerol and DMSO (Dimethylsulfoxide) which were originally used as additives for protecting viable organisms from freezing injury (POLGE et al. 1949; LOVELOCK & BISHOP 1959), were also useful for reducing the ultimate size of ice crystals. Antifreeze agents lowered the melting point, reduced the rate of nucleation and ice crystal growth, but sometimes lowered the recrystallization temperature. This was believed to be due either to the inhibition of aggregation of water molecules that form ice crystals or to the reduction in the total amount of water available to form ice crystals (BOYDE & WOOD 1969). Chloroform as an antifreeze agent had the advantage of not leaving residues on cryofractured tissues. FERNÁNDEZ-MORÁN (1957, 1960) first used antifreeze agents for his cryotechnique (freeze-substitution) in electron microscopy. HAGGIS (1961) applied glycerol to freeze-etching. However, there had been very few reports on morphology showing a relationship between the concentration of glycerol and other freezing conditions, in particular, the rate of cooling (MOOR 1964; STEERE 1969b; NEI, MATSUsaka et al. 1971). Experiments using human erythrocytes showed that an addition of 10 to 20% glycerol was insufficient to prevent the formation of ice crystals during rapid freezing at a rate of approximately 10^4 °C/min (Nei, Matsusaka et al. 1971). The use of glycerol or DMSO is discouraged for freeze-drying because of the low vapor pressure of these coolants retained in the sample. Boyde and Wood (1969) recommended the use of freeze-drying following substitution of water with organic liquids such as amyl acetate. Freeze-drying of specimens frozen in nonpolar organic solvents had the advantage of being very rapid because these solvents had a high vapor pressure below their freezing points. Freeze-drying in the presence of amyl acetate was completed in only half an hour at -75 °C (melting point of amyl acetate is -71 °C) under a vacuum of 5×10^{-3} torr, whereas freeze-drying in the presence of water required up to a week at -70 °C. Further advantages in solvent substitution were slight tissue shrinkage and limited chemical extraction. The use of volatile reagents in place of glycerol was studied by Haggis (1972). He attempted prefixation with 2% glutaraldehyde in a sucrose buffer and impregnation with a mixture of dioxane (49.5%), water (49.5%) and glycerol (1%). This was followed by rapid freezing, fracturing and vacuum drying at -80 °C. Ice crystal formation was minimized by using this procedure.

Glycerol or DMSO had usually been used as a protective additive for long-term preservation of living cells and tissues mainly in concentrations ranging from 10 to 15%. However, considerably higher concentrations (20 to 40%) had generally been used for morphological studies. Buckingham and Staehelin (1969) examined the influence of glycerol on the structure of lecithin bilayer membranes by means of X-ray diffraction and the freeze-etch technique. The effect of additives on the mitotic apparatus was also examined by Rebhun and Sander (1971). Moor (1973) described their harmful effects. An artificial vesiculation of the endoplasmic reticulum cisternae in living cells (Moor 1971) and a swelling of mitochondria in ascites tumor cells were reported. He also emphasized that the way to overcome these problems was to first submit sensitive specimens to a "mild" aldehyde fixation (e.g. 2% glutaraldehyde for one hour), since this reduced the reactivity and improved the permeability without introducing severe fixation artifacts. Sherman and Liu (1973) reported that glycerol was injurious and toxic during pretreatment of rat pancreatic acinar cells. Toxicity, which resembled cryoinjury, was greater at 22 °C than 0 °C pretreatment, showing some morphological alterations.

Holding temperature

Careful attention should be paid to the holding temperatures of specimens during preparation procedures. There was a possibility that ice might recrystallize and grow larger in the vitrified specimens when the temperature was raised above the critical point. The temperature at which recrystallization of ice occurred depended upon the properties of the substances added to the specimen. The freeze-etching technique is beneficial for observing the freezing pattern of the specimen in examining such a problem. As described above, Bank and Mazur (1973) reported the crystal growth in yeast cells suspended in distilled water rapidly frozen at the rate of 1,000 °C/min and rewarmed to -45 °C. Nei (1973b) also observed the recrystallization of human erythrocytes suspended in 30% glycerol saline. Ice free cells, frozen rapidly at a cooling rate of approximately 10^4 °C/min, caused intracellular ice crystal growth on rewarming to -80 °C at which no increase in hemolysis was observed. Although the freeze-drying or freeze-substitution procedure is usually carried out at -70 to -80 °C, an examination should be made to determine whether there is any structural alteration in the frozen specimen kept at these temperatures. Bullivant (1965) obtained good preservation in the morphology of small pieces of mouse pancreas substituted with ethanol at −75 °C for weeks. Van Harreveld and Crowell (1964) compared the morphologies of specimens treated at −25, −50 and −85 °C, and confirmed good results at −85 °C.

Application of Freezing Techniques

Freeze-drying (Meryman 1966; Bullivant 1970; Rebhun 1972)

Altmann (1890) established a freeze-drying technique utilizing vacuum sublimation of water vapor from the frozen state at low temperatures, and applied it to histology under light microscopy as one of the preparation procedures. The main aim of this technique was to prevent shrinkage due to dehydration and also to reduce the release or loss of chemical constituents of the specimen caused by treatment with organic solvents. Similar to other cryotechniques in electron microscopy corresponding to the nature of different specimens, adequate freezing conditions should be chosen to minimize the presence of artifacts caused by ice crystal formation. The frozen specimen is then transferred to a precooled container which is then connected to a freeze-dryer. Sublimation in the frozen specimen should be performed at temperatures as low as possible, usually at −70 °C or below, which depends primarily upon the specimen temperature and size even though this will prolong the duration of drying. Bondareff (1957) and Gersh et al. (1957, 1960) carried out the drying procedure at −30 to −40 °C. Malhotra and Van Harreveld (1965) dried the specimens at −79 °C for 7 days and then at higher temperatures for several more days. Antifreeze agents such as glycerol or DMSO should not be used for freeze-drying, since they are not as volatile as water and their concentration and retention in the specimen during the drying process inhibits further drying. Although freeze-drying is superior to air-drying, it is difficult to completely avoid shrinkage that occurs in the final stage of drying (Nei 1962b). Gersh (1932) introduced a freeze-drying technique for light microscopy. Sjöstrand (1943) first applied this technique to electron microscopic studies. Since then, a considerable amount of work in electron microscopy using the freeze-drying technique had been achieved by these workers with excellent results.

Fundamental conditions in techniques of freeze-drying of biological specimens such as the temperature of the specimen and the cold trap and the vacuum in the system were examined in the early days (Malmstrom 1951; Glick & Malmstrom 1952; Stephenson 1953, 1956, 1960; Gersh & Stephenson 1954). Gersh and Stephenson (1954) introduced standard procedures for animal tissues. In their technique careful attention, especially the temperature rise and frost condensation, was paid to the procedure for specimen transfer after the specimen was rapidly frozen by immersion into isopentane cooled with liquid nitrogen. Dried specimens were fixed prior to embedding. This could be done by introducing vapor fixatives such as OsO_4, formaldehyde, glutaraldehyde and alcohol through a side arm of the drying chamber. After fixation, the degassed embedding medium was poured into the drying chamber under vacuum. They also illustrated the paths of the water molecules after sublimation from ice crystals at the interface between the frozen interior and the dry shell. As drying proceeded, the average thickness of the dry shell increased and that of the undried interior decreased until the tissue was completely dry. Stephenson (1953) estimated the relationship between the specimen size and drying time required for different specimen temperatures and gave a figure for slices of guinea pig liver (Fig. 10). Malmstrom (1951) discussed the factors affecting the rate of drying of tissues; effects of pressure, temperature, partial pressure of water at the trap, the radius of the tube holding the sample and the length of the path from the sample to the trap (Fig. 11).

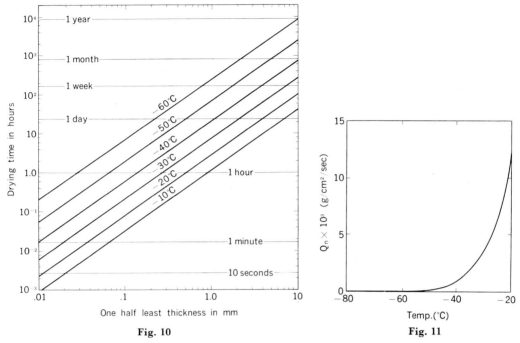

Fig. 10 Drying time of slices of guinea pig liver as a function of temperature and thickness. (From Stephenson J.L. (1953) Bull. Math. Biophys., *15*: 411)

Fig. 11 Rate of evaporation of water as a function of temperature at pressures below 10^{-5} mmHg. (From Malmstrom B.G. (1951) Exptl. Cell Res., *2*: 688)

When the specimen temperature was raised to a certain degree during the drying process, some morphological changes occurred at the sublimation front of the specimen. This was called "collapse" in the course of freeze-drying or freeze-substitution, which was described by MacKenzie (1972). One should take note that the collapse temperature, depending upon the type of solutes, was lower than the eutectic temperature of the same specimen.

Mundkur (1964) modified Gersh's method for a histochemical study of freeze-dried yeast cells. He used difluorodinitrobenzene to stabilize the cells after drying and also used a lower drying temperature in such a way that an ice crystal free preparation was obtained. Sjöstrand and his colleagues (Sjöstrand 1943a, b; Sjöstrand & Baker 1958; Sjöstrand & Elfvin 1964) applied freeze-drying to electron microscopy of tissue cells, but the results were rather inferior in the early stage. Since 1952, many repeated trials were carried out in his laboratory. They usually dried specimens at below $-40\,°C$, particularly at $-70\,°C$,

Table 2 Collapse temperatures. (From MacKenzie A.P. (1972) Freezing, freeze-drying and freeze substitution. *In:* Scanning Electron Microscopy/1972, IIT Research Institute, Chicago)

Sample materials	Freeze-drying	Freeze-substitution	
		in ethanol	in aceton
Dextran T 110	$-10\,°C$		
Ficoll	-19	$-43.5\,°C$	$-36\,°C$
glucose	-41		
ovalbumin	-10	-35	-25
PVP K-30	-23		
Sephadex G-200	-10	-32.5	-23
sucrose	-32		-58

and used OsO_4 vapor to stabilize the dried tissue. MÜLLER (1957) contributed an important approach by designing a method for freeze-drying of plant materials for electron microscopy. The most important problem they discussed was artifacts, mainly regarding ribosomes, which were probably brought about by the conventional chemical treatment. Arguments on the ultrastructure were conducted by several people, particularly on the membrane system in specimens prepared by conventional methods and the freeze-drying method (GRUNBAUM & WELLINGS 1960; SENO & YOSHIZAWA 1960; ELFVIN 1963; SJÖSTRAND & ELFVIN 1964).

One of the ultimate aims of the freeze-drying method is its application to cytochemistry and autoradiography in electron microscopy (ROTH & STUMPF 1969). STIRLING and KINTER (1967) used hamster intestinal tissue and froze a small piece in propane at $-184\,°C$ and then performed the freeze-drying procedure at $-70\,°C$, gradually raising the temperature up to room temperature which required 4 days. NAGATA et al. (1969) devised his own equipment for autoradiographic studies and obtained new findings. STUMPF and ROTH (1967) designed a simple freeze-drying method by means of cryosorption pumping.

Various types of apparatuses for freeze-drying of biological materials, especially for the preparation of tissues, have been designed by many investigators who desired to use them in their laboratories.

Freeze-substitution (PEASE 1967; BULLIVANT 1970; REBHUN 1972)

Freeze-substitution, which SIMPSON (1941) first introduced in this research field, is another dehydration method for frozen specimens. This is much simpler than freeze-drying, which requires special equipment as already described. Specimens are dehydrated by substituting ice with organic solvents at low temperatures. Disadvantages of this method, however, are the long period of time required for substitution and a possible loss of chemical constituents from the biological specimens during treatment with the drying solvents. The most important point in freeze-substitution is probably the maintenance of temperature. Regarding the recrystallization temperature of the specimen, it should be kept as low as possible, as described in the freeze-drying technique. An extremely long exposure for substitution may be required at low temperatures. The ability of organic solvents to dissolve water at a very low temperature is not yet clearly understood. VAN HARREVELD and CROWELL (1964) confirmed the mode of water removal by acetone treatment in their nuclear magnetic resonance studies. They found that at $-50\,°C$ a 14% solution represented saturation and that at $-85\,°C$ the solubility dropped to between 2 and 3%. Thus, at quite low temperatures acetone became a poor substituting medium. FERNÁNDEZ-MORÁN (1952a, b, 1957, 1960) initiated his freeze-substitution experiments for electron microscopy and contributed a considerable amount of work. He advocated the use of glycerol and liquid helium II as a protective additive and a coolant, respectively, and also used a variety of solvents (alcohol, acetone and ethylchloride) with the addition of heavy metal salts. As a result, good preservation was obtained in the lamellar systems of retinal rod outer segments and myelin sheath. BULLIVANT (1960, 1962, 1965) used a simple method, in which liquid helium II and propane were employed and the substitution was carried out at $-75\,°C$. He also confirmed the effectiveness of the addition of glycerol for the protection of cells from ice crystal formation. Good preservation was obtained in a pancreas without additives even in a rare case, but definitely failed with tissue culture cells. REBHUN (1961, 1965, 1972) and REBHUN and SANDER (1971) also carried out many experiments on freeze-substitution using marine eggs and rat pancreas and liver. They examined the effect of a variety of solvents (acetone, ethanol and others) on the morphology of specimens partially dehydrated. The substitution was conducted at $-80\,°C$. Specimens containing 40–50% water seemed to be preferable, since ribosomes were remarkably revealed in the cells with this water content. As described before, VAN HARREVELD and CROWELL (1964) and VAN HARREVELD et al. (1965) attempted to obtain a high rate of cooling by the use of contact with precooled metal. The tissue was substituted in dry acetone with the addition of 2% OsO_4 at $-85\,°C$ for 3 days. Good preservation was obtained to a depth of 10 µm below the surface. His colleagues applied this basic procedure to other tissues (MALHOTRA & VAN HARREVELD 1965a, b; MALHOTRA 1966, 1968). MONROE et al. (1968) made small samples from a pulsating dog heart by shooting a small projectile. PEASE (1967a, b, 1973) had performed many studies involving freeze-substitution of various animal tissues. He examined various freezing conditions. Good preservation in his work was obtained by substituting an eutectic mixture (70%) of ethylene glycol kept at $-50\,°C$, while preservation was a failure with other substituting media such as alcohol and acetone. Arguments on ice crystal formation in freeze-drying and freeze-substitution methods followed by an examination of the freeze-thawed specimen had been presented by several investigators (HANZON et al. 1959; BAKER 1962; SJÖSTRAND & ELFVIN 1964).

Freeze-sectioning (BULLIVANT 1970)

There are two main reasons for carrying out freeze-sectioning. One is that, when a soft specimen is solidified at low temperatures or embedded in ice, cutting can easily be done. The other is to prevent the leakage of tissue constituents by chemical treatment with organic solvents. This technique is usually applied to cytochemical electron microscopy. FERNÁNDEZ-MORÁN and his co-workers (FERNÁNDEZ-MORÁN 1952a, b; FERNÁNDEZ-MORÁN & DAHL 1952; FERNÁNDEZ-MORÁN & FINEAN 1957) first attempted freeze-sectioning. The specimens were either air-dried or freeze-dried immediately after sectioning and subsequently cytochemical studies were carried out directly on the dried sections. Since that time, his method had not been pursued by other workers for more than ten years. Later BERNHARD and his colleagues (BERNHARD & NANCY 1964; BERNHARD & LEDUC 1967; LEDUC et al. 1967; VORBRODT & BERNHARD 1968; ZOTIKOV & BERNHARD 1969; BERNHARD & VIRON 1971) intensively developed this technique. Tissue was fixed in glutaraldehyde embedded in 20% gelatin, partially dehydrated in 50% glycerol, frozen in liquid nitrogen, and sectioned at $-35\,°C$ in 40% DMSO using an ultramicrotome in a deep-freeze. The sections, stained with a variety of heavy metal salts, showed membrane structures similar to those obtained by freeze-drying or freeze-substitution. They attempted some developments in fixation and embedded aldehyde-fixed tissue pieces in a gelatin solution at $37\,°C$ (BERNHARD & VIRON 1971). CHRISTENSEN (1967, 1969) attempted cutting with a glass knife in a nitrogen atmosphere at liquid nitrogen temperature. He developed a simple method that allowed frozen thin sections of fresh-frozen tissue to be cut on a virtually unmodified ultramicrotome kept at room temperature (CHRISTENSEN 1971). Even with this technique, there was still some risks of ice crystal formation. Only 30 µm or thereabout thick surface layers of tissue were favorable. IGLESIAS and his associates (IGLESIAS et al. 1971; BERNIER et al. 1972) modified BERNHARD's technique and used methyl cellulose before freezing. Wide ranges of temperatures were tested and $-50\,°C$ for the knife and $-70\,°C$ for the specimen were found to give the best results. Recently TOKUYASU (1973) reported a simple new technique for ultracryotomy not only for a variety of tissues but also for single cells in suspension. In this technique, sucrose (0.5–2.3 M) was infused into glutaraldehyde-fixed tissue pieces before freezing for the purpose of controlling the sectioning consistency. By choosing the proper combinations of sucrose concentration and sectioning temperature, a wide variety of tissues could be sectioned smoothly.

Freeze-fracturing, etching and replication (BULLIVANT 1970; KOEHLER 1975)

Freeze-fracturing, etching and replication has developed so quickly and dramatically in the past few years and is now widely used as an ultrastructural preparative technique. The reasons for such rapid progress might be due to the following advantages; (1) Fixation and dehydration are not always necessary. This minimized the possibility of the presence of artifacts caused by such treatments. (2) The fracture proceeds along a structurally weak path within frozen membranes. Three-dimensional images in the specimen, particularly the cleaved surface of membranes or organelles, can be obtained. (3) The entire procedure from initial freezing to replication is carried out at about $-100°C$ or at lower temperatures. Recrystallization of ice can hardly occur when compared to freeze-drying or freeze-substitution. (4) Only the replica is observed under an electron microscope, hence no radiation damage. In contrast it is a disadvantage to only observe the replica owing to the difficulty in a correct interpretation of the images obtained. Sectioned specimens, which can be conventionally prepared, should be used concurrently to compensate for this point.

Freeze-etching technique consists of the following steps which involve in sequence; pretreatment, freezing, fracturing, etching, replication and cleaning. This technique has a rather long history, in spite of its recent quick development. A low-temperature replica method of aqueous substances such as a water and silver halide suspension was first tested by HALL (1950) for electron microscopy. HIBI and YADA (1959) also carried out low temperature replication. MERYMAN and KAFIG (1955) devised a technique for electron microscopy of frozen liquids by vacuum evaporated replication. Replicas of the surfaces and fracture planes with or without heat etching were examined using frozen water, ice and other frozen liquids and biological materials. STEERE (1957, 1969a, b) improved the procedure for fracturing the specimen with a cooled scalpel and applied this method initially to his study on plant viruses. HAGGIS (1961) modified Meryman's apparatus and followed Steere's procedure. It was claimed that individual hemoglobin molecules could be seen in human red blood cells to which 20% glycerol was added. MOOR et al. (1961) and MOOR and MÜHLETHALER (1963) devised an excellent but complicated apparatus with which they later did much work on freeze-etching. Specimen temperature during the procedure was kept constant by an electric feedback system governed by a thermister fixed in the stage. Based on these investigations thus far performed, the standard method which is now in current use was almost completed. The glycerinated specimen was frozen rapidly in Freon to avoid ice crystal formation and mounted on a cold specimen stage in the evaporator. Then a high vacuum was obtained and the specimen temperature was raised to about $-100°C$. The surface was fractured with a cooled knife and left for a while to sublime the ice in a vacuum. A carbon platinum replica of the surface was made and then removed from the vacuum for cleaning to dissolve the adhering materials with solvents. Various kinds of equipments had been devised in modification form. Factors affecting good preservation of the specimens had been examined from several points of view. BULLIVANT and AMES (1966) designed a simple device which was later modified. The equipment consisted of a brass block separated into three parts. With this apparatus, contamination deposited on the fractured surface could be minimized, because the specimen was only exposed to the contamination arriving through small tunnels kept at very low temperatures. WEINSTEIN and SOMEDA (1967a, b) and McALEAR and KREUTZIGER (1967) modified the Bullivant's type I and II apparatus, respectively, and obtained good results. Balzer (Moor) and Bullivant types are the two typical freeze-etching equipments which have both advantages and disadvantages.

BRANTON (1966, 1967, 1971) and BRANTON and SOUTHWORTH (1967) emphasized the importance of interpretation of a cleaved membrane surface. It was difficult to determine where the fracture plane proceeded in a membrane, but he applied freeze-etching to various membrane systems and mentioned that fractures proceeded along the inner hydrophobic surfaces of membranes. Surface labelling with a marker was investigated for interpretation of membrane surfaces (PINTO DE SILVA & BRANTON 1970; TILLACK & MARCHESI 1970). Another difficulty in interpretation of images existed in the nature of many small particles seen on the fractured face (STAEHELIN 1968; MEYER & WINKELMANN 1970; BULLIVANT 1973). Etching might leave a granular structure on the surface where ice sublimed away and contamination might also cause similar artifacts. Caution should, therefore, be taken to determine whether the particles were of natural or artificial origin, in particular, by contamination (STAEHELIN 1965; STAEHELIN & BERTAUD 1971).

For confirmation of fine structures, it was useful to prepare double replica specimens and to observe the complementary three-dimensional mirror images produced by fracturing. STEERE and SOMMER (1972), WHERLI et al. (1970), CHALCROFT and BULLIVANT (1970) and SLEYTR (1970) designed adequate devices for this purpose. Skilful technique was required to pick the double replicas up on the grid in the correct position. Recently several investigators attempted to obtain the necessary high rate of cooling by various means, i.e., by droplet (BACHMANN & SCHMITT 1971), oil emulsion (BUCKHEIM 1972) and the sandwich method (WINKELMANN & MEYER 1968; WINKELMANN & WAMMETBERGER 1969; NEI 1974).

Freeze-fracturing, etching and replication technique thus provides a great deal of new information on biological specimens, but it is difficult to interpret the morphology of the ultrastructure only from images obtained by this technique. One has to make an interpretation from both the thin-sectioned and freeze-etched specimens.

Cryotechniques in scanning electron microscopy (BOYDE & WOOD 1969; BOYDE & VESELY 1972; BOYDE & ECHLIN 1973; HAYES 1973; NEI 1974)

Freeze-drying and freeze-substitution methods are used in scanning electron microscopy (SEM) as well as in transmission electron microscopy (TEM). Dried specimens properly treated are placed in an evacuation unit and coated with gold while being revolved on a rotary stage. The cryofracture technique is also useful for SEM, as it has the following advantages; (1) The fracture sometimes takes random directions through the tissue revealing more than that seen when it is sectioned in a single plane. (2) Cryofracture takes place at a much lower temperature than those used for cutting frozen sections. The specimen, fixed in glutaraldehyde, was rapidly frozen in Freon 22 at $-150\,°C$, fractured at $-170\,°C$ with a cold blade, and freeze-dried under vacuum at $-80\,°C$ for 15 hours. In specimens thus prepared, rather large crevices often opened up between cells. Ice particles of various dimensions were observed in cells or solutions (HAGGIS 1970). Nonmechanical cryofracture has been employed for preparing delicate tissues (e.g., inner ear) (LIM 1971). This method reduced the number of artifacts caused by shearing forces. Specimens fixed with fixatives and suspended in ethanol up to 70% were frozen by dropping into liquid nitrogen. Droplets cracked after instant freezing were freeze-dried in a dryer. TANAKA (1972) designed a new technique in which the resin-embedded specimen was solidified and cracked at $-30\,°C$ and then dried by the critical point method for SEM observation.

Direct observation of frozen specimens

Transmission electron microscopy. FERNÁNDEZ-MORÁN (1960) first attempted the application of electron microscopy for the study of ice crystal nucleation and growth at low temperatures. He succeeded in obtaining high-resolution micrographs and electron

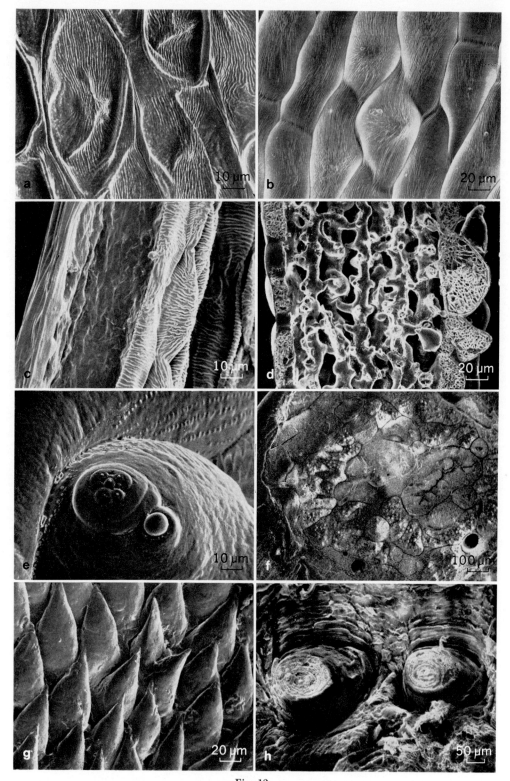

Fig. 12

diffraction patterns by stabilizing the thin ice crystals at appropriate low temperatures. He also observed biological specimens such as cell suspensions as well as fresh and glycerinated thin tissues which were rapidly frozen by spraying or immersing into liquid helium II. Since 1954 when a specimen cooling device for incorporation into the electron microscope was designed for the first time in Japan, Nei (1960) had been attempting the use of this equipment for investigations in cryobiology. He succeeded in his observations of fresh microorganisms without dehydration and of the freeze-drying process of such cells under the electron microscope (Nei 1962a). With a spray freezing device specially designed later (Nei 1966), cell suspensions were instantaneously frozen on a precooled grid under vacuum by spraying with a gun. Frozen specimens were then transferred onto a cold stage installed in the microscope. The process of sublimation of extra- and intracellular ice and the changing cellular morphology of the cells during freeze-drying were recorded by special cinematographic equipment incorporated into the microscope. Rice et al. (1971) tried to observe cryosectioned specimens in the frozen state using a cold stage in a high voltage electron microscope.

Scanning electron microscopy. Attempts had been made to examine frozen specimens without dehydration and metal-coating directly under SEM. Cross (1969) employed SEM for investigation of the surface of evaporating ice and succeeded in observing it directly. It was discovered that ice was a sufficiently good electrical conductor to prevent charging by the electron beam. Echlin et al. (1970) and Echlin (1971) applied this method to biological specimens such as plant and animal cells. Temperature control of the cold stage was obtained by a system analogous to a feedback system. Specimens were mounted on a specimen stub and immediately plunged into Freon 22 maintained at its melting point ($-160\,°C$) by liquid nitrogen. Then the specimen stub was rapidly transferred to the cold stage maintained at $-180\,°C$ with liquid nitrogen, and the microscope column was pumped down to 1×10^{-4} torr. Freon 22 quench-freezing yielded a specimen almost devoid of ice crystal damage. There was no advantage in impregnating the specimens with glycerol or DMSO prior to rapid freezing. The antistatic spray Duran also yielded poor results because it obscured surface details. The best images were obtained by fixation with glutaraldehyde and osmium tetraoxide. If too much water is removed from the specimen, by allowing too great a rise in temperature above $-85\,°C$, the tissue will collapse. A further attempt to coat frozen specimens with carbon and metal was made for higher resolution (Echlin & Moreton 1973). Nei, Yotsumoto et al. (1971, 1973) also conducted direct observations of frozen specimens and concurrently that of the cryofractured specimens using SEM. They employed two types of cold stages, installed in the specimen chamber of the microscope to maintain the specimen holder at $-150\,°C$ during most of the observations and in the pre-evacuation chamber to fracture the frozen specimen with a cold knife at $-100\,°C$ or below, respectively. After the frozen specimen

---

Fig. 12 Direct observation of frozen specimens with a scanning electron microscope. (Nei, Yotsumoto et al. 1971, 1973)
Specimens mounted on the holder are rapidly frozen by abrupt immersion in Freon 22 cooled with liquid nitrogen. After transfer to the cold stage in the microscope, the specimen is observed in the frozen state. Freeze-fracturing can be done on the cold stage with a cold knife installed in the pre-evacuation chamber. **a.** Iris petals, air-dried and coated with gold. ($\times 600$) **b.** Same specimen, directly observed in the frozen state. ($\times 300$) **c.** Chrysanthemum petals, cut and air-dried and then coated with carbon and gold. ($\times 600$) **d.** Same specimen, directly observed in the frozen state. ($\times 300$) **e.** Fruit fly (*Drosophila meganogaster*) larva. Head portion of a frozen specimen. ($\times 600$) **f.** Same specimen. Freeze-fractured face in the frozen state. ($\times 60$) **g.** Papilla of hamster tongue. Frozen state. ($\times 300$) **h.** Hamster tongue. Freeze-fractured face. Papilla and muscle fibre. ($\times 100$)

surface was first observed, the specimen was replaced in the pre-evacuation chamber and fractured with a knife. The fractured surface, still maintained in the frozen state, was then observed. Even in this technique, there remained a problem of antifreeze agents. An addition of glycerol was not desirable because it obscured the details of the fine structure.

Cryo-SEM is a simple and useful technique to observe biological specimens, especially soft tissues.

Conclusion

Various freezing techniques, some of which have long been employed and others which have developed very recently in electron microscopy, are briefly introduced in this general review. There are numerous bibliographies to be referred to this research field, which have been published mostly within the past few years, but limited space does not allow the author to cite all of them.

Although these freezing techniques have their own characteristics and as far as devised, they still have, more or less, many demerits as well as merits theoretically and methodologically from various points of view.

The most fundamental point common to many important problems in these techniques is how to freeze the specimens, in particular, how to make preparations devoid of ice crystals which might introduce cumbersome artifacts in these techniques. More adequate means should be designed for further development of these studies applicable to various types and sizes of specimens to inhibit ice crystal formation.

One of the ultimate aims of electron microscopy is to observe fresh wet specimens without any preparative treatment. For this purpose, freezing of specimens is a useful technique, which can be conducted rather simply. The ultrastructural investigation of biological materials for morphological observation in their native state or living state, in particular, also requires studies to be done concurrently from a physiological point of view. It can be expected that the freezing techniques will share the most important aspect of electron microscopy of living matters in the near future.

ACKNOWLEDGEMENT
The author wishes to thank Dr. A.P. MacKenzie for critical reading of the manuscript.

REFERENCES

[1] Albrecht R. M., Orndorff G. R. and MacKenzie A. P. (1973) Survival of certain microorganisms subjected to rapid and very rapid freezing on membrane filters. Cryobiology, *10:* 233.

[2] Altmann R. (1890) Die Elementarorganismen und ihre Beziehungen zur den Zellen. Leipzig.

[3] Bachmann L. and Schmitt W. W. (1971) Improved cryofixation applicable to freeze-etching. Proc. Nat. Acad. Sci. U.S.A., *68:* 2149.

[4] Baker R. F. (1962) Freeze-thawing as a preparatory technique for electron microscopy. J. Ultrastruct. Res., *7:* 173.

[5] Bank H. (1973) Visualization of freezing damage. II. Structural alterations during warming. Cryobiology, *10:* 157.

[6] Bank H. and Mazur P. (1972) Relation between ultrastructure and viability of frozen-thawed Chinese hamster tissue-culture cells. Exptl. Cell Res., *71:* 441.

[7] Bank H. and Mazur P. (1973) Visualization of freezing damage. J. Cell Biol., *57:* 729.

[8]   Bernhard W. and Leduc E. H. (1967)   Ultrathin frozen sections. I. Methods and ultrastructural preservation.   J. Cell Biol., 34: 757.

[9]   Bernhard W. and Nancy M. T. (1964)   Coupes a congélation ultrafines de tissue inclus la gélatine.   J. Microscopie, 3: 579.

[10]   Bernhard W. and Viron A. (1971)   Improved techniques for the preparation of ultrathin frozen sections.   J. Cell Biol., 49: 731.

[11]   Bernier R., Iglesias J. R. and Simard R. (1972)   Detection of DNA by tritiated actinomycin D on ultrathin frozen sections.   J. Cell Biol., 53: 798.

[12]   Blackman M and Lisgarten N. D. (1958)   Electron diffraction investigations in the cubic and other structural forms of ice.   Adv. Phys., 7: 189.

[13]   Bondareff W. (1957)   Morphology of particulate glycogen in guinea pig liver revealed by electron microscopy after freezing and drying and selective staining en bloc.   Anat. Rec., 129: 97.

[14]   Boyde A. and Echlin P. (1973)   Freeze and freeze-drying. A preparative technique for SEM.   In: Scanning Electron Microscopy/1973, pp. 759–766. IIT Res. Inst., Chicago.

[15]   Boyde A. and Vesely P. (1972)   Comparison of fixation and drying procedures for preparation of some cultured cell lines for examination in the SEM.   In: Scanning Electron Microscopy/1972, pp. 265–272. IIT Res. Inst., Chicago.

[16]   Boyde A. and Wood C. (1969)   Preparation of animal tissues for surface-scanning electron microscopy.   J. Microscopy, 90: 221.

[17]   Branton D. (1966)   Fracture faces of frozen membranes.   Proc. Nat. Acad. Sci., 55: 1048.

[18]   Branton D. (1967)   Fracture faces of frozen myelin.   Exptl. Cell Res., 45: 703.

[19]   Branton D. (1971)   Freeze-etching studies of membrane structure.   Phil. Trans. Roy. Soc. London, B, 261: 133.

[20]   Branton D. and Southworth D. (1967)   Fracture faces of frozen Chlorella and Saccharomyces cells.   Exptl. Cell Res., 47: 648.

[21]   Buchheim W. (1972)   Zur Gefierfixierung wässeriger Lösungen.   Naturwiss., 59: 121.

[22]   Buckingham J. H. and Staehelin L. A. (1969)   The effect of glycerol on the structure of lecithin membranes; a study by freeze-etching and X-ray diffraction.   J. Microscopy, 90: 83.

[23]   Bullivant S. (1960)   The staining of thin sections of mouse pancreas prepared by the Fernández-Morán helium II freeze substitution method.   J. Biophys. Biochem. Cytol., 8: 639.

[24]   Bullivant S. (1962)   Consideration of membranes and associated structures after cryofixation.   In: Electron Microscopy. Breese S. S., Jr. (ed.) Vol. 2, R-2, Academic Press, New York.

[25]   Bullivant S. (1965)   Freeze-substitution and supporting techniques.   Lab. Invest., 14: 1178.

[26]   Bullivant S. (1970)   Present status of freezing techniques.   In: Some Biological Techniques in Electron Microscopy. Parsons D. F. (ed.) pp. 101–146, Academic Press, New York.

[27]   Bullivant S. (1973)   Freeze-etching and freeze-fracturing.   In: Advanced Techniques in Biological Electron Microscopy. Koehler J. K. (ed.) pp. 67–112, Springer-Verlag, New York.

[28]   Bullivant S. and Ames A. (1966)   A simple freeze-fracture replication method for electron microscopy.   J. Cell Biol., 29: 435.

[29]   Chalicroft J. P. and Bullivant S. (1970)   An interpretation of liver cell membrane and junction structure based on observation of freeze fracture replicas of both sides of the fracture.   J. Cell Biol., 47: 49.

[30]   Christensen A. K. (1967)   A simple way to cut frozen thin sections of tissue at liquid nitrogen temperatures.   Anat. Rec., 157: 227.

[31]   Christensen A. K. (1969)   A way to prepare thin sections of fresh tissues for electron microscopy.   In: Autoradiography of Diffusable Substances. Roth L. J. and Stumpf W. E. (eds.) pp. 349–362, Academic Press, New York.

[32]   Christensen A. K. (1971)   Frozen thin sections of fresh tissue for electron microscopy, with a description of pancreas and liver.   J. Cell Biol., 51: 772.

[33] Cowley C. W., Timson W. J. and Sawdye T. A. (1961) Ultra rapid cooling techniques in the freezing of biological materials. Biodynamica, *8:* 317.

[34] Cross J. D. (1969) Study of the surface of ice with a scanning electron microscopy. *In:* Physics of Ice. Riehl N., Bullemer B. and Engelhard H. (eds.) pp. 81–94, Plenum Press, New York.

[35] Derjaguin B. V. (1970) Superdense water. Sci. Amer., *223:* Nov. 52.

[36] Dorsey N. E. (1940) Properties of the ordinary water substance. Amer. Chem. Soc. Monogr. No. 81, Reinhold, New York.

[37] Dowell L. G., Moline S. W. and Rinfret A. P. (1962) A low-temperature X-ray diffraction study of ice structures formed in aqueous gelatin gels. Biochim. Biophys. Acta, *59:* 158.

[38] Dowell L. G. and Rinfret A. P. (1960) Low-temperature forms of ice as studied by X-ray diffraction. Nature, *188:* 1144.

[39] Dupouy G. (1968) Electron microscopy at very high voltages. *In:* Advances in Optical and Electron Microscopy. Barner R. and Cosslett V. E. (eds.) Vol. 2, pp. 168–250, Academic Press, New York.

[40] Dupouy G., Perrier F. and Durriru L. (1960) Microscopie électronique. L'observation de la matière vivante au moyen d'un microscope électroniuqe fonctionnant sous très haute tension. C. R. Acad. Sci., *251:* 2836.

[41] Echlin P. (1971) The examination of biological material at low temperatures. *In:* Scanning Electron Microscopy/1971, pp. 225–232, IIT Res. Inst. Chicago.

[42] Echlin P. and Moreton R. (1973) The preparation, coating and examination of frozen biological materials in the SEM. *In:* Scanning Electron Microscopy/1973, pp. 325–329, IIT Res. Inst. Chicago.

[43] Echlin P., Paden R., Dronzek B. and Wayte R. (1970) Scanning electron microscopy of labile biological material maintained under controlled conditions. *In:* Scanning Electron Microscopy/1970, pp. 49–56, IIT Res. Inst. Chicago.

[44] Eisenberg D. and Kauzmann W. (1969) The Structure and Properties of Water. Oxford University Press, London.

[45] Elfvin L-G. (1963) The ultrastructure of the plasma membrane and myelin sheath of peripheral nerve fibres after fixation by freeze-drying. J. Ultrastruct. Res., *8:* 283.

[46] Fennema O. (1973) Water and ice. *In:* Low-temperature Preservation of Foods and Living Matter. Fennema O. R., Powrie W. D. and Marth E. H. (eds.) pp. 3–100, Marcel Dekker, New York.

[47] Fernández-Morán H. (1952a) Application of the ultrathin freezing sectioning technique to the study of cell structures with the electron microscope. Arkiv. Fysik, *4:* 471.

[48] Fernández-Morán H. (1952b) The submicroscopic organization of vertebrate nerve fibres. An electron microscope study of myelinated and unmyelinated nerve fibres. Explt. Cell Res., *3:* 282.

[49] Fernández-Morán H. (1957) Electron microscopy of nervous tissue. *In:* Metabolism of the Nervous System. Richter D. (ed.) pp. 1–34, Pergamon Press, London.

[50] Fernández-Morán H. (1960) Low-temperature preparation techniques for electron microscopy of biological specimens based on rapid freezing with liquid helium II. Ann. N. Y. Acad. Sci., *85:* 689.

[51] Fernández-Morán H. and Dahl A. O. (1952) Electron microscopy of ultrathin frozen sections of pollen grains. Science, *116:* 465.

[52] Fernández-Morán H. and Finean J. B. (1957) Electron microscopic and low-angle X-ray diffraction studies of the nerve myelin sheath. J. Biophys. Biochem. Cytol., *3:* 725.

[53] Fletcher N. H. (1970) The Chemical Physics of Ice. Cambridge University Press, London.

[54] Frank H. S. (1965) The structure of water. Fed. Proc., *24:* Suppl. 15, S-1.

[55] Frank H. S. (1970) The structure of ordinary water. Science, *169:* 635.

[56] Gersh I. (1932) The Altmann technique for fixation by drying while freezing. Anat. Rec., *53:* 309.

[57] Gersh I., Isenberg I., Stephenson J. L. and Bondareff W. (1957) Submicroscopic struc-

ture of frozen-dried liver specifically stained for electron microscopy. I. Technical. Anat. Rec., *128:* 91.

[58] GERSH I. and STEPHENSON J. L. (1954) Freezing and drying of tissues for morphological and histochemical studies. *In:* Biological Applications of Freezing and Drying. HARRIS R. J. C. (ed.) pp. 329–384, Academic Press, New York.

[59] GERSH I., VERGARA J. and ROSSI G. (1960) Use of anhydrous vapors in postfixation and in staining of reactive groups of proteins in frozen-dried specimens for electron microscopic studies. Anat. Rec., *138:* 445.

[60] GLAESER R. M. (1971) Report of the US-Japan seminar on HVEM, personal communication.

[61] GLICK D. and MALMSTROM B. G. (1952) Simple and efficient freeze-drying apparatus for the preparation of embedded tissue. Exptl. Cell Res., *3:* 125.

[62] GRUNBAUM B. W. and WELLINGS S. R. (1960) Electron microscopy of cytoplasmic structures in frozen-dried mouse pancreas. J. Ultrastruct. Res., *4:* 73.

[63] HAGGIS G. H. (1961) Electron microscope replicas from the surface of a fracture through frozen cells. J. Biophys. Biochem. Cytol., *9:* 841.

[64] HAGGIS G. H. (1970) Cryofracture of biological material. *In:* Scanning Electron Microscopy/1970, pp. 97–104, IIT Res. Inst. Chicago.

[65] HAGGIS G. H. (1972) Freeze-fracture for scanning electron microscopy. Proc. Fifth Europ. Congr. EM. pp. 250–251, Inst. Phys. London.

[66] HALL C. E. (1950) A low temperature replica method for electron microscopy. J. Appl. Phys. *21:* 61.

[67] HANZON V., HERMODSSON L. H. and TOSCHI G. (1959) Ultrastructural organization of cytoplasmic nucleoprotein in the exocrine pancreas cells. J. Ultrastruct. Res., *3:* 216.

[68] HAYES T. L. (1973) Scanning electron microscopy. *In:* Advanced Techniques in Biological Electron Microscopy. KOEHLER J. K. (ed.) pp. 153–214, Springer-Verlag, New York.

[69] HIBI T. and YADA K. (1959) Low temperature replica and its application. J. Electron Microscopy, *7:* 21.

[70] IGLESIAS J. R., BERNIER R. and SIMARD R. (1971) Ultramicrotomy: A routine procedure. J. Ultrastruct. Res., *36:* 271.

[71] KAVANAU J. L. (1965) Structure and Function in Biological Membrane. Vol. 1. Holden-Day Inc., San Francisco.

[72] KLOTZ I. M. (1965) Role of water structure in macromolecules. Fed. Proc., *24:* Suppl. 15, S-24.

[73] KLOTZ I. M. (1970) Polyhedral clathrate hydrates. *In:* The Frozen Cell. WOLSTENHOLME G. E. W. and O'CONNOR M. (eds.) pp. 5–26, J. & A. Churchill, London.

[74] KOBAYASHI K. (1971) Report of the US-Japan seminar on HVEM, personal communication.

[75] KOEHLER J. K. (1972) The freeze-etching technique. *In:* Principles and Techniques of Electron Microscopy. HAYAT M. A. (ed.) Vol. 2, pp. 53–98. Van Nostrand Reinhold Co. New York.

[76] KUNTZ I. D., BRASSFIELD T. S., LAW G. D. and PURCELL G. W. (1969) Hydration of macromolecules. Science, *163:* 1329.

[77] LEDUC E. H., BERNHARD W., HOLT S. J. and TRANZER J. P. (1967) Ultrathin frozen sections. II. Demonstration of enzymic activity. J. Cell Biol., *34:* 773.

[78] LIM D. J. (1971) Scanning electron microscopic observation on non-mechanically cryofractured biological tissue. *In:* Scanning Electron Microscopy/1971, pp. 257–264, IIT Res. Inst. Chicago.

[79] LOVELOCK J. E. and BISHOP M. W. H. (1959) Prevention of freezing damage to living cells by dimethyl sulfoxide. Nature, *183:* 1394.

[80] LUSENA C. V. (1955) Ice propagation in systems of biological interest. III. Effect of solutes on nucleation and growth of ice crystals. Arch. Biochem. Biophys., *57:* 277.

[81] LUSENA C. V. (1960) Ice propagation in glycerol solutions at temperatures below $-40\,°C$. Ann. N. Y. Acad. Sci., *85:* Art. 2, 541.

[82] LUSENA C. V. and COOK W. H. (1953) Ice propagation in systems of biological interest. I.

Effect of membranes and solutes in a model cell system.   Arch. Biochem. Biophys., *46:* 232.
[83] Lusena C. V. and Cook W. H. (1954)   Ice propagation in systems of biological interest. II. Effect of solutes at rapid cooling rates.   Arch. Biochem. Biophys., *50:* 243.
[84] Luyet B. J. (1961)   A method for increasing the cooling rate in refrigeration by immersion in liquid nitrogen or in other boiling baths.   Biodynamica, *8:* 331.
[85] Luyet B. J. (1966)   Anatomy of the freezing process in physical systems.   *In:* Cryobiology. Meryman H. T. (ed.) pp. 115–138, Academic Press, New York.
[86] Luyet B., Kroener C. and Rapatz G. (1958)   Detection of heat of recrystallization in glycerol-water mixtures.   Biodynamica, *8:* 73.
[87] Luyet B. and Rapatz G. (1958)   Patterns of ice formation in some aqueous solutions. Biodynamica, *8:* 1.
[88] MacKenzie A. P. and Luyet B. (1962)   Electron microscope study of the structure of very rapidly frozen gelatin solutions.   Biodynamica, *9:* 47.
[89] MacKenzie A. P. (1972)   Freezing, freeze-drying and freeze-substitution.   *In:* Scanning Electron Microscopy/1972, pp. 273–280, IIT Res. Inst. Chicago.
[90] Malhotra S. K. (1966)   A study of the structure of the mitochondrial membrane system. J. Ultrastruct. Res., *15:* 14.
[91] Malhotra S. K. (1968)   Freeze-substitution and freeze-drying in electron microscopy. *In:* Cell Structure and Its Interpretation. McGee-Russel S. M. and Ross K. F. A. (eds.) pp. 11–21, Edward Arnold Publ. London.
[92] Malhotra S. K. and Van Harreveld A. (1965a)   Some structural features of mitochondria in tissues prepared by freeze-substitution.   J. Ultrastruct. Res., *12:* 473.
[93] Malhotra S. K. and Van Harreveld A. (1965b)   Dorsal roots of the rabbit investigated by freeze-substitution.   Anat. Rec., *152:* 283.
[94] Malmstrom B. G. (1951)   Theoretical considerations of the rate of dehydration by histological freeze-drying.   Exptl. Cell Res., *2:* 688.
[95] Mazur P. (1966)   Physical and chemical basis of injury in single-celled microorganisms subjected to freezing and thawing.   *In:* Cryobiology. Meryman H. T. (ed.) pp. 213–315, Academic Press, New York.
[96] Mazur P. (1970)   Cryobiology: The freezing of biological systems.   Science, *168:* 939.
[97] Mazur P., Leibo S. P., Farrant J., Chu E. H. Y., Hanna M. G. and Smith L. H. (1970) Interaction of cooling rate, warming rate and protective additive on the survival of frozen mammalian cells.   *In:* The Frozen Cell. Wolstenholme G. E. W. and O'Connor M. (eds.) pp. 69–88, J. & A. Churchill, London.
[98] McAlear J. H. and Kreutziger G. O. (1967)   Freeze-etching with radiant energy in a simple cold block device.   Proc. 25th Ann. Meet. Electron Micros. Soc. Amer. pp. 116–117, Claitor's Publishing Division.
[99] Menz, L. J. and Luyet B. J. (1961)   An electron microscope study of the distribution of ice in single muscle fibres frozen rapidly.   Biodynamica, *8:* 261.
[100] Meryman H. T. (1966)   Cryobiology. Academic Press, New York.
[101] Meryman H. T. and Kafig E. (1953)   Migratory recrystallization of ice at low temperature. J. Appl. Phys., *24:* 1416.
[102] Meryman H. T. and Kafig E. (1955)   The study of frozen specimens, ice crystals and ice crystal growth by electron microscopy.   Naval Med. Res. Inst. Rep. NM 000 018. 01. 09., *13:* 529.
[103] Meyer H. W. and Winkelmann H. (1970)   Nachweis der Membranspaltung bei der Gefrierätzpräparation an Erythrozytenghosts und die Beeinflussung der Membranstruktur durch Harnstoff.   Protoplasma, *70:* 233.
[104] Monroe H. G., Gamble W. J., La Farge C. G., Gamboa R., Morgan C. L., Rosenthal A. and Bullivant S. (1968)   Myocardial ultrastructure in systole and diastole using ballistic cryofixation.   J. Ultrastruct. Res., *22:* 22.
[105] Moor H. (1964)   Die Gefrierfixation lebender Zellen und ihre Anwendung in der Elektronenmikroskopie.   Zsch. Zellforsch., *62:* 546.
[106] Moor H. (1971)   Recent progress in the freeze-etching technique.   Phil. Trans. Roy. Soc. London, B, *261:* 121.
[107] Moor H. (1973)   Cryotechnology for the structural analysis of biological material.   *In:* Freeze-Etching. Techniques and applications. Benedetti E. L. and Favard P. (eds.) pp. 11–19, Soc. Franc. Microsc. Electron, Paris.

[108] MOOR H. and MÜHLETHALER K. (1963) Fine structure in frozen-etched yeast cells. J. Cell Biol., *17:* 609.

[109] MOOR H., MÜHLETHALER K., WALDNER H. and FREY-WYSSLING A. (1961) A new freezing-ultramicrotome. J. Biophys. Biochem. Cytol., *10:* 1.

[110] MÜLLER H. R. (1957) Gefriertrocknung als Fixierungsmethoden an Pflanzenzellen. J. Ultrastruct. Res., *1:* 109.

[111] MUNDKUR B. (1964) Electron microscopical studies of frozen-dried yeast. Exptl. Cell Res., *34:* 155.

[112] NAGATA F. and ISHIKAWA I. (1972) Observation of wet biological materials in a high voltage electron microscope. Jap. J. Appl. Phys., *11:* 1239.

[113] NAGATA T., NAWA T. and YOKOTA S. (1969) A new technique for electron microscopic dry-mounting radioautography of soluble-compounds. Histochemie, *18:* 241.

[114] NEI T. (1960) Effects of freezing and freeze-drying on microorganisms. In: Recent Research in Freezing and Drying. PARKES A. S. and SMITH A. U. (eds.) pp. 78–86, Blackwell Sci. Publ. Oxford.

[115] NEI T. (1962a) Electron microscopic study of microorganisms subjected to freezing and drying: Cinematographic observations of yeast and coli cells. Exptl. Cell Res., *28:* 560.

[116] NEI T. (1962b) Freeze-drying in the electron microscopy of microorganisms. J. Electron Microscopy, *11:* 51.

[117] NEI T. (1966) Freezing and drying in electron microscopy: Observations of microorganisms with newly devised cooling equipment. In: Electron Microscopy. UYEDA R. (ed.) Vol. 2, pp. 47–48, Maruzen Co., Tokyo.

[118] NEI T. (1973a) Growth of ice crystals in frozen specimens. J. Microscopy, *99:* 227.

[119] NEI T. (1973b) An attempt at cryocleavage in the freeze-etching technique. J. Electron Microscopy, *22:* 371

[120] NEI T. (1974) Cryotechniques. In: Principles and Methods of Scanning Electron Microscopy. HAYAT M. A. (ed.) Vol. 1, pp. 113–123, Van Nostrand Reinhold Co., New York.

[121] NEI T., ARAKI T. and MATSUSAKA T. (1969) Freezing injury to aerated and nonaerated cultures of *Escherichia coli*. In: Freezing and Drying of Microorganisms. NEI T. (ed.) pp. 3–15, University Tokyo Press, Tokyo.

[122] NEI T., KOJIMA Y. and HANAFUSA N. (1964) Hemolysis and morphological changes of erythrocytes with freezing. Contrib. Inst. Low Temp. Sci. B, *13:* 1.

[123] NEI T., MATSUSAKA T. and ASADA M. (1971) Investigations on the cooling conditions in the freeze-etching technique. Low Temp. Sci. B, *29:* 91.

[124] NEI T., YOTSUMOTO H., HASEGAWA Y. and NAGASAWA Y. (1971) Direct observation of frozen specimens with a scanning electron microscope. J. Electron Microscopy, *20:* 202.

[125] NEI T., YOTSUMOTO H., HASEGAWA Y. and NAGASAWA Y. (1973) Direct observation of frozen specimens with a scanning electron microscope. J. Electron Microscopy, *22:* 185.

[126] NEMETHY G. and SCHERAGA H. A. (1962a) Structure of water and hydrophobic bonding in proteins. I. A model for the thermodynamic properties of liquid water. J. Chem. Phys., *36:* 3382.

[127] NEMETHY G. and SCHERAGA H. A. (1962b) Structure of water and hydrophobic bonding in proteins. II. Model for the thermodynamic properties of aqueous solution of hydrocarbon. J. Chem. Phys., *36:* 3401.

[128] PARSONS D. F., MATRICARDI V. R., SUBJECK J., UYDESS I. and WRAY G. (1972) High voltage electron microscopy of the wet whole cancer and normal cells. Visualization of cytoplasmic structures and surface projections. Biochim. Biophys. Acta, *290:* 110.

[129] PARSONS D. F. and MORETZ R. C. (1970) Microscopy and diffraction of water in the electron microscope. Proc. Intern. Congr. EM. Grenoble, Vol. 1, pp. 497–498.

[130] PEASE D. C. (1967a) Eutectic ethylene glycol and pure propylene glycol as substituting media for the dehydration of frozen tissue. J. Ultrastruct. Res., *21:* 75.

[131] PEASE D. C. (1967b) The preservation of tissue fine structure during rapid freezing. J. Ultrastruct. Res., *21:* 98.

[132] PEASE D. C. (1973) Substitution techniques. In: Advanced Techniques in Biological Electron Microscopy. KOEHLER J. K. (ed.) pp. 35–66, Springer-Verlag, New York,

[133] PINTO DA SILVA P. and BRANTON D. (1970) Membrane splitting in freeze-etching. Covalently bound ferritin as a membrane marker. J. Cell Biol., *45:* 598.

[134] PLATTNER H., FISCHER W. M., SCHMITT W. W. and BACHMANN L. (1972) Freeze etching of cells without cryoprotectants. J. Cell Biol., *53:* 116.

[135] POLGE C., SMITH A. U. and PARKES A. S. (1949) Revival of spermatozoa after vitrification and dehydration at low temperature. Nature, *164:* 666.

[136] RAPATZ G. L., MENZ L. J. and LUYET B. J. (1966) Anatomy of the freezing process in biological materials. *In:* Cryobiology. MERYMAN H. T. (ed.) pp. 139–162, Academic Press, New York.

[137] REBHUN L. I. (1961) Application of freeze-substitution to electron microscope studies of invertebrate oocytes. J. Biophys. Biochem. Cytol., *9:* 785.

[138] REBHUN L. I. (1965) Freeze-substitution: Fine structure as a function of water concentration in cells. Fed. Proc. *24:* Suppl. 15, S-217.

[139] REBHUN L. I. (1972) Freeze-substitution and freeze-drying. *In:* Principles and Techniques of Electron Microscopy. HAYAT M. A. (ed.) Vol. 2, pp. 3–49, Van Nostrand Reinhold Co., New York.

[140] REBHUN L. I. and SANDER G. (1971) Electron microscope studies of frozen-substituted marine eggs. I. Conditions for avoidance of intracellular ice crystallization. Amer. J. Anat., *130:* 1.

[141] RICE R. V., MOSES J. and WRAY G. (1971) High voltage electron microscopy of unstained frozen striated muscle. Read at the US-Japan seminar on HVEM. Personal communication.

[142] RIEHL N., BULLEMER B. and ENGELHARD H. (1969) Physics of Ice. Plenum Co., New York.

[143] RIEHLE U. (1968) Über die Vitrifizierung verdünnter wässeriger Lösungen. Dissertation Nr. 4271, Eidgen. Techn. Hochschule Zürich.

[144] ROTH L. J. and STUMPF W. E. (1969) Autoradiography of Diffusable Substances. Academic Press, New York.

[145] SENO S. and YOSHIZAWA K. (1960) Electron microscope observations on frozen dried cells. J. Biophys. Biochem. Cytol., *8:* 617.

[146] SHERMAN J. K. (1972) Comparison of *in vitro* and *in situ* ultrastructural cryoinjury and cryoprotection of mitochondria. Cryobiology, *9:* 112.

[147] SHERMAN J. K. and KIM K. S. (1967) Correlation of cellular ultrastructure before freezing, while frozen and after thawing in assessing freeze-thaw induced injury. Cryobiology, *4:* 61.

[148] SHERMAN J. K. and LIU K. C. (1973) Ultrastructural cryoinjury and cryoprotection of rough endoplasmic reticulum. Cryobiology, *10:* 104.

[149] SHIMADA K. and ASAHINA E. (1972) Innocuous ice crystallization within ascites tumor cells of rat by rapid freezing. Low Temp. Sci. B, *30:* 65.

[150] SIMPSON W. L. (1941) An experimental analysis of the Altmann technique of freeze-drying. Anat. Rec., *80:* 173.

[151] SJÖSTRAND F. S. (1943a) Electron microscopic examination of tissues. Nature, *151:* 725.

[152] SJÖSTRAND F. S. (1943b) Eine neue Methode zur Herstellung sehr dünner Objektschnitte für die elektronenmikroskopische Untersuchungen von Geweben. Arkiv. Zool., *35:* A-1.

[153] SJÖSTRAND F. S. and BAKER R. F. (1958) Fixation by freezing-drying for electron microscopy of tissue cells. J. Ultrastruct. Res., *1:* 239.

[154] SJÖSTRAND F. S. and ELFVIN L-G. (1964) The granular structure of mitochondrial membranes and of cytomembranes as demonstrated in frozen-dried tissue. J. Ultrastruct. Res., *10:* 263.

[155] SLEYTR U. (1970) Die Gefrierätzung korrespondierender Bruchhäften: Ein neuer Weg zur Aufklärung im Membranstrukturen. Protoplasma, *70:* 101.

[156] STAEHELIN L. A. (1968) The interpretation of freeze-etched artificial and biological membranes. J. Ultrastruct. Res., *22:* 326.

[157] STAEHELIN L. A. and BERTAUD W. S. (1971) Temperature and contamination dependent freeze-etch images of frozen water and glycerol solutions. J. Ultrastruct. Res., *37:* 146.

[158] STEERE R. L. (1957) Electron microscopy of structural detail in frozen biological specimens. J. Biophys. Biochem. Cytol., *3:* 45.

[159] STEERE R. L. (1969a) Freeze-etching simplified. Cryobiology, *5:* 306.
[160] STEERE R. L. (1969b) Freeze-etching and direct observation of freezing damage. Cryobiology, *6:* 137.
[161] STEERE R. L. and SOMMER J. R. (1972) Stereo ultrastructure of nexus faces exposed by freeze-fracturing. J. Microscopie, *15:* 205.
[162] STEPHENSON J. L. (1953) Theory of the vacuum drying of frozen tissues. Bull. Math. Biophys,. *15:* 411.
[163] STEPHENSON J. L. (1956) Ice crystal growth during the rapid freezing of tissues. J. Biophys. Biochem. Cytol., *2:* 45.
[164] STEPHENSON J. L. (1960) Fundamental physical problems in the freezing and drying of biological materials. *In:* Recent Research in Freezing and Drying. PARKES A. S. and SMITH A. U. (eds.) pp. 121–145, Blackwell Sci. Publ. Oxford.
[165] STIRLING C. E. and KINTER W. B. (1967) High resolution radioautography of galactose-^{3}H accumulation in rings of hamster intestine. J. Cell Biol., *35:* 585.
[166] STOWELL R. E., YOUNG D. E., ARNOLD E. A. and TRUMP B. F. (1965) Structural, chemical, physical and functional alterations in mammalian nucleus following different conditions of freezing, storage and thawing. Fed. Proc., *24:* Suppl. 15, S-115.
[167] STUMPF W. E. and ROTH L. J. (1967) Freeze-drying of small tissue samples and thin frozen sections below $-60\,°C$. A simple method of cryosorption pumping. J. Histochem. Cytochem. *15:* 243.
[168] TANAKA K. (1972) Freezed resin cracking method for scanning electron microscopy of biological materials. Naturwiss., *59:* 77.
[169] TILLACK T. W. and MARCHESI V. T. (1970) Demonstration of the outer surface of freeze-etched red blood cell membranes. J. Cell Biol., *45:* 649.
[170] TOKUYASU K. T. (1973) A technique for ultracryotomy of cell suspensions and tissues. J. Cell Biol., *57:* 551.
[171] TRUMP B. F., YOUNG D. E., ARNOLD E. A. and STOWELL R. E. (1965) Effects of freezing and thawing on the structure, chemical constitution, and function of cytoplasmic structure. Fed. Proc., *24:* Suppl. 15, S-144.
[172] VAN HARREVELD A. and CROWELL J. (1964) Electron microscopy after rapid freezing on a metal surface and substitution fixation. Anat. Rec., *149:* 381.
[173] VAN HARREVELD A., CROWELL J. and MALHOTRA S. K. (1965) A study of extracellular space in central nervous tissue by freeze-substitution. J. Cell Biol., *25:* Part I, 117.
[174] VORBRODT A. and BERNHARD W. (1968) Essais de localisation au microscope électronique de l'activité phosphatasique nucléaire dans des coupes à congélation ultrafines. J. Microscopie, *7:* 195.
[175] WEHRLI E., MÜHLETHALER K. and MOOR H. (1970) Membrane structure as seen with a double replica method for freeze fracturing. Exptl. Cell Res., *59:* 336.
[176] WEINSTEIN R. S. and SOMEDA K. (1967a) Freeze-etching of fracture faces of frozen packed red cells with a modified Bullivant-Ames freeze-fracture and replication apparatus. J. Cell Biol., *35:* 190a.
[177] WEINSTEIN R. S. and SOMEDA K. (1967b) The freeze-cleave approach to the ultrastructure of frozen tissues. Cryobiology, *4:* 116.
[178] WHIPPLE H. E. (1965) Forms of water in biologic systems. Ann. New York Acad. Sci., *125:* Art. 2, pp. 249–772.
[179] WILLIAMS R. C. (1954) The application of freeze-drying to electron microscopy. *In:* Biological Applications of Freezing and Drying. HARRIS R. J. C. (ed.) pp. 303–328, Academic Press, New York.
[180] WINKELMANN H. and MEYER H. W. (1968) A routine freeze-etching technique of high effectivity by simple technical means. Part. 1. The principle. Exp. Path., *2:* 277.
[181] WINKELMANN H. and WAMMETBERGER S. (1969) Eine mit einfachen Mitteln durchfürbare Routine-Gefrierätztechnique hoher Effektivität. Teil II. Die technische Anordnung. Exp. Path., *3:* 113.
[182] ZOTIKOV L. and BERNHARD W. (1969) Localisation au microscope électronique de l'activité de certaines nucléases dans des coupes à congélation ultrafines. J. Ultrastruct. Res., *30:* 642.

Membrane Modifications for the Cell Junction in the Mucosal Epithelia of the Rat Small Intestine: A Freeze-Etch Study

Toshi Yuki YAMAMOTO and Hiroshi WATANABE

Tight junction, or zonula occludens, has been noted in the epithelia of a variety of organs including the small intestine to be formed at the most apical regions of the lateral cell surface and to serve as a permeability barrier which separates the intercellular space from the lumen. The freeze-etch technique reveals this type of junction as an area of membrane modification easily discernible from other membrane areas. When fractures expose membrane faces at the tight junctions one can see meshworks of narrow ridges on the A face and complementary patterns of grooves on the B face (Staehelin, Mukherjee & Williams 1969; Chalcroft & Bullivant 1970; Friend & Gilula 1972; Staehelin 1973; Wade & Karnovsky 1974).

This paper reports that such ridges and grooves are not always confined within the narrow apical regions of lateral cell surfaces where tight junction has commonly been known to occur but that they may often diffusely spread on the lateral surfaces of the epithelial cells in the rat small intestine.

Materials and Methods

The duodenum and the oral part of jejunum of adult rats of both sexes were used for this study.

Animals were sacrificed by decapitation and segments of the small intestine were immediately removed. They were cut into small pieces and fixed in 2.5% glutaraldehyde dissolved either in 0.9% sodium chloride or in 0.1 M cacodylate buffer for various periods from 30 min to 3 hr. Then the fixed tissue blocks were immersed in 20 to 40% glycerol in 0.9% sodium chloride for 1 to 24 hr. After rapid freezing in liquid Freon 12, the blocks were plunged into liquid air.

Fracturing, etching, and shadowing were performed using a JEOL-type freeze-etch device. Etching of the fracture faces was carried out for 1 to 2 min at a specimen temperature of about $-100\,°C$. Replicas were formed by shadowing with platinum palladium backed with carbon. A JEM-200A electron microscope was used for examining the replicas.

Results

The typical tight junctions were seen at the most apical zones of the lateral surfaces of the epithelial cells. Here, the A face of cell membrane possessed a series of ridges forming a

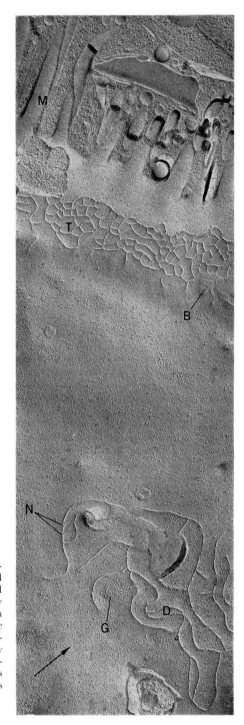

Fig. 1 A lateral cell membrane of a columnar epithelial cell. A typical tight junction (T) is located just below the microvilli (M) layer, and the dispersed junctional ridges and grooves (D) are more basally situated. Note the differences in construction between the two meshworks. The ridges and grooves have many notches (N) on them arranged at varying intervals so that they appear as dashed lines. B, narrow area of the B face where the grooves of the tight junction are formed; G, small aggregation of particles similar to the gap junction. In all figures, the arrows indicate the direction of shadowing. (×26,000)

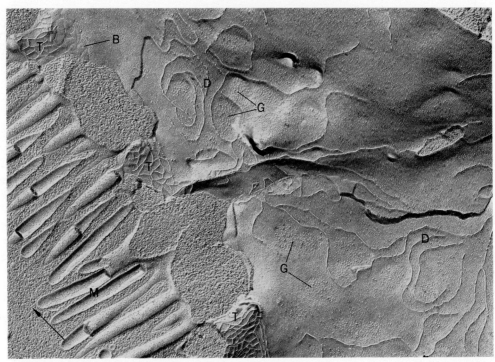

Fig. 2 Another freeze-etch replica showing a typical tight junction (T) and dispersed junctional ridges and grooves (D). B, narrow area of the B face; G, aggregations of small particles; M, microvilli. (×25,500)

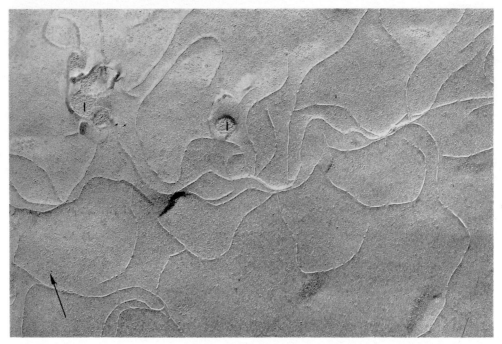

Fig. 3 A typical distribution pattern of the dispersed junctional ridges on the A face of the lateral cell membrane of a columnar epithelial cell. I, crossly fractured interdigitation. (×25,500)

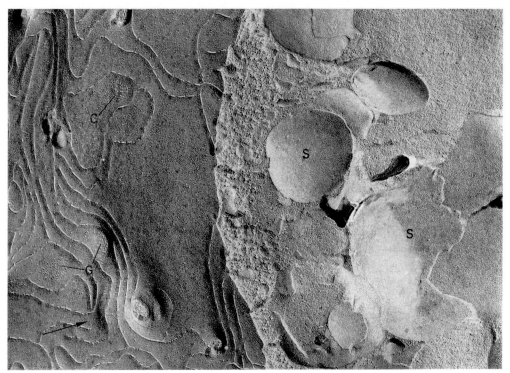

Fig. 4 The junctional ridges dispersed on the A face of the lateral cell membrane of a goblet cell. G, aggregations of small particles; S, secretion granules in the goblet cell. (×31,000)

regular and compact meshwork (Figs. 1 and 2). Diameters of the meshes measured roughly 70 to 250 nm. At higher magnification, the ridges were found interrupted by small notches arranged at irregular intervals (Figs. 1–4). The B face of cell membrane showed corresponding arrays of grooves. The junction constituted a continuous belt, about 0.5 μm in width, which circumferentially surrounded each epithelial cell at the level of the terminal web.

Ridges on the A face and grooves on the B face of the same structures as those just described for the typical tight junctions often were observed to spread to a varying degrees on the lateral cell membranes both of columnar epithelial and of goblet cells (Figs. 1–4). They ran meandering courses and exhibited occasional branchings and anastomoses to form irregular and loose meshworks. Sometimes they were connected upward with the typical tight junctions or extended downward close the basal cell surface. It has been verified by careful observations of freeze-etch replicas, as well as of thin sections (unpublished observation), that the apposed lateral cell membranes of adjacent cells were always joined together along the ridges and grooves with a resultant occlusion of the intercellular space (Figs. 5 and 6).

The gap junctions could be recognized on freeze-etch replicas as areas of dense aggregation of small particles on the A face of cell membrane, and of small depressions on the B face. In our specimens the particles and depressions were seen to form small aggregations with indefinite boundaries here and there on the lateral cell membranes (Figs. 1, 2 and 4). The aggregations occurred occasionally within the meshes of the ridges and grooves. In some places, the ridges and grooves surrounded the roots of cytoplasmic interdigitations

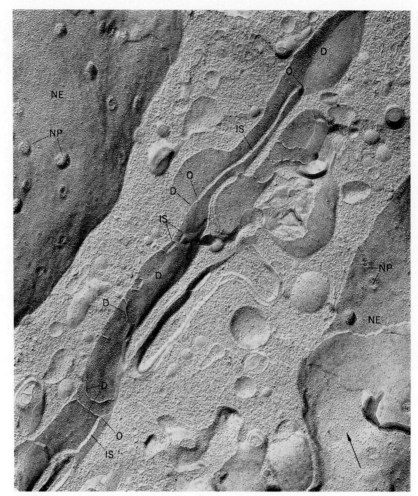

Fig. 5 Apposed cell membranes of two adjacent columnar epithelial cells fractured obliquely. The intercellular space (IS) is occluded along the junctional ridges and grooves (D). NE, fracture face of the nuclear envelope; NP, nuclear pore; O, occluded margin of the intercellular space. (×41,000)

between adjacent cells, however, in other places interdigitations had no connection with the ridges and grooves (Fig. 3).

Discussion

Since FARQUHAR and PALADE (1963) minutely described the junctional complex in the epithelia, many electron microscope studies using thin sections have been devoted to this subject. With these studies it has been established that the localization of the tight junctions in mucosal epithelia is limited to the most apical zone of the lateral cell surface where they form zonules completely encircling the epithelial cell.

Recently the freeze-etch technique has added much to our knowledge of the morphology of junctional modifications of the cell membrane. In particular, this technique has clarified the distinction between tight and gap junctions and also has made it possible to observe the two-dimensional stretches of these junctions on the membrane.

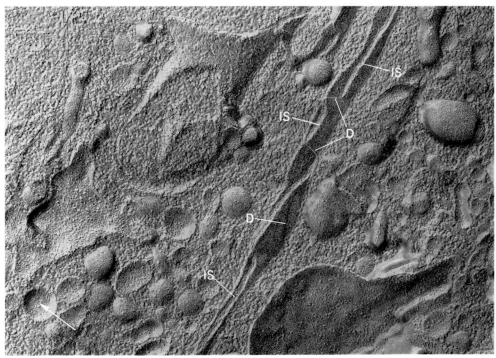

Fig. 6 Obliquely fractured lateral cell membranes. The intercellular space (IS) is occluded along the junctional ridges and grooves (D). (×51,000)

In accordance with most of the previous works (STAEHELIN, MUKHERJEE & WILLIAMS 1969; CHALCROFT & BULLIVANT 1970; FRIEND & GILULA 1970; GOODENOUGH & REVEL 1970; McNUTT & WEINSTEIN 1970; FRIEND & GILULA 1972; STAEHELIN 1973; WADE & KARNOVSKY 1974), the present study confirmed that the tight junctions are represented as areas of meshworks of ridges and grooves in freeze-etch replicas and that these areas form continuous belts or zonules around all epithelial cells. Furthermore, it has been revealed by our study that, in the mucosal epithelium of the rat small intestine, similar ridges and grooves may occur spreading over wide areas of the lateral cell surface. They form loose and irregular meshworks and often show direct continuity to those of the typical tight junctions. The apposed cell membranes of adjacent cells are joined together along these ridges and grooves with a resultant occlusion of the intercellular space. FRIEND and GILULA (1972) also briefly noted the occasional presence of discontinuous or isolated ridges and grooves in the rat intestinal epithelium.

It may be concluded from these findings that the junctional modification fundamentally similar to the tight junction is diffusely formed over the lateral cell surfaces of columnar and goblet cells in the epithelium of rat small intestine. The difficulty in disclosing these junctions on thin sections may chiefly be attributed to their looseness in construction. In typical tight junctions, the meshwork of ridges and grooves are so compact that the apposed cell membranes within the meshes are kept in fairly close contact with each other, while in areas of the dispersed ridges and grooves, the apposed cell membranes in the meshes may be separated by the intercellular space. It is conceivable that the dispersed junctional ridges and grooves are changeable in configuration and distribution depending on the functional state of the cells.

Summary

The mucosal epithelium of the rat small intestine was studied by the freeze-etch method. Ridges on the A face and corresponding grooves on the B face spread to varying degrees on the lateral surfaces of both columnar epithelial and goblet cells. These ridges and grooves are similar in structure to those of the typical tight junction and often show direct continuity with them. Meshworks of the dispersed ridges and grooves are irregular and loose. Apposed cell membranes always come into close contact with each other along these ridges and grooves with a resultant occlusion of the intercellular space.

REFERENCES

[1] CHALCROFT J. P. and BULLIVANT S. (1970) An interpretation of liver cell membrane and junction structure based on observation of freeze-fracture replicas of both sides of the fracture. J. Cell Biol., *47:* 49.

[2] FARQUHAR M. G. and PALADE G. E. (1963) Junctional complexes in various epithelia. J. Cell Biol., *17:* 375.

[3] FRIEND D. S. and GILULA N. B. (1970) Cell junctions of the rat epididymis. J. Cell Biol., *47:* 66a (Abstract).

[4] FRIEND D. S. and GILULA N. B. (1972) Variations in tight and gap junctions in mammalian tissues. J. Cell Biol., *53:* 758.

[5] GOODENOUGH D. A. and REVEL J. P. (1970) A fine structural analysis of intercellular junctions in the mouse liver. J. Cell Biol., *45:* 272.

[6] McNUTT N. S. and WEINSTEIN R. S. (1970) The ultrastructure of the nexus. A correlated thin-section and freeze-cleave study. J. Cell Biol., *47:* 666.

[7] STAEHELIN L. A. (1973) Further observations on the fine structure of freeze-cleaved tight junctions. J. Cell Sci., *13:* 763.

[8] STAEHELIN L. A., MUKHERJEE T. M. and WILLIAMS A. W. (1969) Freeze-etch appearance of tight junctions in the epithelium of small and large intestine of mice. Protoplasma, *67:* 165.

[9] WADE J. B. and KARNOVSKY M. J. (1974) The structure of the zonula occludens. A single fibril model based on freeze-fracture. J. Cell Biol., *60:* 168.

IV.

Quantitative Methods in Electron Microscopic Cytology

Cytometric Analysis of Thymic Small Lymphocytes, Studied by a Stereological Method in Electron Microscopy

Takashi ITO *and* Kazuhiro ABE

As is generally known, the thymus is a major organ of lymphocyte production. Lymphocytes proliferate actively in this organ and differentiate from large through medium to small lymphocytes within the cortex. Small lymphocytes thus produced in the cortex move into the medulla, from which they leave this organ as so-called thymus-derived lymphocytes eventually to join the peripheral circulation as a recirculating lymphocyte pool which populates certain thymus-dependent areas in the peripheral lymphatic tissues. Lymphocytes are thought to be functionally "processed" in the thymus, because lymphocytes acquire immunological competence within the environment of this organ. However, small lymphocytes are considered to be immunologically incompetent in the cortex, while they are probably immunologically competent in the medulla (for reviews see Everett & Tyler 1967; Miller & Osoba 1967; Yoffey & Courtice 1970; Elves 1972). If so, are thymic small lymphocytes accompanied by any morphological changes as well as functional during movement from the cortex to the medulla? In fact, some differences between thymic cortical and medullary small lymphocytes have been observed by previous authors in certain behavioral and morphological aspects (Trowell 1961; Ito & Hoshino 1962; Kostowiecki & Sherman 1963; Saint-Marie & Leblond 1964; Weber 1966).

In previous papers, we also have confirmed by qualitative and quantitative analyses with electron microscopy that cortical small lymphocytes are morphologically discernible from medullary small lymphocytes in the thymus of adult and neonatal mice (Abe & Ito 1970, 1972; Abe et al. 1973b) During the course of these studies on thymic small lymphocytes at the electron microscopic level, a stereological approach was needed for the determination of the precise size of spherical objects at the three-dimensional level, because lymphocytes are generally spherical in shape. Stereological analysis has recently been attempted for the determination of the size of spherical objects in electron microscopy as well as in light microscopy (for reviews see Weibel 1969; Elias et al. 1971). The methods previously used, however, seem to be rather complicated and troublesome in practice. In this respect, the stereological method used in our studies was relatively simple and convenient for practical use, and it seems to be useful for the above purpose. In this paper, the principle of the stereological method which we have proposed is first considered theoretically, and the cytometric results obtained by this method on thymic small lymphocytes are presented.

Material and Methods

Normal thymuses from 4- and 35-day-old dd-mice were fixed in 5% formalin in 0.05 M

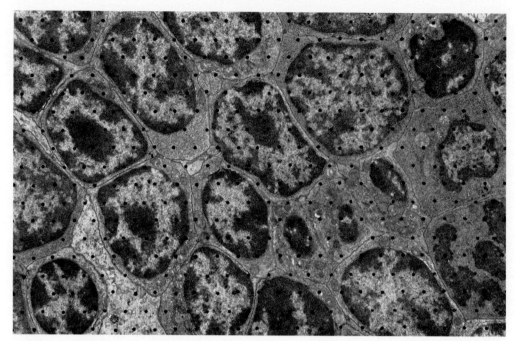

Fig. 1 A cortical portion of the thymus of a 35-days-old mouse. The electron micrograph is superimposed by a regular lattice of points for the measurement of nuclear and cellular sectional areas. ($\times$ 6,000)

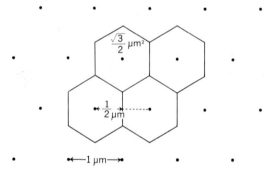

Fig. 2 One point of the lattice for measurement represents a hexagonal area of $\sqrt{3}/2$ μm^2. ($\times 6,000$)

phosphate buffer (pH 7.5) at 4 °C for 48 hrs and postfixed in 2% OsO$_4$ in the same buffer for 1.5 to 2 hrs. After dehydration in the usual way the tissues were embedded in Epon. Ultrathin sections were cut with glass knives on an LKB ultramicrotome and doubly stained with uranyl acetate and lead citrate. They were examined with a Hitachi-electron microscope (HS-8). Areas of nuclear and cellular profiles of lymphocytes in both the cortex and medulla were determined by a point-counting method (WEIBEL et al. 1966) on electron micrographs enlarged to a final magnification of 6,000$\times$. For the point-counting a transparent plastic sheet with a regular lattice of 540 points was superimposed on the micrographs (Fig. 1), and the numbers of points lying on the nuclei and cell bodies were counted. On the sheet each point was 6 mm from its nearest neighbor, and the points were 1 μm apart when superimposed on micrographs enlarged 6,000$\times$. In this case, as seen in Fig. 2, the area occupied by one point can be regarded as representing a hexagonal area of $\sqrt{3}/2$ μm^2. For the statistical calculation a microcomputer (SOBAX 2700) was used in the stereological procedures as described in detail below.

Stereological Principle and Procedures

Sampling

As pointed out by WEIBEL (1969), each step in a stereological study demands random sampling to avoid possible biases. For example, the steps include choice of the animals used, selection of the tissue blocks and recording and analyses of the electron micrographs. However, complete random sampling is not always carried out in the strict sense and samples used are systematically selected under a constant and regular system decided in advance, depending on the characteristics of the objects. Such sampling is called systematic random sampling (EBBESSON & TANG 1967; WEIBEL 1969; TANG & EBBESSON 1972).

With thymic cortical and medullary small lymphocytes, it should be assumed that in the thymus, lymphocytes are distributed without anisotropy throughout the cortex and medulla and that no individual variation is present in the size distribution for cortical and medullary small lymphocytes. Several blocks were selected at random for each animal from an adequate number of each block pool from the thymus. Then the blocks were trimmed and examined by light microscopy to decide whether they were from the cortex or medulla. Electron micrographs were systematically sampled under a magnification of 3,000× from a single section of either the cortex or medulla for each block. Thus the sampled electron micrographs were considered to contain random profiles of lymphocytes. All profiles of lymphocytes on the micrographs were measured for quantitation, and almost equal numbers of profiles were collected in each case.

Volumetry

In this study, the volumetric ratio of nucleus to cell in small lymphocytes was obtained from the total numbers of points lying on areas of the sections of the nucleus and the cell body in the electron micrographs. The value thus obtained was one of the probable values. In such point-counting volumetry, a sampling error should be considered for evaluation of the reliability of the values obtained. The relative error in the ratio obtained was thought to depend on the total numbers of test points counted. This error would be reduced in practice when the total numbers of test points counted were appropriately larger. The number of test points for counting would increase, if the points were densely distributed. In the lattice of points for counting, however, the points should be coarsely enough distributed for efficient point-counting volumetry. In relation to this problem, HILLIARD and CAHN (1961) and WEIBEL (1969) stated that the optimal density of points depended on the size of the objects, and that the points should be so spaced that no more than one point would be included in the individual profile. Under such conditions the sampling error is smallest in a given total number of points for volumetry. Thus the sampling error (m) in point-counting volumetry can be obtained by the following formula,

$$m = \sqrt{\frac{X(1-X)}{P}} \qquad (1)$$

where P expresses either the number of test points counted when points are distributed coarsely enough or the number of profiles counted when points are distributed more densely and X represents the ratio obtained. As indicated by this formula, the value of m represents the standard error, and the value of X falls within a range of $X \pm m$. The statistical significance between two ratios is evaluated by the following formula in Student's t test.

$$t = \frac{X_1 - X_2}{\sqrt{m_1^2 - m_2^2}} \qquad (2)$$

The number of samples necessary for the measurements is dependent on the range of m. If it is expected that the error (m) of X is below $e\%$, the number of points or profiles (Pe) necessary for point-counting volumetry is determined by the following formula

$$Pe = \frac{1-X}{10^{-4}e^2 X} \qquad (3)$$

Size distribution of spherical objects

In stereological microscopy, a three-dimensional interpretation of two-dimensional images on sections is needed. Thus, the stereological determination of the size of spherical objects is achieved on the basis of size distribution of the sectional profiles on slices. The basic problems related to the characterization of the size distribution have been considered in detail by WEIBEL (1969). If spheres of a diameter of $2R$ are randomly sectioned, the sectional profiles appear as circles of varying diameter $2r$ $(R>r)$. As shown in Fig. 3, if the profile circles on sections are grouped according to the radius ranges r_0 to r_1, r_1 to r_2, ..., r_{i-1} to r_i, ..., r_{n-1} to r_n, the distribution frequency within each range is dependent on the distance between two sectional circles limiting each range (Fig. 4). Therefore, the frequency (F_i) of sectional circles with a range between two diameters of r_{i-1} and r_i can be written as

$$F_i = C\left(\sqrt{R^2 - r_{i-1}^2} - \sqrt{R^2 - r_i^2}\right) \qquad (4)$$

where C is constant.

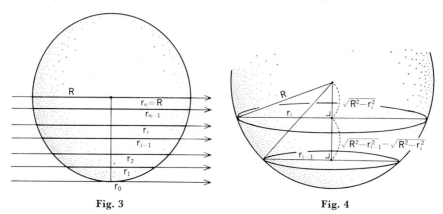

Fig. 3 The sectional circles of a sphere with a diameter $2R$ are classified into n classes of limiting ranges $r_0, r_1, r_2, \ldots, r_i, \ldots, r_n$.

Fig. 4 Distance between two sectional circles of a sphere.

In general, size distribution of spherical objects is obtained by measuring the diameters of the sectional profile circles on sections. If the circles on sections are classified regularly into n classes according to the radii, r_i is written as

$$r_i = \frac{iR}{n}$$

In this situation, if r_i is substituted into equation (4), the frequency F within the nth range is expressed as

$$F_i = C\left(\sqrt{n^2 - (i-1)^2} - \sqrt{n^2 - i^2}\right)\frac{R}{n} \qquad (5)$$

The frequency distribution is highest where $i=n$ in the range between r_{n-1} and r_n. The shape of the frequency distribution curve is variable depending on the value of C, although it exhibits a similarity. If one gives 1 to the peak value as a matter of convenience for calculation, the frequency F_i is obtained as follows:

$$F_n = 1$$
$$F_i = \frac{\sqrt{n^2 - (i-1)^2} - \sqrt{n^2 - i^2}}{\sqrt{2n-1}} \qquad (6)$$

F_i in this formula is variable depending on n. In practice, however, the calculation is relatively complex and troublesome. In addition, a precise measurement of diameters is often attended with some difficulties, because the objects presumed to be spherical in biological specimens do not always occur as perfectly circular profiles on sections. For this reason, we have proposed to use areas of the circular profiles instead of the diameters. In our stereological method, the procedures proved to be easier and more convenient in practice. The principle and procedures will be mentioned below.

If the circular areas on sections are classified into n classes in which the circular area of the largest limiting range is πR^2 and the circular area of the ith limiting range is $i/n \cdot \pi R^2$, the area limiting the nth range is written as

$$\pi r_i^2 = \frac{i}{n} \pi R^2$$

Hence

$$r_i^2 = \frac{i}{n} R^2$$

In this situation, if r_i^2 is substituted into equation (4), the frequency F_i within the ith range is expressed as

$$F_i = C \left(\sqrt{n - i + 1} - \sqrt{n - i} \right) \frac{R}{\sqrt{n}} \qquad (7)$$

The frequency distribution is greatest where $i = n$ in the range between r_{n-1} and r_n. If one gives 1 to the value at the peak, one obtains the frequency F_i as follows

$$F_n = \frac{CR}{\sqrt{n}} = 1$$
$$F_i = \sqrt{n - i + 1} - \sqrt{n - i} \qquad (8)$$

Thus the frequency distributions in turn are obtained successively as

$$F_n = \sqrt{1} - \sqrt{0}$$
$$F_{n-1} = \sqrt{2} - \sqrt{1}$$
$$F_{n-2} = \sqrt{3} - \sqrt{2}$$
$$\vdots$$

Thus the frequency distributions obtained show a regular declination from the largest to the next smaller classes independent of the number of classes (Fig. 5), and the frequency F_i is obtained from the next formulas (9 or 10), if the classes are expressed in turn from the largest to the smallest class in terms of $0, 1, 2, 3, \ldots i, \ldots$ or $1, 2, 3, 4, \ldots i, \ldots$

$$F_i = \sqrt{i+1} - \sqrt{i} \quad i = 0, 1, 2, 3, \ldots\ldots, i, \ldots\ldots \qquad (9)$$
or
$$F_i = \sqrt{i} - \sqrt{i-1} \quad i = 1, 2, 3, 4, \ldots\ldots, i, \ldots\ldots \qquad (10)$$

A total of F_i is obtained from the formula (10) as follows

$$\sum_{i=1}^{n} F_i = \sqrt{n} \qquad (11)$$

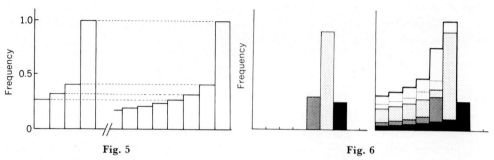

Fig. 5 The frequency distribution of circular areas on sections show a regular declination from the largest to the smaller classes independently of the number of classes.

Fig. 6 If there is a mixture of three populations of spheres with different sizes (left), the size distribution of the circular areas of the random sections of the spheres would represent the heavy line on the right, because the three frequency distributions of sectional circles for the respective populations of spheres are superimposed.

When the value A of the frequency at the peak is determined by actual measurement, the frequency distribution in each class can be derived directly from the formula (9) as follows:

$$AF_0 \ (F_0 = 1)$$
$$AF_1$$
$$AF_2$$
$$\vdots$$
(12)

If the objects comprise a mixture of two or more populations of spheres of different diameters, each population will give rise to a size distribution of the profile areas. Thus the curve actually obtained will form a compound distribution curve, because each frequency distribution for each population is superimposed in the actual curve (Fig. 6). When the frequency distributions determined by measurements of the profile areas are $A_0, A_1, A_2, A_3, \ldots, A_i, \ldots$, the three-dimensional size distributions $B_0, B_1, B_2, B_3, \ldots, B_i, \ldots$ of spheres are successively deduced from Table 1 in the same way as done by the earlier authors (BAUDHUIN & BERTHET 1967; ELIAS & HENNING 1967; COUPLAND 1968; WEIBEL 1969).

Table 1

$$A_0 - 0 = B_0 F_0$$
$$A_1 - B_0 F_1 = B_1$$
$$A_2 - B_0 F_2 - B_1 F_1 = B_2$$
$$A_3 - B_0 F_3 - B_1 F_2 - B_2 F_1 = B_3$$
$$A_4 - B_0 F_4 - B_1 F_3 - B_2 F_2 - B_3 F_1 = B_4$$
$$A_5 - B_0 F_5 - B_1 F_4 - B_2 F_3 - B_3 F_2 - B_4 F_1 = B_5$$
$$\cdots\cdots\cdots\cdots\cdots\cdots\cdots\cdots\cdots\cdots\cdots$$
$$A_i - B_0 F_i - B_1 F_{i-1} - B_2 F_{i-2} - B_3 F_{i-3} - \cdots\cdots - B_{i-2} F_2 - B_{i-1} F_1 = B_i$$
$$\cdots\cdots\cdots\cdots\cdots\cdots\cdots\cdots\cdots\cdots\cdots$$

$i = 0, 1, 2, 3, \ldots, i, \ldots \quad F_i = \sqrt{i+1} - \sqrt{i} \quad F_0 = 1$

If the largest circular area of the spheres is expressed by S, the diameter $(2R)$ can be determined from

$$2R = 2\sqrt{\frac{S}{\pi}} \qquad (13)$$

The stereological method proposed is thought to be useful in determining the size of spherical objects. The procedures may sound rather complicated, but they are relatively easy and simple in practice, if the computational procedures are performed with the use of appropriate computer programs. Furthermore, even if the profiles are not perfectly circular, measurements of the areas would be possible by means of a suitable method such as point-counting. When the diameters ($2r$) of the circles are measured, the above procedures would also be available if the areas of the circles are expressed in terms of πr^2.

Effect of section thickness

When thicknesses of the sections are obviously larger than the width of each class of the frequency distribution, the distributions obtained on the basis of the above principle are influenced by the section thickness. If spheres with a diameter $2R$ are randomly sectioned, the profiles projected on slices are circles of a diameter $2R$ or $2r$ ($R>r$) (Fig. 7). Relation

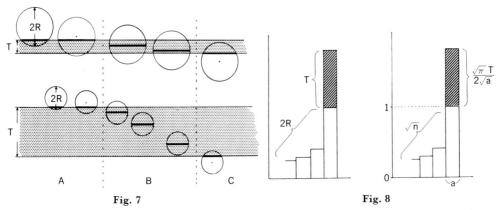

Fig. 7 Fig. 8

Fig. 7 When spheres of a diameter $2R$ are sectioned with thickness T, the projected circles have either a diameter smaller than $2R$ (A and C) or a diameter of $2R$ (B). Heavy lines in the spheres reveal diameters of projected circles.

Fig. 8 If the frequency of circles of smaller than $2R$ in diameter (A and C in Fig. 7) is proportional to $2R$, the frequency of circles of $2R$ in diameter (B in Fig. 7) is proportional to T (left). If I is given to the peak value of the largest class in the frequency distribution of smaller circles than $2R$ in diameter, the size distribution of the projected circles classified into n classes in section thickness T is shown on the right.

between the frequency F_r of circles smaller than $2R$ in diameter (A and C in Fig. 7) and the frequency F_R of circles $2R$ in diameter (B in Fig. 7) is written as follows:

$$\frac{F_R}{F_r} = \frac{T}{2R} \quad (T \text{ is section thickness})$$

Therefore, F_r is proportional to $\sum F_i$, if the formula (11) is substituted into the above.

$$F_R = \frac{T\sqrt{n}}{2R} \quad (14)$$

In this situation, the size of the circles is presented according to the profile areas. From this we obtain

$$\pi R^2 = na$$

Where a is the width of each class.
Hence

$$F_R = \frac{T}{2}\sqrt{\frac{\pi}{a}} \quad (15)$$

Then, the frequency distribution is expressed in combination with the effect of the section thickness in the formulas (10, 15) as follows:

$$F_1 + \frac{T\sqrt{\pi}}{2\sqrt{a}} \quad (F_1 = 1)$$
$$F_2$$
$$F_3$$
$$\vdots$$

(16) (Fig. 8)

In practice, the principle mentioned above is also available even when the section thickness is taken into consideration.

Another problem related to the effect of section thickness occurs when the profiles are poorly recognizable as with spheres that are sectioned near the pole. In such a case, the angle, which is indicated by θ in Fig. 9, should be considered. Such an effect of section thickness was also discussed by BACH (1967), ELIAS and HENNIG (1967) and SATO and YONEMARU (1973). In any case, the size distribution of spheres stereologically obtained shifts to the frequency distribution of circles measured directly, as the section thickness relative to the size of the spherical object increases.

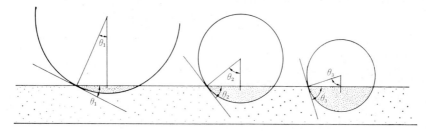

Fig. 9 When the spheres are sectioned near the pole, the profile of the projected circles would become less identifiable as the angle θ decreases. If the circles with the same diameter are projected, their margin is poorly defined on the left and well defined on the right.

For stereological quantitation the statistical calculation and data proccessing may be performed more easily with simpler computor programs. Nevertheless, in fact, stereological measurements are still laborious. The recent development of a computer system lined directly with the electron microscope may enable the automatic measurement, but we cannot yet profit by the automatic image analyzer for stereological quantitation of many objects in electron microscopy, because the computer system is available merely for objects that give simple and well contrasted images.

Cytometric Analysis of Thymic Small Lymphocytes

Cytometric analysis of thymic small lymphocytes in neonatal and adult mice was performed by the stereological procedures mentioned above.

Volumetric ratio of nucleus to cell in small lymphocytes

The nuclens: cell ratios of small lymphocytes in the cortex and medulla were volumetrically obtained by counting the numbers of test points lying on the nuclei and whole cells as follows:

$$Nucleus : cell\ ratio = \frac{\sum N_i}{\sum C_i}$$

Where N_i and C_i represent the number of points lying on the nuclear and cellular profile areas, respectively. In electron microscopy, since small lymphocytes could be identified by the nuclear characteristics (ABE & ITO 1970; ABE et al. 1973b), measurements were performed on small lymphocytes which contained the nuclear profiles on sections.

In the thymus of 4-day-old mice, 1,828 cortical small lymphocytes and 1,260 medullary small lymphocytes were measured. The ratio (mean$\pm$S.E.) was 0.72 ± 0.01 for cortical small lymphocytes and 0.56 ± 0.01 for medullary small lymphocytes, respectively. The difference between the two ratios was statistically significant ($p<0.001$). In the thymus of 35-day-old mice, 2,459 cortical small lymphocytes and 2,328 medullary small lymphocytes were measured. The ratio was 0.75 ± 0.01 for cortical small lymphocytes and 0.55 ± 0.01 for medullary small lymphocytes, respectively. The two ratios were also significantly different ($p<0.001$). In addition, the ratios of cortical and medullary small lymphocytes were not significantly different between 4- and 35-days-old mice.

Nuclear and cellular sizes of thymic small lymphocytes

Neonatal mice. The frequency distribution curves of sectional nuclear areas were obtained from measurements of 975 cortical lymphocytes and 748 medullary lymphocytes in the thymus of 4-day-old mice (Fig. 10). From each of the curves in this figure, the three-dimensional size distribution was statistically calculated by the stereological procedure mentioned above. The curves obtained are presented in Fig. 11. The practical procedures of calculation have been referred to in a previous paper (ABE & ITO 1972).

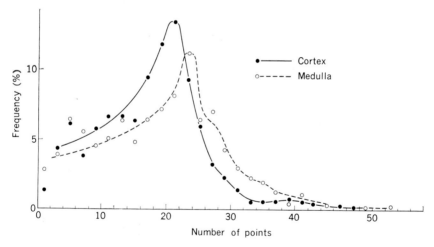

Fig. 10 The frequency distributions of the nuclear profiles of cortical and medullary lymphocytes in the thymus at 4 days of age. The results were obtained from actual measurements on sections.

As seen in Fig. 11, the nuclei of thymic small lymphocytes fell within a range between 4 and 6 μm in diameter, and their peak diameter was 4.9 μm for cortical lymphocytes and 5.1 μm for medullary lymphocytes, respectively. Since the curve obtained from the thymic cortex showed an almost normal distribution in profile pattern, the cortex was thought to contain a single population of typical small lymphocytes. On the other hand, the curve obtained from the thymic medulla indicated that the medulla constaned a proportionately a fair number of lymphocytes with larger nuclei, which, however, were within the size range of small lymphocytes.

The three-dimensional distribution of cellular size was calculated from the frequency

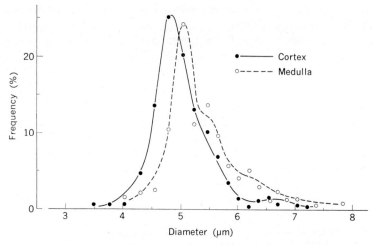

Fig. 11 The three-dimensional size distibutions of nuclei of cortical and medullary thymic lymphocytes at 4 days of age. They were obtained from calculation by the stereological method proposed.

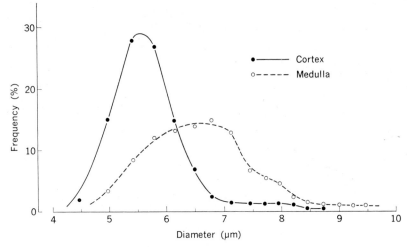

Fig. 12 The three-dimensional distributions of cellular sizes of cortical and medullary thymic lymphocytes at 4 days of age.

distribution of cellular profile areas of 801 cortical and 537 medullary lymphocytes (Fig. 12). As seen in this figure, the two curves were different in shape. A comparison of profile patterns of the distribution curves indicated that medullary lymphocytes were about 1 μm larger in diameter than cortical lymphocytes. As seen in Fig. 12, cortical small lymphocytes were 4 to 7 μm and medullary small lymphocytes were 5 to 8 μm in diameter. The diameters of lymphocytes encountered most frequently fell between 5.4 and 5.8 μm in the cortex and 6.2 and 7.0 μm in the medulla, respectively. The flatness at the peak of the size distribution curve of medullary lymphocytes seemed to be caused by an irregularity in cell shapes.

Adult mice. The nuclear size distribution curves of thymic lymphocytes in 35-day-old mice were obtained by calculation from the distribution of nuclear profile areas of 1,247 cortical and 1,249 medullary lymphocytes (Fig. 13). As seen in this figure, the two curves were almost similar in the position of the peak. Thus the majority of small lymphocytes

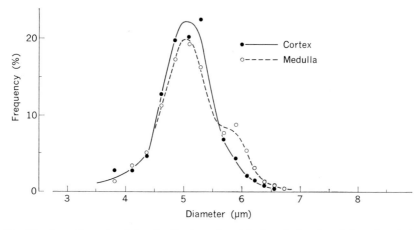

Fig. 13 The three-dimensional size distributions of nuclei of cortical and medullary thymic lymphocytes at 35 days of age.

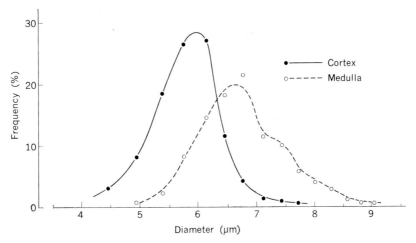

Fig. 14 The three-dimensional distributions of cellular sizes of cortical and medullary thymic lymphocytes at 35 days of age.

in both the cortex and medulla had a diameter of about 5 μm. However, the two curves were different in their slopes on the right side. Such a difference appeared to be caused by the presence of a fair number of small lymphocytes with larger nuclei (5.5 to 6.0 μm in diameter) in the medulla. As mentioned above, such small lymphocytes with larger nuclei were numerous in the thymic medulla of neonatal mice. Thus, small lymphocytes with larger nuclei were still contained in a relatively small proportion even in the thymic medulla of 35-day-old mice.

The cellular size distribution of thymic lymphocytes in 35-day-old mice was calculated from the distribution of cellular profile areas of 1,092 cortical and 928 medullary lymphocytes (Fig. 14). The curves indicated that medullary small lymphocytes were larger than cortical small lymphocytes and that the diameter was different by about 1 μm between cortical and medullary small lymphocytes.

In conclusion, cytometric analysis reveals significant quantitative differences between thymic cortical and medullary small lymphocytes. The nuclei are generally the same in size in the cortex and medulla of neonatal and adult mice. They are 4 to 6 μm, and the most frequently encountered nuclei are about 5 μm, in diameter. However, small lymphocytes with larger nuclei of diameters between 5.5 to 6.0 μm are more numerous in the medulla than in the cortex. Such small lymphocytes are very numerous in the medulla of neonatal mice, but are also present in a small proportion in the medulla of adult mice. The cellular size of small lymphocytes is significantly different between the cortex and the medulla in both 4- and 35-day-old mice. The cell diameter of cortical small lymphocytes is between 4 and 7 μm while that of medullary small lymphocytes is between 5 and 8 μm. Thus medullary small lymphocytes are larger by about 1 μm in diameter. For the above reason, the volumetric ratio of nucleus to cell in small lymphocytes in significantly larger in cortical small lymphocytes than that in medullary small lymphocytes.

As mentioned above, thymic small lymphocytes are cytometrically different between the cortex and medulla. In light of recent concepts of the thymic function, it seems reasonable to assume that the morphometric difference between thymic cortical and medullary small lymphocytes represents a reflexion of functional maturation of lymphocytes within the environment of this organ (for reviews see EVERETT & TYLER 1967; MILLER & OSOBA 1967; YOFFEY & COURTICE 1970; ELVES 1972). As is generally accepted, medullary small lymphocytes emigrate from the thymus and join the peripheral circulation. Since emigrant lymphocytes from the thymus leave the organ via the medulla, they are considered to present cytological features such as do medullary small lymphcytes. If so, lymphocytes in the thymus-dependent areas of the peripheral lymphoid organs would be similar in cytological details to medullary small lymphocytes, because the areas are populated with thymus-derived lymphocytes. In fact, as observed in our previous studies (ABE & ITO 1970; ABE et al. 1973b), small lymphocytes in the thymus-dependent areas of the spleen are similar in cytological and cytometric details to thymic medullary small lymphocytes.

One of the interesting findings obtained was that the thymic medulla contained small lymphocytes characterized by having larger nuclei than those of typical cortical small lymphocytes and that such peculiar small lymphocytes were predominant in neonatal mice, although they were still present in small proportions even in adults. The occurrence of such peculiar small lymphocytes in the medulla was dealt with in detail in our previous karyometric study at the light microscopic level (ABE et al. 1973a).

Summary

We have proposed a new stereological method for determining the size of spherical objects in electron microscopy. The principle and procedures were dealt with in some detail in this article. Then the method was applied to a cytometric analysis of small lymphocytes in the thymus of neonatal and adult mice.

The cytometric results indicated that thymic cortical and medullary small lymphocytes are similar in nuclear size distribution in both neonatal and adult mice, and that the cellular size distributions of small lymphocytes are significantly different between the cortex and medulla. The morphometric differences between cortical and medullary small lymphocytes are considered a reflexion of functional maturation of small lymphocytes within the thymus.

REFERENCES

[1] ABE K. and ITO T. (1970) Fine structure of small lymphocytes in the thymus of the mouse: qualitative and quantitative analysis by electron microscopy. Z. Zellforsch., *110:* 321.

[2] ABE K. and ITO T. (1972) A new stereological method for determination of the size of spherical objects in electron microscopy: its application to small lymphocytes of the thymus. Arch. histol. jap., *34:* 203.

[3] ABE K., SASAKI K. and ITO T. (1973a) Peculiarity of small lymphocytes in the thymic medulla in neonatal mice. Z. Anat. Entw. Gesch., *140:* 203.

[4] ABE K., SASAKI K. and ITO T. (1973b) Comparative ultrastructure and cytometric analysis of small lymphocytes in haemopoietic organs of neonatal mice. J. Anat., *115:* 393.

[5] BACH G. (1967) Kugelgrößenverteilung und Verteilung der Schnittkreise; ihre wechselseitigen Beziehungen und Verfahren zur Bestimmung der einen aus der anderen. *In:* Quantitative Methods in Morphology. WEIBEL E. R. and ELIAS H. (eds.) pp. 23–45, Springer-Verlag, Berlin-Heidelberg-New York.

[6] BAUDHUIN P. and BERTHET J. (1967) Electron microscopic examination of subcellular fractions. II. Quantitative analysis of the mitochondrial population isolated from rat liver. J. Cell Biol., *35:* 631.

[7] COUPLAND R. E. (1968) Determining sizes and distribution of sizes of spherical bodies such as chromaffin granules in tissue sections. Nature, *217:* 384.

[8] EBBESSON S. O. E. and TANG D. B. (1967) A comparison of smapling procedures in a structured cell population. *In:* Stereology. ELIAS H. (ed.) pp. 131–132, Springer-Verlag, New York.

[9] ELIAS H. and HENNIG A. (1967) Stereology of the human renal glomerulus. *In:* Quantitative Methods in Morphology. WEIBEL E. R. and ELIAS H. (eds.) pp. 130–166, Springer-Verlag, Berlin-Heidelberg New York.

[10] ELIAS H., HENNIG A. and SCHWARZ D. E. (1971) Stereology: applications to biomedical reaserch. Physiol. Rev., *51:* 158.

[11] ELVES M. W. (1972) The lympohcytes, 2nd ed., Lloyd-Luke Medical Books Ltd., London.

[12] EVERETT N. B. and TYLER R. W. (1967) Lymphopoiesis in the thymus and other tissues: functional implications. Internat. Rev. Cytol., *22:* 205.

[13] HILLIARD J. E. and CAHN J. W. (1961) An evaluation of procedures in quantitative metallography for volume-fraction analysis. Transactions of the Metallurgical Society of AIME, *221:* 344.

[14] ITO T. and HOSHINO T. (1962) Histological changes of the nouse thymus during involution and regeneration following administration of hydrocortisone. Z. Zellforsch., *56:* 445.

[15] KOSTOWIECKI M. and SHERMAN J. (1963) Thymocyte and lymphocyte differentiation studies by means of acridine orange fluorescence microscopy. Z. Zellforsch., *61:* 605.

[16] MILLER J. F. A. P. and OSOBA D. (1967) Current concept of immunological function of the thymus. Physiol. Rev., *47:* 437.

[17] SAINTE-MARIE G. and LEBLOND C. P. (1964) Cytologic features and cellular migration in the cortex and medulla of thymus in the young adult rat. Blood, *23:* 275.

[18] SATO M. and YONEMARU M. (1973) Proposed methods for estimation of number and size of cells and cytoplasmic granules. J. Electron Microscopy, *22:* 173.

[19] TANG D. B. and EBBESSON S. O. E. (1972) A comparison of a systematic sampling method method with complete random sampling for estimating total numbers of nerve fibers. Anat. Rec., *174:* 495.

[20] TROWELL O. A. (1961) Radiosensitivity of the cortical and medullary lymphocytes in the thymus. Int. J. Radiat. Biol., *4:* 163.

[21] WEBER W. T. (1966) Difference between medullary and cortical thymic lymphocytes of the pig in their response to phytohaemagglutinin. J. Cell Physiol., *68:* 117.

[22] WEIBEL E. R., KISTLER G. S. and SCHERLE W. R. (1966) Practical stereological methods for morphometric cytology. J. Cell Biol., *30:* 23.

[23] WEIBEL E. R. (1969) Stereological principles for morphometry in electron microscopic cytology. Internat. Rev. Cytol., *26:* 235.

[24] YOFFEY J. M. and COURTICE F. C. (1970) Lymphatics, Lymph and the Lymphomyeloid Complex. Academic Press, London & New York.

Quantitative Analysis of Secretory Granules of the STH-cell in the Rat Hypophysis

Mitsuru SATO, Masayuki YONEMARU and Shiro SONODA

Methods for estimating the number and size of spherical structures were proposed by several investigators (Weibel et al. 1966; Baudhuin & Berthet 1967; Weibel 1969; Abe & Ito 1972). Most of the methods were based on the assumption that the section thickness was negligible as compared with the size of the structures. In a method proposed by Kawamura (1954a, b), on the other hand, the section thickness was positively introduced into the stereological principle. This method was theoretically reinforced by Sato and Yonemaru (1973).

Hedinger and Farquhar (1957), who differentiated acidophilic cells of the anterior lobe of the hypophysis into two cell types, reported that the cells of one type contained granules of about 350 nm in their maximal diameters and corresponded to the secretion of growth hormone. Bloom and Fawcett (1962) reported that the specific alpha granules varied in size even in the same cell. Considerable variation in the size of secretory granules of STH-cell was also noted by Kurosumi (1968).

In the present paper, numbers of secretory granules in the STH-cell were statistically estimated and their size distributions were discussed.

Material and Methods

The hypophysis of an adult male rat (Wistar strain) was fixed in 2% osmium tetroxide in phosphate buffer and embedded in Epon. Sections of about 100 nm in thickness were cut on a Porter-Blum ultramicrotome, stained with uranyl acetate and lead citrate and examined with a JEM-7 electron microscope. Electron micrographs (including their copies) of 20,000 magnifications were obtained from six different sections through a STH-cell, and the number and size of the profiles of the secretory granules were repeatedly measured nine times.

Results

The principle of a method for the estimation of the number and size of spherical structures had been reported by Sato and Yonemaru (1973). When a sphere with a diameter $2R$ is cut into sections of a thickness t, and the profile sizes are classified into n classes, the frequency F_i of profiles belonging to Class i is given as

$$F_i = \sqrt{R^2 - (i-1)^2} - \sqrt{R^2 - i^2} \tag{1}$$

and that of profiles belonging to Class n as

$$F_n = \sqrt{R^2 - (n-1)^2} + 1 \tag{2}$$

where
$$n - 1 \leq R \leq n \quad (3)$$

and the class interval s equals the section thickness t.

With the assumption that the object spheres are evenly distributed within a range where the radius R of each sphere lies within the qualification (3), the average frequencies $\bar{F}_i$ and $\bar{F}_n$ are calculated as follows:

$$\begin{aligned}
\bar{F}_i &= \frac{1}{n-(n-1)} \int_{n-1}^{n} F_i dR = \int_{n-1}^{n} (\sqrt{R^2 - (i-1)^2} - \sqrt{R^2 - i^2})\, dR \\
&= \frac{1}{2}\Big[(R\sqrt{R^2 - (i-1)^2} - (i-1)^2 \log|R + \sqrt{R^2 - (i-1)^2}|) \\
&\quad - (R\sqrt{R^2 - i^2} - i^2 \log|R + \sqrt{R^2 - i^2}|)\Big]_{n-1}^{n} \quad (4)
\end{aligned}$$

$$\begin{aligned}
\bar{F}_n &= \frac{1}{n-(n-1)} \int_{n-1}^{n} F_n dR = \int_{n-1}^{n} (1 + \sqrt{R^2 - (n-1)^2})\, dR \\
&= 1 + \frac{1}{2}\Big[R\sqrt{R^2 - (n-1)^2} - (n-1)^2 \log|R + \sqrt{R^2 - (n-1)^2}|\Big]_{n-1}^{n} \quad (5)
\end{aligned}$$

A calculating table for estimating the number and size of spherical structures prepared with the use of these equations is shown in Table 1. When $t \neq s$, the average frequency $\bar{F}_{n'}$ of Class i and that $\bar{F}_{i'}$ of Class n are obtained as follows:

$$\bar{F}_{i'} = \bar{F}_i / a \quad (6)$$
$$\bar{F}_{n'} = (\bar{F}_n - 1)/a + 1 \quad (7)$$

where t is settled as $t = as$.

Though an algorithm is explained in the paper of SATO and YONEMARU (1973), the following equation may be more serviceable for the use of a computer.

$$\sum_{i=1}^{n} F_{ij} N_i = P_j \quad (j = 1, 2, \ldots, n) \quad (8)$$

where F_{ij} shows frequencies of profiles of Class j per one sphere of Class i (Table 2).

Measurements of profiles of secretory granules of a STH-cell are shown in Table 3. Numbers of granules estimated from the data of each measurement repeated nine times are plotted in Fig. 1. Some variations can be noticed among the repeated measurement. Nevertheless, the average distribution appears to be bimodial. The distribution curve presented by the mean values of each of the nine points was adjusted to polynomial approximation. The smoothened curve presents two apparent peaks (Fig. 2). This curve suggests that two varieties of secretory granules, i.e. small and large granules, may be mixed in the STH-cell. If it can be supposed that each variety of granules is normally distributed, the distribution curve can be decomposed into two curves with the method of optimal solution. The mean diameters m_s (of small granules) and m_l (of large granules), the standard deviations σ_s and $\sigma_{l'}$, and the distribution ratio S/L fluctuated, and the optimal solution was obtained as follows:

$$m_s = 162.5 \text{ nm}, \qquad \sigma_s = 37.5 \text{ nm}$$
$$m_l = 290.0 \text{ nm}, \qquad \sigma_l = 61.7 \text{ nm}$$
$$S/L = 31/59$$

On the basis of these values, the decomposed distribution curves are described in Fig. 2.

From the above analysis, the following matters may be induced. Two small and large varieties of secretory granules are contained in the STH-cell. The larger granules are more abundant than the smaller ones. The size distribution of the larger granules is broader than that of the smaller ones.

Table 1 Calculating table for estimating spherical structures (Where section thickness t equals class interval s).

Class number of the size of structures	Class number of the size of profiles														
	1	2	3	4	5	6	7	8	9	10	11	12	13	14	15
1	.500+1														
2	.426	1.073+1													
3	.212	.858	1.429+1												
4	.147	.487	1.152	1.712+1											
5	.113	.358	.684	1.389	1.954+1										
6	.091	.285	.516	.843	1.591	2.170+1									
7	.077	.238	.419	.645	.980	1.772	2.366+1								
8	.067	.204	.355	.531	.757	1.101	1.935	2.547+1							
9	.059	.179	.308	.453	.627	.856	1.210	2.087	2.716+1						
10	.052	.160	.273	.397	.540	.714	.946	1.311	2.228	2.875+1					
11	.047	.144	.245	.354	.475	.617	.792	1.029	1.405	2.360	3.025+1				
12	.043	.131	.223	.320	.426	.546	.687	.864	1.106	1.494	2.486	3.169+1			
13	.040	.121	.204	.292	.386	.491	.610	.752	.932	1.179	1.577	2.606	3.306+1		
14	.037	.111	.188	.268	.354	.446	.550	.669	.813	.995	1.247	1.656	2.720	3.438+1	
15	.034	.104	.175	.249	.326	.410	.502	.605	.725	.870	1.055	1.313	1.732	2.830	3.565+1

When $t=as$, these values must be modified by multiplying by $1/a$.

Table 2 Symbols for Equation 8.

Class number of the size of spheres	Number of spheres	Class number of profile size					
		1	2	...	j	...	n
		Number of profiles					
		P_1	P_2	...	P_j	...	P_n
1	N_1	F_{12}	F_{12}	...	F_{1j}	...	F_{1n}
2	N_2	F_{21}	F_{22}	...	F_{2j}	...	F_{2n}
.	.	.	.		.		.
.	.	.	.		.		.
.	.	.	.		.		.
i	N_i	F_{i1}	F_{i2}	...	F_{ij}	...	F_{in}
.	.	.	.		.		.
.	.	.	.		.		.
n	N_n	F_{n1}	F_{n2}	...	F_{nj}	...	F_{nn}

N.B. $F_{ij}=0$, where $j>i$.

Table 3 Results of measurements: Frequencies of profiles on sections of a STH-cell.

Person in charge of measurement	Time of the measurement	Range of Profile size (nm)											Total
		50|100	100|150	150|200	200|250	250|300	300|350	350|400	400|450	450|500	500|550	550|600	
M. Sato	1	48	455	545	531	521	372	193	66	21	1	1	2754
	2	70	464	507	502	479	365	230	98	30	8	1	2754
	3	89	463	513	521	490	388	194	70	23	3		2754
M. Yonemaru	4	101	435	531	498	484	410	205	66	22	1	1	2754
	5	96	437	543	495	495	398	202	63	23	1	1	2754
	6	99	433	546	499	474	413	203	60	25	1	1	2754
S. Sonoda	7	137	441	527	469	497	391	194	69	26	2	1	2754
	8	127	451	474	459	471	394	233	107	31	11		2754
	9	155	504	513	462	483	377	184	58	15	3		2754

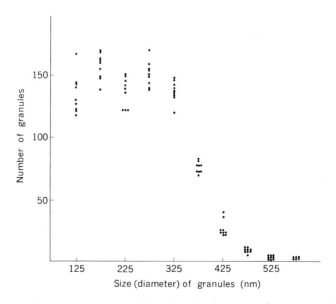

Fig. 1 Estimated numbers and sizes through nine repeated measurements.

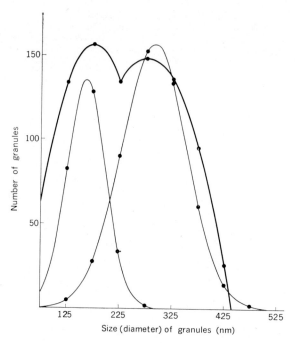

Fig. 2 Distribution curves of the secretory granules of a STH-cell. The thick line represents the curve smoothened by polynomial approximation and the thin lines represent the decomposed curves.

Discussion

The need for quantitative analysis of the cells and cytoplasmic components has become increasingly important in the field of electron microscopy. Methods for the estimation of number and size of spherical and multiform structures have already been proposed by SATO and YONEMARU (1973). The methods may be applicable especially to randomly oriented structures, for example, endocrine cells, lymphoid cells, tumor cells, secretory granules, lysosomes, vacuoles and vesicles, certain mitochondria, and so forth. We proposed an algorithm approximating the equation presented in this paper.

Considerable variation in the size of secretory granules of STH-cell was noted by BLOOM and FAWCETT (1962) and KUROSUMI (1968), nevertheless, further details of the size distribution of the granules remained obscure. In this paper, it has been clearly demonstrated that the size distribution of the granules is bimodial, and that the distribution curve could be decomposed into two normal distribution curves. This fact implies that two varieties of granules might be present in the STH-cell, and these varieties could correspond to the secretion of two different hormones.

Summary

Secretory granules in the STH-cell of the anterior pituitary of a male rat were analysed quantitatively.

The size distribution of the granules was bimodial, and the distribution curve was divided into two components, with the assumption that each variety of granules is normally distributed in size.

The mean diameter of the smaller granules was estimated at 162.5 nm with a standard deviation of 37.5 nm and that of the larger granules was 290.0 nm with a standard deviation of 61.7 nm. The distribution ratio of the smaller to the larger granules was obtained as 31 to 59.

REFERENCES

[1] Abe K. and Ito T. (1972) A new stereological method for determination of the size of spherical objects in electron microscopy: Its application to small lymphocytes of the mouse thymus. Arch. histol. jap., *34:* 203.

[2] Baudhuin P. and Berthet J. (1967) Electron microscopic examination of subcellular fractions, II. Quantitative analysis of the mitochondrial population isolated from rat liver. J. Cell Biol., *35:* 631.

[3] Bloom W. and Fawcett D. W. (1962) A textbook of Histology. 8th ed., pp. 339, W. B. Saunders Company, Philadelphia & London.

[4] Hedinger C. E. and Farquhar M. G. (1957) Elektronenmikroskopische Untersuchungen von zwei Typen acidophiler Hypophysenvorderlappenzellen bei der Ratte. Schweiz. Z. Path. Bakt., *20:* 766.

[5] Kawamura M. (1954a) Quantitative-histological studies on the thymus, especially on the Hassall's bodies. Med. Bull. Kagoshima Prefect. Univ., *6:* 211.

[6] Kawamura M. (1954b) On the determination of the number and size of the glomerules in human fetus kidney. Med. Bull. Kagoshima Prefect. Univ., *6:* 367.

[7] Kurosumi K. (1968) Functional classification of cell types of the anterior pituitary gland accompanied by electron microscopy. Arch. histol. jap., *29:* 329.

[8] Sato M. and Yonemaru M. (1973) Proposed methods for estimation of number and size of cells and cytoplasmic granules. J. Electron Microscopy, *22:* 173.

[9] Weibel E. R. (1969) Stereological principles for morphometry in electron microscopic cytology. *In:* International Review of Cytology. Bourne G. H. and Danielli J. F. (eds.) Vol. 26, pp. 235–302, Academic Press, New York & London.

[10] Weibel E. R., Kistler G. S. and Scherle W. F. (1966) Practical stereological methods for morphometric cytology. J. Cell Biol., *30:* 23.

Significance of Morphometry in the Cytophysiology of Secretion: Its Application to the Endocrinological Studies of the Rat Hypophysis

Kazumasa KUROSUMI

Results of counting pituitary cell numbers at various functional states have been reported by some authors (Dingemans 1969; Ishikawa & Yoshimura 1969), using the light and/or electron microscopes. These studies are crude in view of the ultrastructural level even though electron microscopes were used. They are not comparable to the delicate functional fluctuations of this labile endocrine organ which is always affected by both the hypothalamic control associated with the influence of other parts of the central nervous system and the peripheral hormonal secretion through the so-called feedback mechanism. The measurement of size and number of secretory granules is very useful for cell morphology as well as cell physiology of glandular organs especially the hypophysis and hypothalamus. The volume of the cell organelles, such as the rough endoplasmic reticulum, Golgi apparatus and mitochondria may be intimately concerned with the rate of secretory activity, especially the synthetic process of secretion and the production of secretory granules. The point-counting method of Weibel (Weibel 1969; Weibel & Bolender 1973) is the most useful and reliable technique for measuring areas of scattered or concentrated portions of cell organelles

Quantitative studies were attempted on the neurosecretory system of the hypothalamus and posterior pituitary, and on the corticotrophs and gonadotrophs of the anterior pituitary under various experimental conditions. The measurements of size and number of secretory granules, and that of the volume (area on sections) of the cell body, endoplasmic reticulum and Golgi apparatus were performed.

Measurement of Granule Size (Hypothalamic Neurosecretory System)

It is commonly known that the so-called posterior pituitary hormones (vasopressin or ADH and oxytocin) are not produced in the posterior pituitary, but that they are synthesized in certain nerve cells and transported through the processes (axons) of these hormone-producing neurons, and are finally stored and released by the posterior pituitary itself. Two different groups of nerve cells (neurosecretory nuclei) producing the posterior pituitary hormones are recognized in the mammalian hypothalamus, and they are called the supraoptic and paraventricular nuclei. The precise localization for the production site of each hormone has not yet been determined, though some authors argue that the supraoptic nucleus may produce vasopressin while the paraventricular nucleus may synthesize oxytocin (Hild & Zetlaer 1952).

Not only these two well-known posterior pituitary hormones, but also many other active substances which stimulate or inhibit the secretion of the anterior pituitary hormones are also produced in the hypothalamus. They are either called relaesing or inhibiting hormones and accumulate in the nerve terminals facing the blood capillaries of a specific circulation, that is, the hypophyseal portal system. The releasing or inhibiting hormones are discharged into the blood stream of this specific circulation and thus irrigate the anterior pituitary tissue at a relatively high concentration. They stimulate the corresponding cell types in the anterior pituitary under the so-called key and lock relationship, for instance, the CRH (corticotropin-releasing hormone) only stimulates the corticotrophs that secretes ACTH (corticotropin) (Fig. 1).

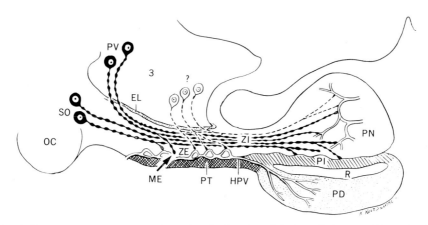

Fig. 1 A diagrammatic illustration of the mid-sagittal section of the hypothalamus and hypophysis of the rat. OC: optic chiasm, SO: supraoptic nucleus, PV: paraventricular nucleus, 3: third ventricle, ?: production site of releasing hormones of the anterior pituitary hormones, EL: ependymal layer, ME: median eminence, ZE: zona externa, ZI: zona interna, PT: pars tuberalis, HPV: hypophyseal portal veins, PD: pars distalis (anterior pituitary), PI: pars intermedia, PN: pars nervosa (posterior pituitary), R: residual cavity of Rathke's pouch.

Both the posterior pituitary and the releasing or inhibiting hormones are oligopeptides and are synthesized in the nerve cells of the hypothalamus and transported through their axons to the site of discharge; that is, the posterior pituitary for the former, and the median eminence for the latter. These neurohormones may be packed in spherical granules in the Golgi apparatus of the neurosecretory cells, like the secretory substance of other protein or peptide-secreting cells. Electron microscopy demonstrates the presence of small round secretory granules in the cell body of nerve cells in the supraoptic as well as the paraventricular nuclei. Cisternae of the Golgi apparatus contain rather small granules, being suggestive of a new formation of granules in the Golgi apparatus (Fig. 2). Full-grown granules are often located within the axons both in the hypothalamus and the posterior pituitary (Fig. 3).

DE ROBERTIS and his co-workers (GERSCHENFELD et al. 1960; ZAMBRANO & DE ROBERTIS 1966) measured the size of neurosecretory granules from various portions of the hypothalamo-hypophyseal system of the toad and rat, and concluded that the neurosecretory granules increase in size during transport through the axons. They said that the material synthesized in the perikaryon might be the carrier or an inactive protein and that along the axon the active peptides could be formed and added. The granulometry by DE ROBERTIS and his school was not performed on the median eminence, dividing it especially into the internal and external layers with both the toad and the rat.

We made a similar but more detailed study on the rat hypothalamus and neurohypophy-

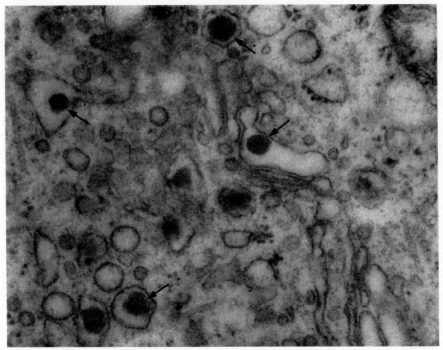

Fig. 2 The Golgi apparatus of a neurosecretory cell in the supraoptic nucleus of a rat. Newly formed neurosecretory granules (arrows) are seen in the vesicles and slightly dilated cisternae of the Golgi apparatus. ($\times 50,000$)

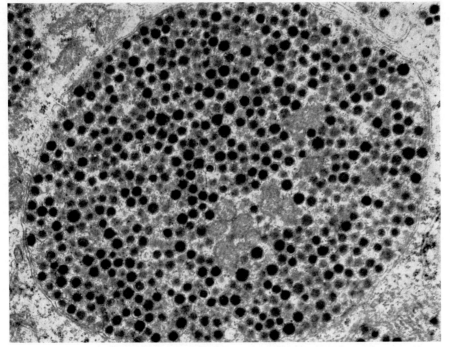

Fig. 3 A cross section of the swollen part of a neurosecretory axon filled with abundant neurosecretory granules and some mitochondria. The posterior pituitary contains a large number of such axons. ($\times 20,000$)

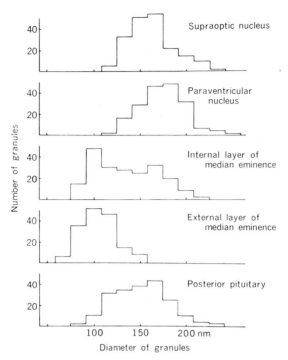

Fig. 4 Size distribution of neurosecretory granules at various regions of the rat hypothalamus and hypophysis. There are two types of granule, each belonging to a different neurosecretory system.

sis, and measured the size of the neurosecretory granules at five points. These were the perikarya in the supraoptic and paraventricular nuclei, internal and external layers of the median eminence and the posterior pituitary glands. The result is shown in Fig. 4. The mean granule size of the supraoptic nucleus is about 150 nm diameter, while that of the paraventricular nucleus is about 180 nm. Two peaks are observed in the distribution curve of the granule size in the internal layer of the median eminence; one is about 100 nm and the other 170 nm, suggesting the presence here of two different types of neurosecretory granules. On the other hand, the external layer of the median eminence contains only one type of granule, the mean of which is about 100 nm in diameter. Finally, there is only one peak in the distribution curve of granule size in the posterior lobe of the hypophysis that is about 170 nm. The granule size is not altered during transport although the granules in the paraventricular nucleus is slightly larger than those in the supraoptic nucleus. The mean diameter of the granules in these two nuclei is about 160–170 nm and this value is continuously observed in the internal layer of the median eminence and the posterior pituitary.

The diameter of the neurosecretory granules of the hypothalamo-hypophyseal system does not change during transport, but quite a different type of secretory granule is added at the median eminence. The latter granules are about 100 nm in diameter and occur in both the internal and external layers of the median eminence. None of the larger type of granules can be observed in the external layer. In sections of the internal layer, two different axons are observed; one is cut transversally and the other longitudinally. The former contains larger granules of about 170 nm in diameter, while the latter contains granules of about 100 nm (Fig. 5). The former fibers are those of the hypothalamo-hypophyseal system conveying the so-called posterior pituitary hormones, and advance

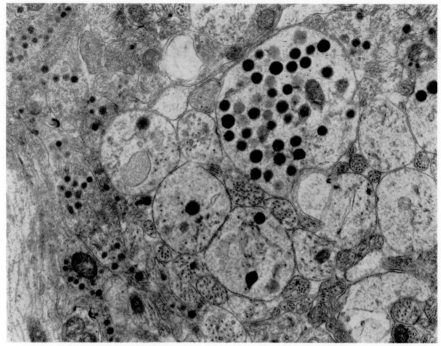

Fig. 5 The internal layer (zona interna) of a rat median eminence. Axons containing large granules are cut transversely (right), while those with small granules are cut longitudinally (left). (×17,000)

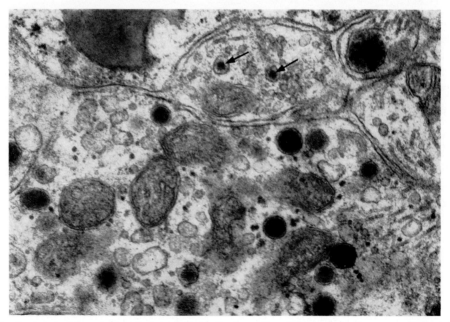

Fig. 6 A part of the posterior pituitary of a rat. Sections of two different types of axons are illustrated. The lower axon contains large neurosecretory granules, while the upper contains small granules (cored vesicles) as indicated with arrows. (×36,000)

posteriorly into the posterior lobe. The latter fibers containing smaller granules may convey the hypothalamic releasing (and inhibiting) hormones which are liberated into the blood stream of the hypophyseal portal system in the external layer of the median eminence. Therefore, the latter fibers are bent at right angles in the internal layer and proceed downward toward the external layer. This is the reason why two different groups of nerve fibers containing neurosecretory granules are arranged perpendicularly to each other in the internal layer of the median eminence. Only a few of the smaller granules are transported into the posterior pituitary (Fig. 6).

GERSCHENFELD et al. (1960) and ZAMBRANO and DE ROBERTIS (1966) both measured granule size in various areas of the hypothalamus and hypophysis of the toad and the rat, and concluded that the granules grew larger as they moved from the nucleus to the hypophysis. They considered that the synthesis of the neurosecretory substance might occur even in the axons and that some substance could be added to the moving granules. However, neither ribosomes nor the endoplasmic reticulum were observed in the neurosecretory axons both in the median eminence and the posterior pituitary. Therefore, synthesis of the secretory material at the axons could be denied. Our results of granulometry indicated no size differences between the granules in the perikarya and those in the fibers and terminals. It is difficult to explain why the data of the DE ROBERTIS' school are so disagreed with that of ours. They did not distinguish two different systems of neurosecretion in the hypothalamus. One is the hypothalamo-phypophyseal system producing ADH and oxytocin, while the other is the hypothalamo-infundibular system producing the releasing or inhibiting hormones of the adenohypophysis. They must have confused the granules of two different systems, and first probably measured the smaller granules of the hypothalamo-infundibular system and then measured the larger granules of the hypothalamo-hypophyseal system at the stalk or posterior lobe. Such a result could bring about the conclusion that the granules grew larger during transport. Detailed and systematic granulometry could indicate the true feature of granule formation and transportation in this complicated system of neurohormone production.

Counting Granule Numbers (Corticotroph Granules after Stress)

In glandular cells of an endocrine organ secreting a proteinaceous hormone, the number of secretory granules in the cell may indicate to some extent the hormonal content within the cell. If the production of the hormone is accelerated or the release of the hormone is suppressed, the number of granules in the cell as well as the hormonal content could increase, and vice versa.

The cellular source of ACTH (adrenocorticotropic hormone or corticotropin) has long been the focus of debate in the field of cytophysiology of the adenohypophysis. Since morphology of the secretory granules is one of the most reliable criteria for cell identification of the anterior pituitary, the ACTH-secreting cell (corticotroph) must also be characterized by the morphology of the secretory granules. The author and his co-workers (KUROSUMI & KOBAYASHI 1966; KUROSUMI & OOTA 1966) discovered a new cell type in the anterior pituitary of normal and experimentally treated rats. The secretory granules of this cell type are pleomorphic: either vesicles with or without central cores, or solid granules of varing electron density. Cored vesicles are most characteristic among the various forms of granule. Granules increase in number after adrenalectomy and solid dense granules become predominant. Along with changes in other cellular structures such as the rough endoplasmic reticulum and Golgi apparatus, the above mentioned alter-

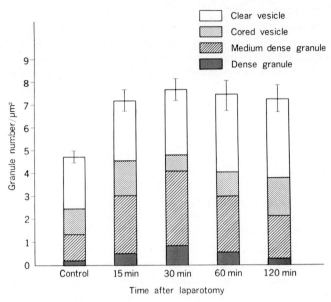

Fig. 7 Fluctuation in the population density of secretory granules of corticotroph after somatic stress (laparotomy). An experiment on rats of the Wistar strain.

ations in granular morphology after adrenalectomy clearly indicate that this cell may be the source of ACTH.

This tentative corticotroph is the cell with a relatively large cell body and irregular processes, which are often extended towards the wall of the blood capillaries. We can observe closely packed secretory granules within the tips of such cytoplasmic processes, most of which are cored vesicles. Immunocytochemistry using peroxidase-labeled antibody method indicated that the ACTH-secreting cell at the light microscopy level extends relatively numerous cytoplasmic processes mainly towards the blood capillaries (NAKANE 1970). This finding is quite similar to that obtained with electron microscopy (KUROSUMI 1968). Direct proof of ACTH cell by a method using enzyme-labeled antibody on ultrathin sections fixed with osmium tetroxide and embedded in epoxy resin was hitherto impossible. There have been some divergent notions on the submicroscopic identification of the ACTH cells (SIPERSTEIN & MILLER 1970; BOWIE et al. 1973). Such divergence could be due in part to the differences in preparation technique for the electron microscopic samples, that is, one of the morphological characteristics, the secretory granules in the form of cored vesicles only appears after osmium fixation, but glutaraldehyde fixation always fails to demonstrate the cored vesicular granules (KUROSUMI 1968). Such a peculiar form of the secretory granule, cored vesicle, may be an artefact, but this feature is very specific for this cell type. Therefore, fixation with glutaraldehyde and osmium, either successive or mixed, is not suitable for study of the corticotroph. Such confusion in the morphophysiology of ACTH-secreting cell might be settled by using methods of quantitative electron microscopy.

It is well known that ACTH is actively liberated by exposure to somatic or emotional stress. We tried stress experiments upon rats using a surgical operation (laparotomy under ether anesthesia) as the somatic stress. Before and after stress, numbers of secretory granules of the corticotrophs were counted on the electron micrographs at 10,000 times in magnifiation. Numbers of each granular type, that is the solid dense, medium dense,

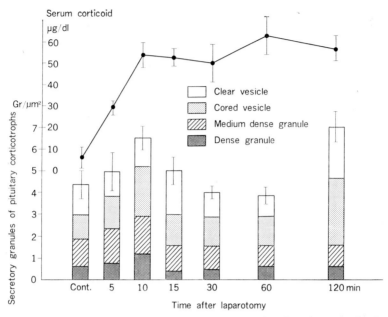

Fig. 8 Fluctuation in the population density of secretory granules of corticotroph after laparotomy. An experiment on rats of the Holzmann strain. Influence of laparotomy in the serum corticoid level is also shown in the upper part of the diagramm.

cored and empty vesicles were counted. The results are shown in Figs. 7 and 8. These figures indicated the population densities of corticotroph granules (granule number per square micrometer of cytoplasm) in each stage after laparotomy. Before exposure to stress majority of the granules are cored or empty vesicles but there are a few solid granules (Fig. 9). Even after several minutes the total number of granules in a unit area of the corticotroph cytoplasm increases. This increase in total granule number is brought about by an increase of solid granules especially very dense granules (Fig. 10). These dense granules are thought to be newly formed.

The first experimental result shown in Fig. 7 was performed on rats of the Wistar strain, while the second experiment shown in Fig. 8 was carried out on rats of the Holzmann strain. The serum content of corticoids at every stage is also shown in Fig. 8, and clearly indicates that ACTH secretion is accelerated after stress. Increase in granule number occurred promptly after the stress exposure: 15 min in the first experiment and 5 min in the second. Extremely dense as well as medium dense granules markedly increased in number after stress. In the first group the peak of the granule increase occurred at 30 min, while in the second, the peak was at 10 min. In the Holzmann rats, the reaction occurred earlier and also stopped earlier than in the Wistar strain.

At 120 min in the second experiment, the total number of secretory granules had markedly increased. The extremely dense and medium dense granules were few at this stage and vesicles with and without cores were rather numerous. Empty and cored vesicles which were thought to be old granules increased, while dense or medium dense granules probably newly formed were less numerous. The analysis of 4 types of granular morphology clearly indicated that the increase in corticotroph granules at 120 min might be explained since the release of hormone was highly suppressed probably due to the high level of serum corticoids through the so-called negative feedback mechanism. The results of quantitative electron microscopy on the corticotroph granules after stress as well as the

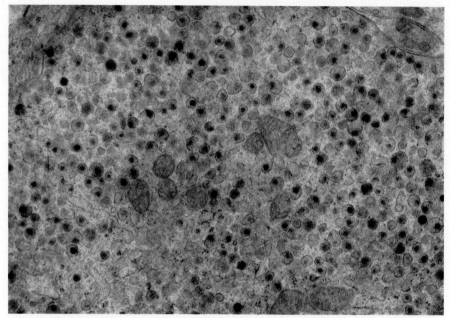

Fig. 9 A part of the cytoplasm of a corticotroph in a normal rat. Most secretory granules are cored vesicles, but only a few dense granules and clear vesicles are mixed. (×20,000)

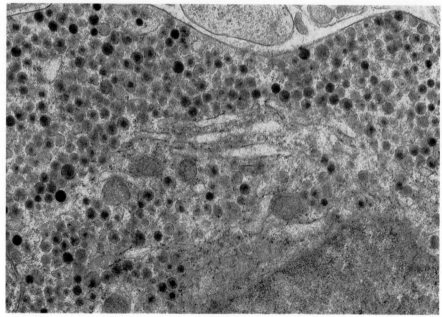

Fig. 10 A part of the cytoplasm of a corticotroph in a rat 30 minutes after stress exposure (laparotomy). Highly dense or medium dense granules are increased, but vesicles with or without cores are few in number. (×20,000)

results of purely morphological observation after adrenalectomy clearly demonstrated the possibility of this newly found cell type as the source of ACTH in the anterior pituitary.

Volumetry of Cell Organelles (Gonadotrophs after Castration)

Weibel (1969) and Weibel and Bolender (1973) introduced the point-counting planimetry originally demonstrated in 1933 by Glagoleff, a Russian geologist. In the present study this method was applied to the gonadotrophic cells after castration with the cross-points of a square grid as markers, because it had been repeatedly shown that castration induces a great change in morphology, especially of the endoplasmic reticulum of the gonadotrophs (Farquhar & Rinehart 1954; Yoshimura & Harumiya 1965). Prior to the final experiment, a series of preparatory experiments were performed. Points of the rectangular lattice superimposed on the profiles of cells, nuclei, cell organelles and any other structures, whose volumes (area in a plane of the section) were desired to be measured were counted. Sum of the counts of all of the points appearing upon the structure in question was parallel to the total volume of the structure.

Table 1 Comparison between the results of planimetry using the planimeter and those by the point-counting method.

Cell (Code number)	Area of whole cell (μm^2)		Error		Area of nucleus (μm^2)		Error	
	Planimeter	Point counting	μm^2	%	Planimeter	Point counting	μm^2	%
70A1572	122.0	118.0	−4.0	3.3	29.6	29.0	−3.6	12.2
70A1577	261.0	260.5	−0.5	0.2	19.6	16.5	−3.1	15.8
4523A	295.6	298.5	2.9	0.9	36.0	37.8	1.8	4.8
68A1969	181.0	181.5	0.5	0.3	15.3	16.8	1.5	9.5
4539A	240.3	229.3	−11.1	4.6	45.0	40.0	−5.1	11.2
70A1567	359.3	361.3	2.0	0.5	38.0	36.8	−1.3	3.2
70A1573	149.3	150.0	0.7	0.5	—	—	—	—
Mean	229.8	228.4	3.1	1.5	30.6	29.0	2.7	9.5

The results of measuring with a planimeter were compared with those obtained with point-counting on the same structure (Table 1). If the measured area was small, errors became larger; but the error was negligible if the measuring area was large enough. A standard line was set on an electron micrograph of a castration cell (Fig. 11), and the deviation of a point on the cell contour from the standard line was measured by point-counting and recorded. Such a measurement of deviation was scanned at regular (unit) interval on the grid for point-counting. From such coordinate data, the cell contour was reconstructed in a stairway-like pattern (Fig. 12). Freehand adjustment of this pattern was performed keeping in mind that the internal area should be altered as little as possible. The result is shown in Fig. 12 with a solid curve line. The broken curve line in this figure is the actual cell contour drawn by superimposing the figure upon the original photograph. The deviation between these two curve lines is very slight as shown in this figure.

The area of the rough endoplasmic reticulum (ER) was measured by the point-counting method as one of the preliminary works. The cell type used for this trial was the gonadotroph type 1 or FSH cell of Barnes (Barnes 1962) from rats either normal or castrated. The endoplasmic reticulum in the normal state is divided into enormously large numbers

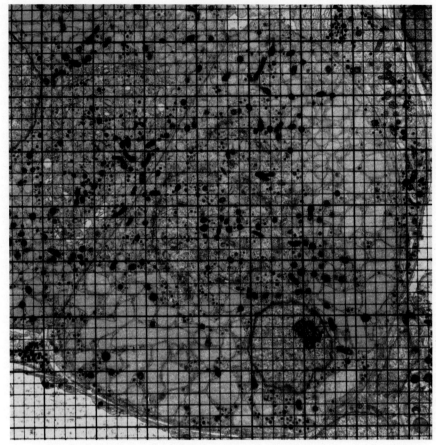

Fig. 11 An electron micrograph of a gonadotroph at 2 months after castration was superimposed with a square grid whose cross-points were used as markers for point-counting planimetry. (×6,000)

Table 2 Results of a preliminary measurement of the ER areas of normal and castrate gonadotrophs.

Experimental condition	Cell (Code number)	Ratio of area occupied by ER to the entire cytoplasmic area (%)	Illustration*
Normal	4523A	19.9	
Normal	68A1969	22.4	Fig. 13
Normal	4539A	62.2**	Fig. 14
Castration			
2 month	70A1577	41.4	Fig. 15
2 month	70A1572	50.8	Fig. 16
2 month	70A1573	61.3	Fig. 17
3 month	70A1567	61.3***	Fig. 18

* Magnifications of figures are not the same as micrographs used for measuring.
** Very rare case of normal gonadotroph.
*** So-called signet ring cell.

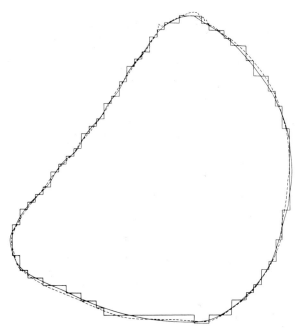

Fig. 12 A stairway-like outline of a castrate gonadotroph, which is recorded by point-counting. Reconstructed outline of a cell adjusted from a stairway-like pattern is illustrated by a solid curve line, and is not grossly deviated from the actual outline of the cell as shown by the broken curve line.

of cisternae which are randomly scattered throughout the cell. Exact measurement of all the area of the cut surface of the cisternae is very difficult, because the numbers of cisternae are too great, and each cisterna is too small to be measured with the planimeter. Point-counting is the most energy saving and reliable method for planimetry. The result of a preliminary measurement of ER of the gonadotrophs is shown in Table 2 and Figs. 13–18.

In the final experiment, the pituitaries of castrated rats 15, 30, 60, and 90 days after orchidectomy were removed and fixed with 1% OsO_4 adjusted at pH 7.4 with veronal buffer and made isotonic with sucrose. The pituitaries from normal control rats were also exised and likewise processed. Epon embedded sections were stained with uranyl acetate and lead hydroxide and observed with an electron microscope. Micrographs were taken at a magnification of $2,500\times$ and enlarged to $10,000\times$. Areas of the entire cell body, nucleus, rough ER and Golgi apparatus were measured by the point-counting method on sections (micrographs) of the gonadotrophs type 1 (FSH cells) at every stage after castration. The areal percentage of the ER or Golgi apparatus to the whole cytoplasm was calculated. The total number of secretory granules in a given cell was counted and the population density of the granules was calculated as follows:

$$D = \frac{G}{Z}$$

where $Z = C - (N+E)$, D: population density of granules of a given cell, G: total number of granules appearing in a section of the cell, Z: area of the cytoplasmic matrix (μm^2), C: area of the whole cell (μm^2), N: area of the nucleus (μm^2), E: area of the rough ER (μm^2). The results are shown in Fig. 19.

The most remarkable and characteristic change in the gonadotroph type 1 after castra-

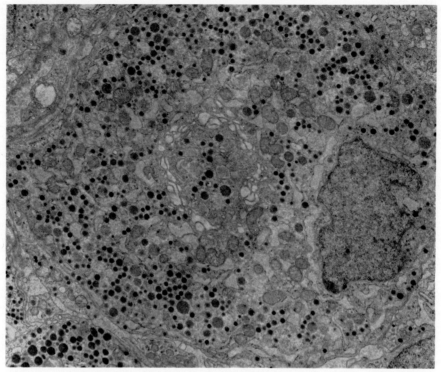

Fig. 13 A gonadotroph of a normal control rat. The ratio of the cut area of ER to the cytoplasmic area is 22.4%. (×7,500)

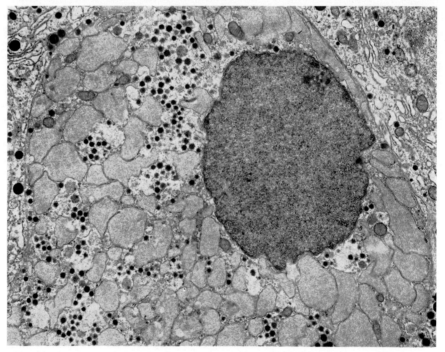

Fig. 14 A rarely occurring gonadotroph of a normal rat with dilated cisternae of the ER. ER/cytoplasm ratio is 62.2%. (×7,500)

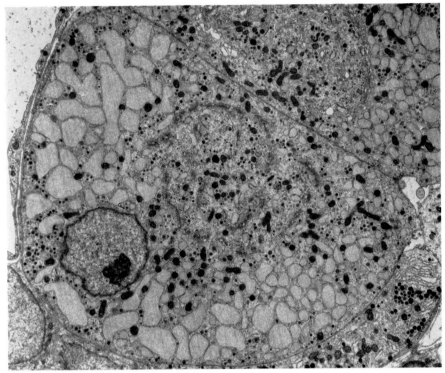

Fig. 15 A gonadotroph of a rat 2 months after castration. ER/cytoplasm ratio is 41.4%. (×6,000)

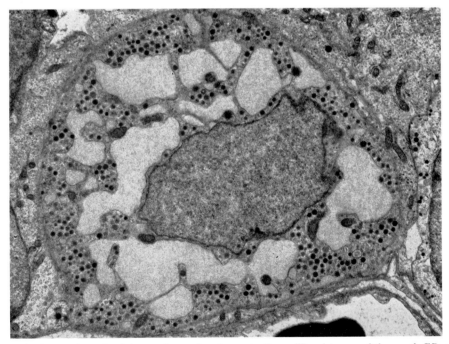

Fig. 16 A gonadotroph of the rat 2 months after castration. The cisternae of the rough ER are strongly dilated and partly fused to one another. ER/cytoplasm ratio is 50.8%. (×8,000)

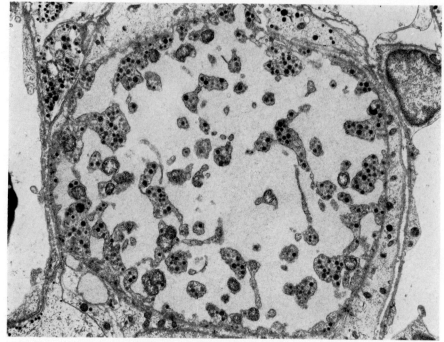

Fig. 17 A gonadotroph of a rat 2 months after castration. Dilatation and fusion of the cisternae of the rough ER are remarkable. ER/cytoplasm ratio is 61.3%. (×7,000)

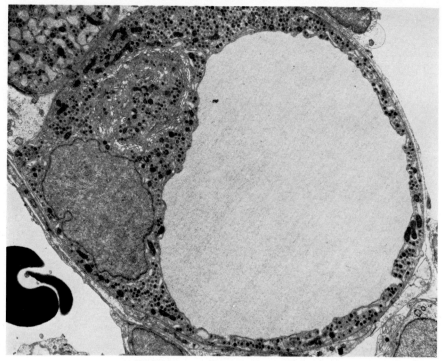

Fig. 18 A signet-ring cell of a rat anterior pituitary 3 months after castration. A large cavity formed by the fusion of ER cisternae is observed. The ratio of such an extremely dilated ER to the entire cytoplasmic area is 61.3%. (×7,000)

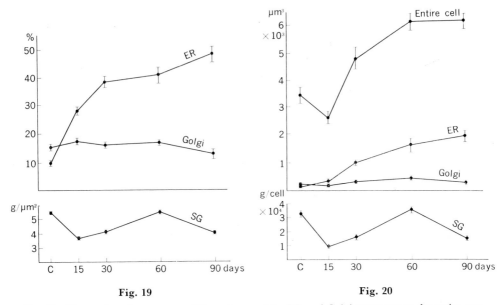

Fig. 19 Fluctuation of percentage of the cut area of the ER and Golgi apparatus to the entire cytoplasmic area, and population density of secretory granules of the gonadotrophs after castration.

Fig. 20 Estimated three-dimensional volume fluctuation of the entire cell body, ER and Golgi as well as that of the average number of secretory granules per cell after castration.

tion is the tremendous increase in volume of the rough ER. This may be caused by the accumulation of substance within the cavity of the cisternae of the rough ER. Protein synthesis in this cell may be accelerated and the newly synthesized substance may be accumulated in the ER cavity. However, the population density of the granules decreases at 15 days after castration. It is questioned whether this decrease is due to the hypertrophy of the cell body or to the actual decrease of granules. In order to solve this problem, it is necessary to compare the entire cell volume of the castrated cells with that of the normal control. In electron micrographs of gonadotrophs at each stage in this experiment, the major and minor diameters of the cell body were measured. To avoid tangential sections through the cell periphery and transverse sections perpendicular to the long axis of the cell, only cell profiles in which both the nucleus and Golgi apparatus appeared were used for the measurements, because the gonadotrophs were elliptical in shape and the nucleus and Golgi apparatus were situated along the major axis of the cell body. The estimated volume of the entire cell body was calculated as an ellipsoid rotated about its major axis.

$$V = \frac{4}{3}\pi \cdot \frac{a}{2} \cdot (\frac{b}{2})^2$$

Where V: volume of the entire cell body (μm^3), a: major diameter of the cell (μm), b: minor diameter of the cell (μm).

Estimated volumes of the ER and the Golgi areas were also calculated, through conversion of two-diamensional (areal) percentage to a three-dimensional (volume) percentage. The calculation was performed by the following formula:

$$V_E = \frac{V \cdot P_E}{100}$$

where $P_E = (\sqrt{R})^3/10$, V_E: volume of ER (or Golgi apparatus) (μm^3), V: volume of the entire cell body (μm^3), P_E: three-dimensional percentage of the volume of the ER (Golgi apparatus) to the entire cell volume, R: two-dimensional percentage of the area of the ER (Golgi apparatus) to the entire cell area.

It was assumed that the granules were almost evenly distributed throughout the cytoplasmic matrix (cytoplasmic field excluding ER) and the total number of granules per cell was calculated by the following formula:

$$G = (\sqrt{D})^3 \cdot V_0$$

where $V_0 = V \cdot P_M/100$, G: estimated number of granules in a given cell, D: two-dimensional population density of granules in the cytoplasmic matrix, V_0: volume of the cytoplasmic matrix (μm^3), V: volume of the entire cell body (μm^3), P_M: three-dimensional percentage of the volume of the cytoplasmic matrix, $P_M = (\sqrt{R_M})^3/10$, R_M: two-dimensional percentage of the area of the cytoplasmic matrix to the entire cell area.

These results are demonstrated in Fig. 20. Though the calculated values were crudely estimated, it had been clearly demonstrated that both the entire cell volume and the total number of granules per cell decreased at 15 days, but they increased soon after until 60 days after castration. The decrease in population density of the granules at 15 days shown in Fig. 19 might be understood not to be the result of hypertrophy of the cell body but due to an actual decrease in granule number at this stage. Measurement of the entire cell volume demonstrated that the gonadotrophic cells themselves became smaller during the early stage of castration. This could be accomplished by an acute discharge of secretory granules at this stage, as the decrease of estimated granule number per cell suggested. Volume of the ER continuously increased but that of the Golgi apparatus was not changed as markedly. The entire cell volume at 90 days remained rather unchanged as compared with 60 days. This could be caused from the fact that the granules were decreased again in number at 90 days. Expansion of the rough ER became extreme at this stage, and pushed away the Golgi apparatus which had decreased in size, and consequently forced the cell function of the secretory granule formation to a rather suppressed condition and decreased the granule number.

Morphometry of the entire cell body as well as the cell organelles and counting secretory granule numbers clearly demonstrated the kinetics of the secretory activity, which had been vague when the electron microscopic images of the glandular cells at various functional states were only simply observed without quantitative consideration.

REFERENCES

[1] BARNES B. G. (1962) Electron microscope studies on the secretory cytology of the mouse anterior pituitary. Endocrinology, 71: 618–628.

[2] BOWIE E. P., WILLIAMS G., SHIINO M. and RENNELS E. G. (1973) The corticotroph of the rat adenohypophysis: A comparative study. Am. J. Anat., 138: 499–520.

[3] DINGEMANS K. P. (1969) On the origin of thyroidectomy cells. J. Ultrast. Res., 26: 480–500.

[4] FARQUHAR M. G. and RINEHART J. F. (1954) Electron microscopic studies of the anterior pituitary gland of castrate rats. Endocrinology, 54: 516–541.

[5] GERSCHENFELD H. M., TRAMEZZANI J. H. and DE ROBERTIS E. (1960) Ultrastructure and function in neurohypophysis of the toad. Endocrinology, 66: 741–762.

[6] HILD W. and ZETLER G. (1952) Vorkommen der Hypophysenhinterlappenhormone in Zwischenhirn einiger Säugetiere. Dtsch. Z. Nervenhk., 167: 205–214.

[7] ISHIKAWA H. and YOSHIMURA F. (1969) Factors affecting the increase and decrease in number of acidophils in adenohypophysis. Gunma Symp. Endocrin., 6: 267–295.

[8] Kurosumi K. (1968) Functional classification of cell types of the anterior pituitary gland accomplished by electron microscopy. Arch. histol. jap., 29: 329–362.
[9] Kurosumi K. and Kobayashi Y. (1966) Corticotrophs in the anterior pituitary glands of normal and adrenalectomized rats as revealed by electron microscopy. Endocrinology, 78: 745–758.
[10] Kurosumi K. and Oota Y. (1966) Corticotrophs in the anterior pituitary glands of gonadectomized and thyroidectomized rats as revealed by electron microscopy. Endocrinology, 79: 808–814.
[11] Kurosumi K. and Fujita H. (1974) An Atlas of Electron Micrographs: Functional Morphology of Endocrine Glands. Igaku Shoin Ltd., Tokyo.
[12] Nakane P. K. (1970) Classifications of anterior pituitary cell types with immunoenzyme histochemistry. J. Histochem. Cytochem., 18: 9–20.
[13] Siperstein E. R. and Miller K. J. (1970) Further cytophysiological evidence for the identity of the cells that produce adrenocorticotropic hormone. Endocrinology, 86: 451–486.
[14] Weibel E. R. (1969) Stereological principles for morphometry in electron microscopic cytology. Int. Rev. Cytol., 26: 235–302.
[15] Weibel E. R. and Bolender R. P. (1973) Stereological techniques for electron microscopic morphometry. In: Principles and Techniques of Electron Microscopy, Biological Applications. Hayat M. A. (ed.) Vol. 3, Van Nostrand Reinhold Co., New York.
[16] Yoshimura F. and Harumiya K. (1965) Electron microscopy of the anterior lobe in normal and castrated rats. Endocrinol. Jap., 12: 119–152.
[17] Zambrano D. and De Robertis E. (1966) The secretory cycle of supraoptic neurons in the rat. A structural-functional correlation. Z. Zellforsch., 73: 414–431.

Quantitative and Three-dimensional Analyses of Dendritic Cells and Keratin Cells in the Human Epidermis Revealed by Transmission and Scanning Electron Microscopy

Yutaka MISHIMA *and* Masahiro MATSUNAKA

The disadvantage of electron microscopic investigation as compared with biochemical approaches has been the difficulty of quantitation which prevented a well correlated study between these two types of research. In this paper we will describe two major quantitation methods. The first method using transmission electron microscopes has been developed to determine the total population density of particular types of cells such as Langerhans cells or melanocytes composing the human epidermis, since they cannot be totally detected by light microscopy using cytochemical methods such as the ATPase and dopa reactions due to the presence of enzymically inactive cells. The second method using scanning electron microscopes quantitates three-dimensional changes of surface ultrastructures of the human epidermis occurring in keratinization disorders. One part is concerned with a determination of the area and pattern of particular surface ultrastructures using photo-densitometric analyses of scanning electron micrographs. The other is to determine the height of particular surface ultrastructures calculated from stereo-paired scanning electron micrographs using a goniometer.

Quantitation of Langerhans Cells, Melanocytes and α-Dendritic Cells in the Normal and Abnormal Human Epidermis using a Transmission Electron Microscope

Quantitation of three types of dendritic cells in the human epidermis using a transmission electron microscope becomes necessary in order to clarify the pathogenesis of depigmentary disorders. The primary defect in vitiligo vulgaris was thought to be the replacement of active melanocytes by Langerhans cells in the basal cell layers based on qualitative electron microscopic findings (Birbeck et al. 1961). However, lately Breathnach et al. (1968) demonstrated that Langerhans cells did not belong to the melanocyte ontogeny due to the discovery of Langerhans cells in mouse skin experimentally deprived of its neural crest component. Further, Brown et al. (1967) reported no change in the total Langerhans cell population in vitiligo epidermis as demonstrated by the light microscopic ATPase reaction. On the other hand it has been reported that the ATPase reaction was not specific for Langerhans cells but could reveal even melanocytes and keratinocytes. Furthermore, there has been some evidence that the ATPase negative Langerhans cells, similar to the dopa reaction negative melanocytes (Mishima & Widlan 1967) were present in the normal skin. Current studies based on enzymic histochemistry and qualitative electron microscopic observation have thus yielded conflicting conclusions, leading us to the development of the following method of quantitating these dendritic cells by direct electron microscopic counting resulting in a newer precise disclosure of cell kinetics in the disease process. The example of such quantitating methods of cell kinetics at the ultrastructural level might further be applied to many other fields of pathobiology dealing with organs composed of relatively compact cell compartments.

Classification of cell type

The classification of dendritic cell types is according to the terminology of the Sixth International Pigment Cell Conference (Fitzpatrick et al. 1966).

Electron microscopy reveals two cell types in human epidermis, keratinocytes and non-keratinocytes. The latter can be subdivided into melanocytes characterized by the presence of melanosomes and various stages of premelanosomes, Langerhans cells characterized by the synthesis of specific Langerhans cell granules, and α-dendritic cells (Fig. 5) similar to other dendritic non-keratinocytes in many aspects including the absence of tonofilaments and desmosomes but differing in that they totally lack specific granule synthesis as well (Silberberg & Mishima 1966). Since these cells could not be classified as either Langerhans cells or melanocytes, we have called them "α-dendritic cells" (Silberberg & Mishima 1966; Mishima & Kawasaki 1969). Although these cells may be related to the Langerhans cell or melanocyte ontogeny, or may even be inactive members of one of these groups, their distinct lack of observable specific granules makes a separate classification necessary for this type of study. Occasionally small non-keratinocytes are found in the epidermis, which are apparently blood cells that have migrated from the dermis. These cells are excluded from the count.

Counting procedure

In order to obtain an accurate dendritic cell count per mm² of the epidermal surface, the following procedures were performed in selecting and preparing representative samples for counting. Samples studied were obtained from normal skin as well as various hypopigmented and depigmented lesions, and were prepared by a routine electron microscopic tissue preparation procedure. A minimum of 10 trimmed pyramids were taken from each block (Fig. 1). Since only cells whose nuclei were included in a section were to be

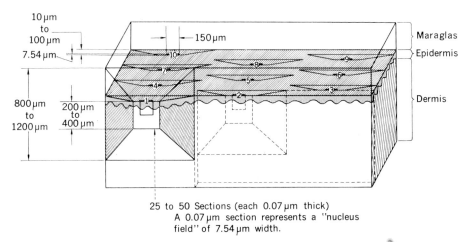

Fig. 1 A three-dimensional diagram of the sampling procedure for electron microscopic quantitation showing the manner in which blocks of tissue were cut. The pyramids are a minimum of 10 μm thick in order to prevent the appearance of the nucleus of one cell in more than one section. Ten or more pyramids were cut from each biopsied sample in the manner shown in the diagram.

counted, the interval between pyramids was made at least 10 μm in order to guarantee that no nuclei would appear in more than one section. The pyramids' width and length were in the 200–400 μm range. Grids of 150-mesh possessing square counting fields of 150 × 150 μm were used. Sections exhibiting all layers of the epidermis were cut and mounted to be completely included in these counting fields. Three to four grids, each having more than 30 gold to silver sections, were sampled from each pyramid. The number of cells per grid in each pyramid was estimated by counting 3 or more fields on each of the grids and averaging the number of cells of each type per field. The average number of each cell type for each pyramid was added together to yield the total cell count for each case.

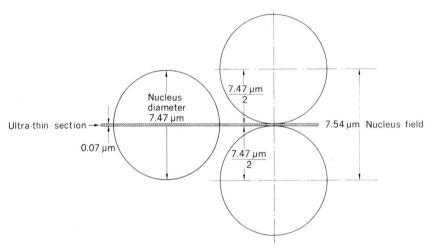

Fig. 2 A diagram showing the principle of a nucleus field. Since each thin section contains at least a portion of every nucleus which is located within a nuclear radius from either side of the surface of the section, each thin section represents a section of tissue equal to the diameter of the nucleus plus the thickness of the thin section. This thickness is designated as a nucleus field.

Calculation of cells per mm² of epidermis

ABERCROMBIE's method (1946) for calculating the number of cells per mm² of epidermal surface observed in vertical sections was originally devised for light microscopy. However, the method can also be employed for electron microscopy.

The basic principle of the method is shown here as a modification of ABERCROMBIE's original diagram (Fig. 2). Based upon measurements with light microscopy the average diameter of the nuclei counted was 7.47 μm, and the average thickness of an ultrathin section was 0.07 μm. Each thin section contains at least a portion of every nucleus, the center of which is located on either side of the section within a distance of one nuclear radius (7.47 μm/2) from the proximal surface. Therefore, each thin section represents a section of tissue 7.54 μm thick which is designated a "nucleus field".

$$F = \frac{T \times 10^6}{L \times N \, (m + s)}$$

where F: cells/mm² of epidermal surface, T: total cell count for the case, L: mean length of thin section counted = 150 μm, m: mean diameter of cell nuclei of the tissue counted = 7.47 μm, s: mean thickness of the thin section = 0.07 μm, N: number of trimmed pyramids.

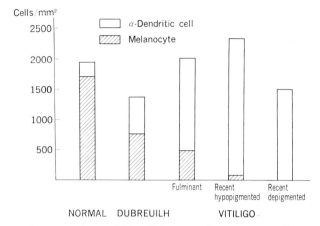

Fig. 3 Melanocytes decrease and α-dendritic cells increase when we progress from normal to hypopigmented and finally depigmented lesions. The total number of α-dendritic cells plus melanocytes remains relatively constant except for the Dubreuilh's regressing lesion in which a degradation of melanocytes is observed.

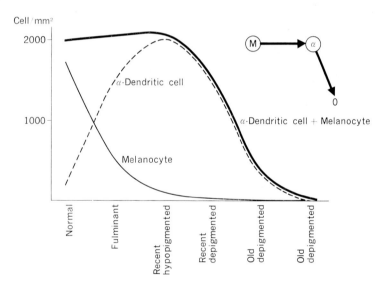

Fig. 4 Suggested cytopathogenesis of vitiligo. According to the stages of the depigmenting process, a reciprocal relationship of melanocyte and α-dendritic cell population is seen in the early stages of vitiligo, which disappears later with both cell types finally reaching zero. A transformation of melanocytes to α-dendritic cells can be speculated.

For verification of this method, the number of keratinocytes derived by this method was compared with the light microscopic quantitation previously reported by Pinkus (1952) yielding essentially identical results.

Population density of dendritic cells in the human epidermis obtained by electron microscopic quantitation

Though details of our results obtained with the above method can be found elsewhere (Mishima & Kawasaki 1969), we will attempt to exemplify the significance of such a quantitation method. After analyzing the population densities of three types of dendritic

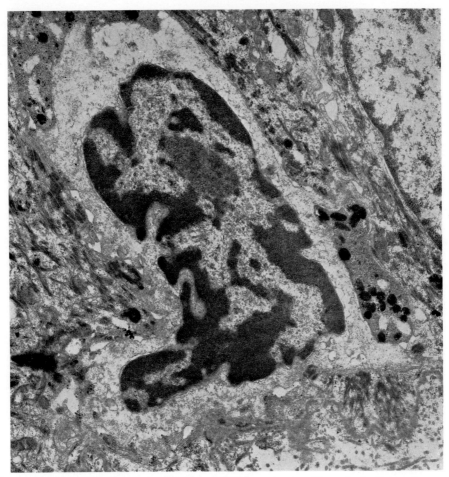

Fig. 5 α-Dendritic cells seen in the basal layer of vitiligo vulgaris. No synthesis of premelanosomes nor Langerhans granules is seen. The dopa reaction is negative. Fixed with glutaraldehyde and postfixed with OsO_4 after the dopa reaction. Stained with uranyl acetate and Reynold's lead citrate. ($\times 17,000$)

cells, we have observed a close quantitative relationship between the density dynamics of melanocytes and α-dendritic cells in various stages of the depigmentation processes as shown in Figs. 3 and 4. In vitiligo of a relatively short duration, the number of melanocytes in the basal layer gradually decreases with the age of the lesion while the number of α-dendritic cells (Fig. 5) increases with their sum remaining fairly constant. Later, when the number of melanocytes approaches zero, the α-dendritic cells also begin to decrease in number. It is only then that Langerhans cells show a change in distribution. Langerhans cells are now observed to increase in the basal layer but their total number in the entire thickness of the epidermis remains constant (MISHIMA & KAWASAKI 1969). The events in leucoderma acquisitum centrifugum are similar (MISHIMA & KAWASAKI 1969). In regressing areas of Dubreuilh's precancerous melanosis we observe distinct evidence of cytoplasmic degradation in the neoplastic melanocytes together with an increase of α-dendritic cells (MISHIMA & KAWASAKI 1969).

It can be concluded that, with the exception of the regressing lesion of Dubreuilh's melanosis which shows distinct melanocyte degradation, the total number of melanocytes

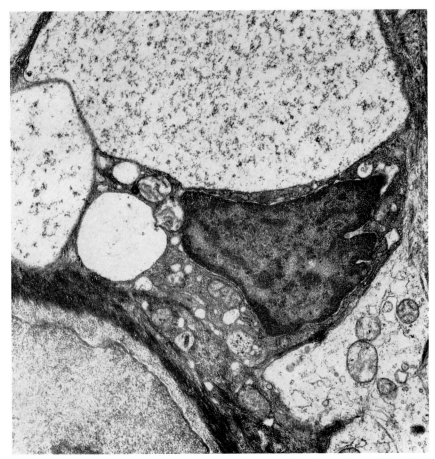

Fig. 6 A large cytoplasmic vacuole formation occurring around a Langerhans cell in a lesion of vitiligo vulgaris. Fixed with OsO$_4$ and stained with uranyl acetate and Reynold's lead citrate. ($\times$ 23,000)

plus α-dendritic cells remains relatively constant in the depigmenting diseases studied. As can be seen in Figs. 3 and 4, the melanocytes show a decrease as we progress from the normal to the hypopigmented lesion of vitiligo and finally to the fully depigmented lesion. Reciprocal to this decrease, α-dendritic cells increase from the normal to the hypopigmented zone.

Following our report (SILBERBERG & MISHIMA 1966; MISHIMA & KAWASAKI 1969) ZELICKSON and MOTTAZ (1970) also used a different electron microscopic quantitative method and reported that an increase of melanocytes and a decrease in the number of epidermal indeterminate cells which correspond to α-dendritic cells, were seen in human skin 48 hours after a single exposure to ultraviolet irradiation. If we are to speculate further by plotting the various vitiligo lesions according to the various stages of the disease process as in Fig. 4, the following hypothesis can be made. Active melanocytes are transformed to α-dendritic cells, with or without the intermediate stage of inactive melanocytes, although the possibility that some α-dendritic cells are derived from a different ontogeny such as from Langerhans cells cannot be excluded. These α-dendritic cells would then undergo excretion or degradation in the later stages of the disease process. Thus, as the numbers of melanocytes decrease in the early stages of vitiligo the numbers of α-dendritic cells reciprocally increase, while their total remains relatively constant. In older lesions, when the melanocyte population is approaching zero, the population of α-dendritic cells begins to decrease and the sum total of these cells reaches zero. Although as mentioned earlier, the total number of epidermal Langerhans cells remains unchanged during progressive depigmentations, there is an increase of Langerhans cells in the basal cell layer resulting in a redistribution of epidermal Langerhans cells. It can be stated that this redistribution is related to an immunological abnormality in the lesion or some other stimuli which may be related to a decrease of melanocytes and an increase of α-dendritic cells in a progressive depigmentation. We often find degenerative changes such as large cytoplasmic vacuole formation around Langerhans cells (Fig. 6), inactive melanocytes, and α-dendritic cells. We can also find phagocytic and lysosomal catabolic activity (SILBERBERG, BAER & ROSENTHAL 1974) in epidermal Langerhans cells. Therefore, the possibility that the increased Langerhans cells in the basal layer is related to a loss of melanogenic activity such as the premelanosome formation in vitiligo melanocytes and to a subsequent increase of α-dendritic cells cannot be excluded.

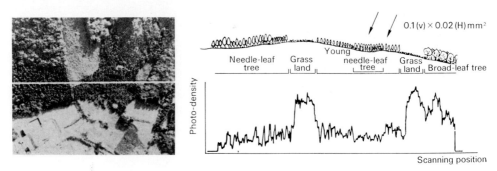

Fig. 7 A photodensitometric analysis of an aerial photograph of a land surface.

Quantitative Analyses of Scanning Electron Micrographs

Photodensitometric analysis

It has been found by scanning electron microscopy (MISHIMA et al. 1973; ORFANOS et al. 1973) that the villus-like projections and the pore-like depressions were fairly characteristic three-dimensional surface ultrastructures of a psoriatic epidermis. Furthermore these unique surface structures seem to occur parallel to the severity of the disease. Thus a quantitation of the extensiveness of these structural changes in three-dimensional surface morphology can be used for the evaluation of the disease process. It has been shown that an aerial photograph of the ground surface can be made into scanning density-distribution curves using a photodensitometer (MISHIMA et al. 1973). This permits an analysis of the type and amount of trees or grass from their characteristic patterns (Fig. 7). Concerning the quantitation of scanning electron micrographs, factors influencing the densities (mV) of scanning electron micrographs are not only by the height or depth of surface structures but also by (1) the ratio of change of the electron beam at every reflexion angle; (2) the facing angle in relation to the incident angle; and (3) the physical properties of the specimen surface related to the absorption and reflexion of electron beams. In addition, the degree of darkness of the printed micrograph is very difficult to control precisely. Therefore, densitometric analyses of scanning electron micrographs (Fig. 8) is not aimed at meas-

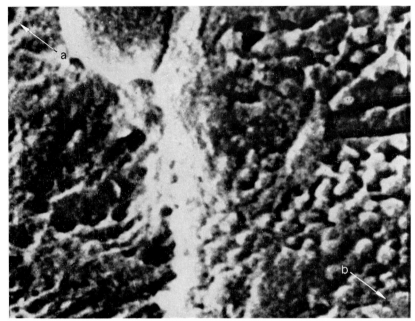

Fig. 8 A scanning micrograph of psoriatic epidermis showing both villus-like projections and pore-like depressions. (×11,000)

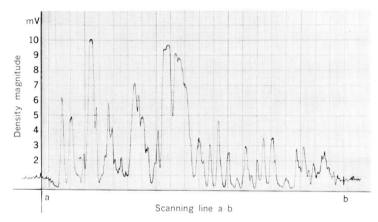

Fig. 9 A photodensitometric curve of Fig. 8 along scanning line "a" to "b".

uring absolute densities but at obtaining (1) densitometric pattern analyses (Fig. 9) (cycles); (2) differences of densities between normal levels and pathological projections or depressions (amplitude); and (3) extent of the counting areas (Fig. 10, Table 1) at various mV (height or depth) in reference to the average densities of a normal surface area. A scanning electron micrograph of villus-like projections (Fig. 11) and one of the corresponding photodensitometric curves obtained along lines 1, 2 and 3 of Fig. 11 is shown in Fig. 12. Fig. 13 is a representative scanning electron micrograph of pore-like depressions seen in psoriatic epidermis, which is subjected to scanning by a photodensitometer along lines A, B, and C. Fig. 14 represents a photodensitometric curve obtained by measurement along line B showing a corresponding periodic pattern. In comparison to this, photodensitometric curves obtained by densitometric analyses of a scanning micrograph (Fig. 15) of the

Table 1 Counting areas at various density levels (mV). Density distribution of a scanning electron micrograph measured by photodensitometer D-250-1H.*

	Density level (mV)	Accumulated count	(average)	Counting area (mm²)	Ratio (%)
1	2	178999 180807			
			179903	1799.03	37.5
2	3	123127 124246			
			123686.5	1236.865	25.8
3	4	97558 95070			
			96314	963.14	20
4	5	68943 66124			
			67533.5	675.335	14.1
5	6	46354 46042			
			46198	461.98	9.63
6	7	31968 31339			
			31653.5	316.535	6.6
7	8	21133 21220			
			21176.5	211.765	4.4
8	9	8518 8351			
			8434.5	84.345	1.76

Slit; 0.5(Y)×0.2(X)m/m, Y axis scanning speed; 0.5m/m, Scanning area; 60×80=4800 m/m², 1 count/unit area; 0.5×0.02=0.01 mm².
* Ohyo Denki Research Institute Co., Tokyo.

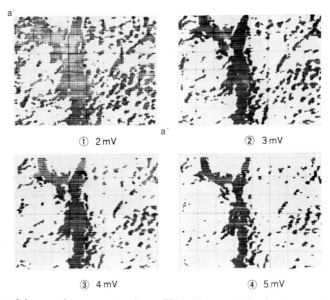

① 2 mV ② 3 mV
③ 4 mV ④ 5 mV

Fig. 10 Extent of the counting areas at various mV in reference to the density magnitude of scanning micrographs.

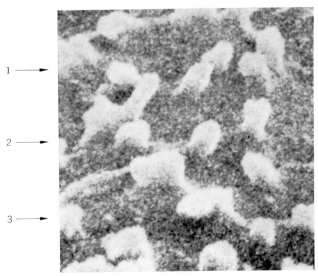

Fig. 11 A scanning electron micrograph of villus-like projections seen on the surface of psoriatic epidermis. Arrows 1, 2 and 3 indicate scanning lines for densitometric analysis. (×28,000)

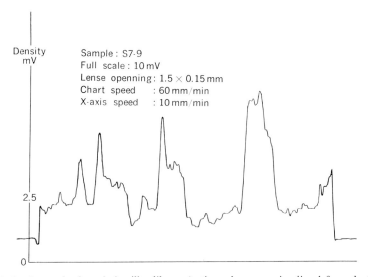

Fig. 12 A densitograph of psoriatic villus-like projections along scanning line 1 from the left to right edge of Fig. 11 having a magnification of 28,000.

normal epidermis reveal distinctly lower amplitude levels of three-dimensional surface patterns (Fig. 16). Summarizing graphs of amplitude and cycles measured from densitometric curves obtained from scanning lines of 2,500 nm of both normal and psoriatic human epidermis indicate that a quantitative evaluation of pathological changes occurring in a three-dimensional cell surface morphology is possible (Fig. 17 and Table 2).

Stereopaired analysis of the height of three-dimensional surface structures

Using a scanning electron microscope with a goniometer stage, we have attempted to determine the height of particular three-dimensional surface structures such as the villus-

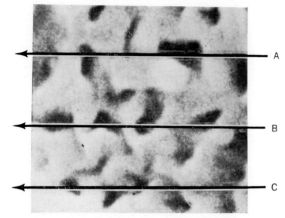

Fig. 13 A representative scanning electron micrograph of psoriatic pore-like depressions, subjected to a photodensitometric analysis along lines A, B, and C. ($\times 17{,}000$)

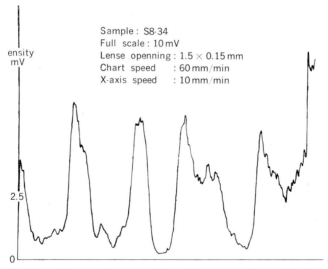

Fig. 14 Photodensitometric curves corresponding to the pattern of pore-like depressions which are present along line B from the right to the left edge in Fig. 13 having a original magnification of 20,000.

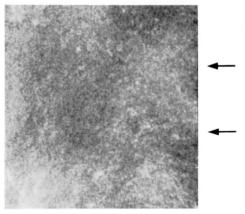

Fig. 15 A scanning micrograph of normal epidermis. ($\times 4{,}000$)

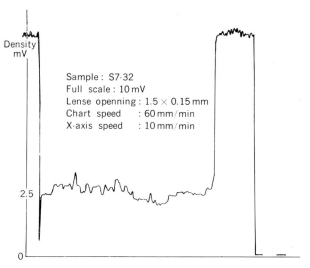

Fig. 16 Photodensitometric curves obtained by densitometric analysis of a scanning electron micrograph, Fig. 15 having a magnification of 4,000 of normal human epidermal surface along the upper arrow line.

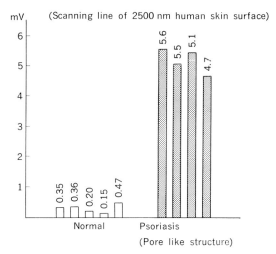

Fig. 17 Photodensitometric patterns revealed by the amplitude and cycles of densitometric curves obtained from scanning electron micrographs of normal and psoriatic epidermis.

Table 2 Salient findings of densitometric analysis of psoriatic pore-like structures.

	Psoriasis (Pore-like structure)	Normal skin
Electron micrograph	×20,000	×4,000
Scanning line of 50mm (EM)	2,500nm	12,500nm
Scanning line of 2,500nm (skin)	50mm (EM)	10mm (EM)
Amplitude (Average)	5.2mV	0.31mV
Cycle	4	5
	(T74–8)	(T74–12)

EM: electron micrograph,   Skin: human skin surface

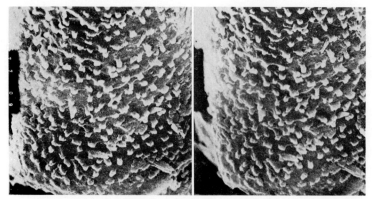

Fig. 18 Stereopaired scanning micrograph of psoriatic villus-like projections. The right hand picture is tilted at a 5° angle at the dividing axis between the two pictures. (×5,000)

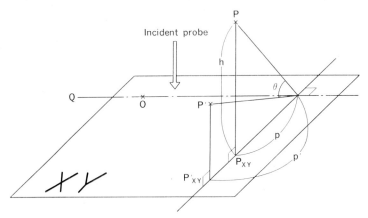

Fig. 19 A diagram of principle of stereomicrographic analysis (OSHIMA et al. 1970).

like projections, prominently occurring in a psoriatic epidermis. Fig. 18 is a paired scanning micrograph of these structures.

Stereomicrographic analyses with paired scanning micrographs had been advanced by OSHIMA et al. (1970) and also by HOWELL and BOYDE (1972). Based on the principle of stereomicrographic analysis observed by OSHIMA et al. (Fig. 19), we have studied the height of three-dimensional surface ultrastructures of the human epidermis. According to their equation we observed stereopaired scanning micrographs of villus-like projections seen on the undersurface of the keratin layer of the palm skin obtained from a psoriasis patient clinically appearing to be normal as represented in Fig. 20a and b. One of the villus-like projections seen between the arrows in the stereopaired micrographs, Fig. 20a and b is measured and the height is calculated as follows:

$$h = \frac{P'}{\sin \theta} - \frac{P}{\tan \theta}$$

$$h = \frac{27.8 \text{ mm}}{\sin 20°} - \frac{25.0 \text{ mm}}{\tan 20°} = \frac{27.8 \text{ mm}}{0.3420} - \frac{25.0 \text{ mm}}{0.3640}$$

$$= 81.2865 \text{ mm} - 68.6813 \text{ mm} = 12.6052 \text{ mm}$$

P' and P are seen on the scanning micrograph enlarged 15,840 times.

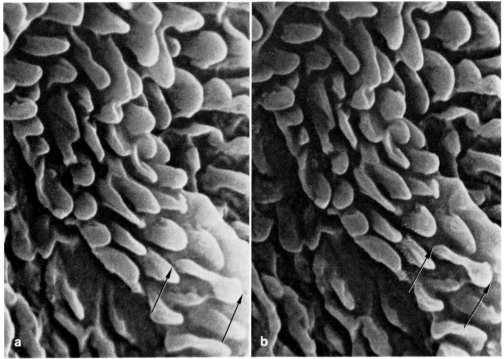

Fig. 20 Stereopaired scanning electron micrographs of villus-like projections observing the undersurface keratin layer of the skin of a human palm which clinically appears to be normal. It was obtained from a psoriasis patient. Left hand micrograph (a) is tilted 20° at the dividing axis between the two pictures. (a: ×11,600, b: ×11,600)

$$\therefore \quad h = \frac{12.6052\ mm}{15,840} = 795\ nm\ \text{height}$$

Although some distortion of the picture increases in with very low magnifications (×100 or less) or very high magnifications (×10,000 or more), due to a pin cushion distortion by the scanning of the incident probe on the specimen in the former, and due to a disturbed scanning by stray magnetic fields in the latter, the scanning electron microscope has a far better depth of focus than that of the transmission electron microscope. Thus stereoscanning micrographs can be taken more precisely to reveal the surface topography of a specimen at the ultrastructural level.

It has been demonstrated that the above stereopaired analytical method together with a densitometric analysis can provide more quantitative data of real three-dimensional fine structures.

REFERENCES

[1] ABERCROMBIE M. (1946) Estimation of nuclear population from microtome sections. Anat. Rec., *94:* 239.

[2] BIRBECK M. S., BREATHNACH A. S. and EVERALL J. D. (1961) An electron microscope study of basal melanocyte and high level clear cell (Langerhans cells) in vitiligo. J. Invest. Dermatol., *37:* 51.

[3] BREATHNACH A. S., SILVERS W. K., SMITH J. and HEYNER S. (1968) Langerhans cells in mouse skin experimentally deprived of its neural crest component. J. Invest. Dermatol., *50:* 147.

[4] Brown J., Winkelmann R. K. and Wolff K. (1967)　Langerhans cells in vitiligo: A quantitative study.　J. Invest. Dermatol., *49:* 386.

[5] Fitzpatrick T. B., Quevedo W. C., Jr., Levene A. L., McGovern V. J., Mishima Y. and Oettle A. G. (1966)　Terminology of vertebrate melanin-containing cells: 1965.　Science, *152:* 88.

[6] Howell P. G. T. and Boyde A. (1972)　Comparison of various methods for reducing measurements from stero-pair scanning electron micrographs to "real 3-d data".　*In:* Scanning electron microscopy/1972, Proc. of the 5th Ann. Scan. Electron Micros. Symposium. Johari O. and Corvin I. (eds.) pp. 234–240, IIT Research Institute, Chicago.

[7] Mishima Y. and Kawasaki H. (1969)　Dendritic cell dynamics in progressive depigmentation. Read before the VII Internat. Pigment Cell Conf., Seattle, Sept. 2–6, (1970) J. Invest. Dermatol., *54:* 83; (1972)　Arch. Dermat. Forsch. *243:* 67.

[8] Mishima Y., Matsunaka M. and Nagao S. (1973)　Subcellular changes in keratin and granular cells of keratin-stripped and psoriatic epidermis revealed by scanning and transmission electron microscopy.　Acta Dermatovener., *53:* 83.

[9] Mishima Y. and Widlan S. (1967)　Enzymically active and inactive melanocyte populations and ultraviolet irradiation: Combined dopa-premelanin reaction and electron microscopy. J. Invest. Dermatol., *49:* 273.

[10] Orfanos C., Schaumburg-Lever G., Mahrle G. and Lever W. (1973)　Alterations of cell surfaces as a pathogenetic factor in psoriasis.　Arch. Dermatol., *107:* 38.

[11] Oshima T., Kimoto S. and Suganuma T. (1970)　Stereomicrography with a scanning electron microscope.　Photogrammetric Engineering. Aug.: 874.

[12] Pinkus H. (1952)　Examination of the epidermis by the strip method.　J. Invest. Dermatol., *19:* 431.

[13] Silberberg I., Baer R. I. and Rosenthal S. A. (1974)　The role of Langerhans cells in contact allergy. I. An ultrastructural study in actively induced contact dermatitis in guinea pigs.　Acta Dermatovener., *54:* 321.

[14] Silberberg I. and Mishima Y.: (1966)　Subcellular characterizations of three distinct disturbances of melanization in man.　J. Appl. Physics, *37:* 3943.

[15] Zelickson A. S. and Mottaz J. (1970)　The effect of sunlight on human epidermis. A quantitative electron microscopic study of dendritic cells.　Arch. Dermatol., *101:* 312.

A Study on the Computerization of Quantitative Electron Microscopy: Application of the Distance Function to Ribosome Count*

Kensuke BABA *and* Teiji OKAYASU

With recent advances in techniques of electron microscopy, we can now obtain many good quality electronmicrographs without much difficulty. A number of investigators have started to quantitatively analyze figures of good electronmicrographs either with or without the morphometric method which has already been comprehensively reviewed by WEIBEL and ELIAS (1967). Tools and devices specially designed for morphometry are now available for making simple measurements of length, diameter, circumference, distance, size, area, darkness and regularity, as well as taking counts of figures represented in the picture. Although scientists may use the tools and devices for morphometry and the morphometric methods, they still have to expend tremendous effort in statistical analyses.

However, most investigators have access to a digital computer without much trouble, if they do not request a large system or special processors. The computer technique in the field of picture processing primarily had its development in pattern recognition of characters in improvement of a poor image and in automatic diagnosis of a cytological specimen with the aid of specially designed devices and/or with a huge computer system. Not many attempts have been made in the utilization of the popularized digital computer for morphometry.

The authors have made studies on quantitative morphology with the popularized computer and applied the well known and the newly designed algorithms in morphometry, in order to spare the extra trouble of going through a statistical treatment of innumerable photos which is needed prior to establishing a computerized morphometric system.

Using the popular computer we have undertaken, as the simplest example, counts of ribosomes from electron micrographs. An algorithm such the principle of spatial (digital) filtering (BABA et al. 1973) has already been proposed for the extraction of dot-like figures such as ribosomes.

In the following, we have applied the well known distance function after ROSENFELD and PFALTZ (1968) in counting the ribosomes. This paper demonstrates the procedure and program for counting ribosomes with due consideration to the practical problems of the popularized computer which is to permit the wide use of procedures and programs that are written in FORTRAN without any coding or correction.

Material and Method

An electron microscopic picture of a portion of the cytoplasm of a normal rat pancreatic

---

* This study was supported in part by a grant-in-aid for cancer research from the Ministry of Health and Welfare and a grant-in-aid for scientific research from the Ministry of Education of Japan.

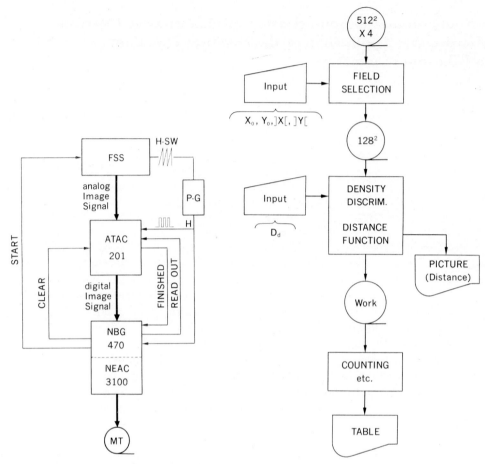

Fig. 1 A functional block diagram of the signal in-put system.

Fig. 2 Processing flow for the ribosome count.

acinar cell was printed on film at a magnification of 50,000×. The density distribution in the area which measured 7.5 centimeter square on the picture was scanned and converted into electrical signals with a flying spot densitometer (FSS-TKF 102-2/Ikegami Tsusin) that divides one frame into 512 lines. The electrical image signals were translated into 12 bit digital values at a rate which yields 512 samples per line with an analogue to digital converter (slightly modified ATAC 201/Nihon-Kohden). In this way, the image information recorded on the picture in the form of its optical density was translated into a computable digital matrix composed of 512×512 elements per frame. This meant that the size of an element in the matrix was equivalent to 30×30 A on the cytological specimen. The digital image signals were temporarily stored in a digital computer (NEAC 3100/NEC) that had a 16 K word core memory and 4 magnetic decks via a data input device (NBG 470/NEC, specially designed for the calculation of dose distribution of radiation) that had been modified to serve as an interface between the NEAC computer and other devices. The FSS flying spot densitometer and the ATAC A/D converter were controlled by the NBG interface with a pre-loaded program in the NEAC computer. The functional block diagram of the signal input system (BABA et al. 1973) is illustrated in Fig. 1. Digitalized

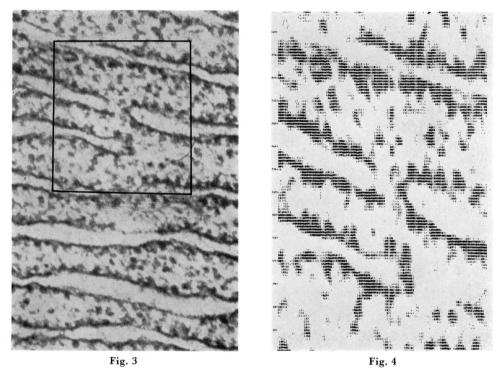

Fig. 3

Fig. 4

Fig. 3 A sample of the original electron micrograph. The data in Figs. 4–8 were obtained from the squared-out field on Fig. 3.

Fig. 4 A perceptual form of density distribution displayed on a line-printer.

image data were successively recorded on magnetic tapes in the NEAC computer frame by frame.

After the data in a binary form on the magnetic tape were rewritten on an IBM compatible 9 track tape in the unpacked form, the image data computable with FORTRAN were processed as follows with a larger capacity computer (HITAC 8350/Hitachi) having 390 K byte core memory. The main portion was processed by a program written in FORTRAN. The process flow chart is shown in Fig. 2. The program is presented in the appendix.

The size of the objective field (]x[×]y[) was designated from a card reader of the computer. In the present study, both]x[and]y[were fixed at 128 (this value is fixed at 128 but it may be revised by statements in steps 5 to 8 of the main program). The frame number and the co-ordinate for the origin in objective field (x_0, y_0) of the frame was designated from a console typewriter (both x_0 +]x[and y_0 +]y[should not be greater than 512). The image data on the selected field were transmitted from the magnetic tape to a work tape. The field selection and data transmission were performed by statements in steps 10 to 36 of the main program. For better understanding we have shown the analyzed original in Fig. 3. The density distribution in the field squared-out in Fig. 3 was displayed on a line printer in the form of a numerical density map and/or in the perceptual form as shown in Fig. 4.

The density level at which the respective elements in the selected field was distinguished as black or white, as shown in Fig. 5, was designated from a console typewriter. In Fig. 5,

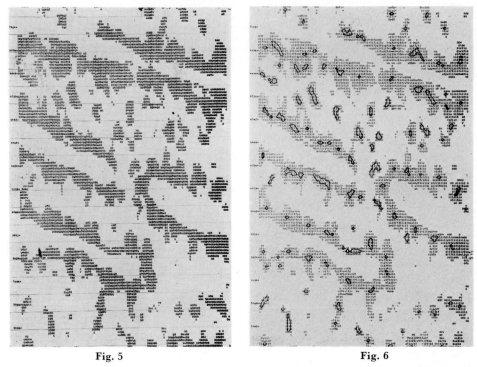

Fig. 5 A discriminated black-or-white picture. Here, the reference level for discrimination for the darkest density on the picture is set at 30/64.

Fig. 6 A city block distance map. Each number indicates the city block distance of the black area from the black point nearest the white area. The cluster constructed of maximum distance is traced out with a pencil mark. This map was obtained from the discriminated black-or-white picture on Fig. 5.

the reference criteria was set at 30 against the density value of 64 at the darkest point in the selected field. To this binary representation (black or white) of the selected field in the internal memory, the algorithm of the city block distance function after ROSENFELD and PFALTZ (1968) was applied without using external memory. The city block distances of the black points from the black point nearest to the white area were distributed as shown in Fig. 6. The above was processed by a subroutine program in steps 347 to 537 of the program attached in the appendix.

The element clusters composed of the local maximum (traced out in Fig. 6) in the city block distance domain corresponded very well with the positions of ribosomes. To search for the clusters that are valued locally as maximum we designed an algorithm which was fundamentally based upon the cluster counting linear algorithm and which was widely applied in the hardware of the rapid feature analyzer such as the Omnicon or Classimat. Location, size and the maximum value of each cluster were listed in the form of a table that is only partially shown in Fig. 7. The last part of the processing was performed by statements corresponding to steps from 39 to 340.

Similarly, to find the difference in the selectivity of ribosomes between two distance functions, we also obtained a distribution of square distances and a counting table for the same field.

NO.	X-MIN	MAX	Y-MIN	MAX	MAX	X	Y	N
1	17	17	2	2	1	1	1	1
2	81	82	3	3	2	2	1	2
3	103	103	3	4	2	1	2	2
4	28	29	6	6	2	1	1	1
5	90	90	8	8	2	1	1	1
6	52	57	9	10	2	6	2	8
7	114	114	8	8	2	1	1	1
8	61	61	11	11	2	1	1	1
9	83	83	15	15	5	1	1	1
10	9	15	13	17	3	7	5	13
11	63	65	11	12	2	3	2	3
12	28	29	17	17	4	2	1	2
13	93	101	15	19	3	9	5	12
14	123	124	18	20	4	2	3	4
15	4	5	15	15	3	2	1	2
16	34	35	16	16	2	2	1	2
17	110	111	20	20	3	2	1	2
18	40	41	21	22	3	2	2	3
19	48	52	22	24	4	5	3	7
20	13	13	21	21	2	1	1	1
21	23	25	21	23	3	3	3	5
22	28	30	21	22	3	3	2	4
23	113	113	21	21	3	1	1	1
24	81	82	25	26	4	2	2	2
25	62	62	25	25	5	1	1	1
26	2	4	24	25	1	3	2	3
27	7	11	26	27	1	5	2	8
28	32	36	29	34	2	5	6	12
29	96	100	29	30	3	5	2	8
30	103	108	30	34	3	6	5	10
31	24	26	32	34	2	3	3	5
32	115	115	31	31	1	1	1	1
33	124	124	34	34	4	1	1	1
34	4	6	36	38	3	3	3	4
35	78	80	33	35	1	3	3	6

Fig. 7 Upper part of the table on the count of ribosome.
1st column (NO.): Number for each ribosome figures. 2nd colum (X-MIN): The abscissa reading for the left end of the (maximum) elements traced out in Fig. 6. 3rd column (MAX): The abscissa reading for the right end of the (maximum) elements traced out in Fig. 6. 4th column (Y-MIN): The ordinate reading for the top end of the (maximum) elements traced out in Fig. 6. 5th column (MAX): The ordinate reading for the bottom end of the (maximum) elements traced out in Fig. 6. 6th column (MAX): The reading for distance maximum of the elements traced out in Fig. 6. 7th column (X): Horizontal projection of the elements constructed from the maximum distance. 8th column (Y): Vertical projection of the elements constructed from the maximum distance. 9th column (N): The number for the elements constituting the maximum distance cluster.

Results

In principle the count of ribosome by the city block distance function was greater than the count by the square distance function as partly indicated in Fig. 8a and b. However, the clusters located at the upper left corner and at the top of Fig. 9b were counted as two and three respectively by the square distance function while the same clusters were counted out as one and one by the city block distance function as shown in Fig. 9a. The count was not influenced greatly by the choice between the two distance functions, rather it was affected more by the reference level for the gray density where the optical information was discriminated as black or white.

When the reference level for classifying the gray density into black or white was altered, the count of the ribosomes varied greatly and the peak that was half way out in the horizontal direction shifted slightly to the origin as seen in Fig. 10. The peak count for the ribosome was considerably greater than the manual result, when our count was limited to the figures that were clearly distinguishable as ribosomes. On the other hand, it was markedly smaller than the count that included the probable figures. However, two experts on electron microscopy who independently counted ribosome in the usual way reported very similar results to us which agreed well with the peak count obtained by the distance function.

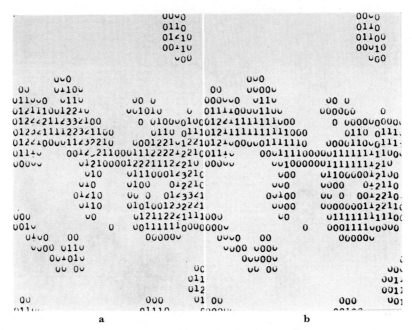

Fig. 8 a. City block distance, a part of the map. **b.** Square distance, a part of the map.

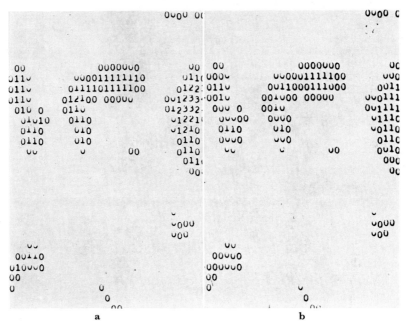

Fig. 9 a. City block distance, a part of the map. **b.** Square distance, a part of the map.

Thus, we came to the conclusion that a count at the peak might be allied to the manual count and, consequently, by relying on the peak we could obtain a count of ribosomes effectively by the distance function without a particular consideration of photographic standardization and/or normalization, hence, free ourselves from the knotty problems in the field of analytical image processing with automatic identification of pattern recognition.

The study required 260 seconds per frame corresponding to 512×512 elements, to transform the optical image information on an electron micrograph read by the FSS flying spot densitometer into digital image signals and to store this in a magnetic tape of a NEAC computer, and approximately 10 minutes to count up ribosome figures in a selected and discriminated image field of 128×128 elements.

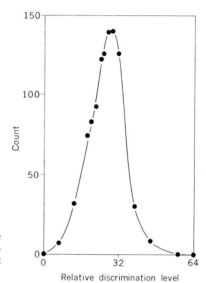

Fig. 10 The change in count with a shift in reference level for discrimination. The number 64 on the abscissa is a comparable value for expressing the darkest density in the field.

Discussion

The number of ribosomes registered for the peak fluctuates only slightly, even when contamination in a scanned film image is great and produces a false maximum. We processed the image without any special consideration in the treatment of the contrast of the original image because we thought that the peak value of the count, i.e., discrimination criteria would not be influenced much by this treatment. We were freed of restraints from the delicate treatment of contrast and from the problem of black contamination that could occur in the process because there was little chance of the peak count being influenced by the peak position which usually shifts on contamination and/or treatment of photography for contrast and brightness.

We primarily examined how the relative magnification of an electron micrograph influences the efficiency of the distance function in counting of ribosome figures. In this examination we have already remarked that the size corresponding to an image element might be less than 35 A on the specimen. On the other hand, the time required for one processing was approximately proportional to the square of the length of an image element on the specimen, in other words, inversely proportional to the square of the magnification of the film scanned. To trade-off problems about the reliability of the analysis and about the speed of processing we fixed the length at 30 A.

ACKNOWLEDGMENT

The authors wish to express our appreciations to Dr. YONOSUKE WATANABE, Associate Professor, Department of Pathology, Medical School, Keio University for the electron micrographs used in the analysis.

REFERENCES

[1] WEIBEL E. R. and ELIAS H. (1967) Quantitative Method in Morphology. Springer-Verlag, Berlin-Heidelberg-New York.
[2] BABA K. AMARI H. and WATANABE Y. (1973) Extraction of Ribosome feature by digital filter. *In:* Applied Electron-microscopy for Medical Science and Biology. YAMADA Y. et al. (eds.) pp. 529–535, Ishiyaku-shuppan, Tokyo. (in Japanese)
[3] ROSENFELD A. and PFALTZ J. L. (1968) Distance functions on digital pictures. Pattern Recognition (Pergamon Press), *1:* 33–61.

Appendix

```
FOR131   EY100 SOURCE PROGRAM

    1           PROGRAM COUNT2
    2           IMPLICIT INTEGER*2(I-N)
    3           DIMENSION ISIGN(500),IXMIN(500),
    4          1IXMAX(500),IYMIN(500),IYMAX(500),
    5          2IDMAX(500),ID(130,2),LP(132),
    6          3IDMAXT(500),IDP(130),IDPT(130),
    7          4ILINE(500),ICOUNT(500),IE(512)
    8           COMMON LD(130,130)
    9           DATA BLANK,ISTRSK/' ','*'/
   10           READ(5,100)IX1,IY1
   11           REWIND 14
   12   2030    CONTINUE
   13           REWIND 12
   14           REWIND 13
   15           READ(1,60)ICO,IP,MX,MY,JD
   16           IF(ICO.EQ.2) GO TO 5010
   17    60     FORMAT(2I1,2I3,I1)
   18           IF(ICO.EQ.1) GO TO 5000
   19   100     FORMAT (2I5)
   20           IPP=IP-1
   21    61     IF(IPP)68,64,62
   22    62     DO 63 I=1,512
   23    63     READ(12,66)(IE(J),J=1,512)
   24           IPP=IPP-1
   25           GO TO 61
   26    64     IF(MY.EQ.0) GO TO 70
   27           DO 65 I=1,MY
   28    65     READ(12,66)(IE(J),J=1,512)
   29    66     FORMAT(4(128I5))
   30    70     IEND=MX+IX1
   31           IST=1+MX
   32           DO 67 I=1,IY1
   33           READ(12,66)(IE(J),J=1,512)
   34    67     WRITE(13,101)(IE(K),K=IST,IEND)
   35           REWIND 13
   36           REWIND 12
   37    68     CONTINUE
   38   5010    CALL DISTNC(IX1,IY1,IP,MX,MY,JD)
   39   1035    IX=IX1+2
   40           IY=IY1+2
   41           IY2=IY+1
   42           NOY=0
   43           NOD=0
   44           DO 1 I=1,IX
   45           IDP(I)=0
   46     1     ID(I,1)=0
   47     2     NOY=NOY+1
   48           IF(NOY.GT.IY) GO TO 50
   49           READ(14,101)(ID(I,2),I=1,IX)
   50   101     FORMAT(60I5)

FOR131   EY100 SOURCE PROGRAM   COUNT2   PROGRAM

   51   1031    DO 3 I=1,IX
   52     3     IDPT(I)=ID(I,2)
   53           DO 45 I=1,IX
   54           IF(ID(I,2).EQ.0) GO TO 45
   55           IF(I.EQ.1) GO TO 4
   56           IF(ID(I-1,2).NE.0) GO TO 7
   57           IF(ID(I-1,1).NE.0) GO TO 6
   58     4     IF(ID(I,1).NE.0) GO TO 8
   59           IF(I.EQ.IX) GO TO 5
   60           IF(ID(I+1,1).NE.0) GO TO 9
   61     5     NOD=NOD+1
   62           IDMAX(NOD)=ID(I,2)
   63           ISIGN(NOD)=0
   64           IXMIN(NOD)=I
   65           IXMAX(NOD)=I
   66           IYMIN(NOD)=NOY
   67           IYMAX(NOD)=NOY
   68           IDMAXT(NOD)=ID(I,2)
   69           ID(I,2)=NOD
   70           ILINE(NOD)=1
   71           ICOUNT(NOD)=1
   72           GO TO 45
   73     6     NO=ID(I-1,1)
   74           GO TO 10
   75     7     NO=ID(I-1,2)
   76           GO TO 10
   77     8     NO=ID(I,1)
   78           GO TO 10
   79     9     NO=ID(I+1,1)
   80    10     IF(IDMAXT(NO)-ID(I,2))22,12,11
   81    11     DO 111 K=1,3
   82           IT=I+K-2
   83           IF(IT.LT.1.OR.IT.GT.IX) GO TO 111
   84           IF(IDP(IT).EQ.0) GO TO 111
   85           IF(IDP(IT).NE.ID(I,2)) GO TO 111
   86           NOT=ID(IT,1)
   87           IF(IDMAX(NOT).GE.IDMAX(NO))
   88          5GO TO 113
   89           IF(IDMAX(NOT).EQ.IDP(IT)) GO TO 14
   90   113     ID(IT,1)=NO
   91   111     CONTINUE
   92           IDMAXT(NO)=ID(I,2)
   93   112     ILINE(NO)=0
   94           ISIGN(NO)=1
   95           ID(I,2)=NO
   96           GO TO 45
   97    12     IF(IDMAX(NO).LE.ID(I,2)) GO TO 124
   98           DO 121 K=1,3
   99           IT=I+K-2
  100           IF(IT.LT.1.OR.IT.GT.IX) GO TO 121
```

COMPUTERIZATION OF QUANTITATIVE EM

```
FOR131  EY100 SOURCE PROGRAM  COUNT2  PROGRAM

101           IF(IDP(IT).EQ.0) GO TO 121
102           IF(IDP(IT).NE.ID(I,2)) GO TO 121
103           NOT=ID(IT,1)
104           IF(IDMAX(NOT).GE.IDMAX(NO))
105         ¥GO TO 121
106           IF(IDP(IT).EQ.IDMAX(NO)) GO TO 14
107           IF(ID(IT-1,2).GT.IDP(IT))
108         ¥GO TO 121
109           IF(ID(IT,2).GT.IDP(IT)) GO TO 121
110           IF(ID(IT+1,2).GT.IDP(IT))
111         ¥ GO TO 121
112           IF(IDP(IT).EQ.IDMAX(NOT)) GO TO 14
113    121 CONTINUE
114           GO TO 112
115    124 DO 13 K=1,2
116           IT=I+K-1
117           IF(IT.GT.IX) GO TO 13
118           IF(ID(IT,1).EQ.0) GO TO 13
119           IF(IDP(IT).NE.ID(I,2)) GO TO 13
120           NOT=ID(IT,1)
121           IF(IDMAX(NO).LE.IDMAX(NOT).AND.
122         6ID(I,2).EQ.IDMAX(NO)) GO TO 125
123           IF(IDMAX(NO).GE.IDMAX(NOT).AND.
124         7ID(I,2).EQ.IDMAX(NOT)) GO TO 125
125           IF(IDMAX(NO).NE.IDMAX(NOT))
126         8GO TO 13
127           IF(ID(I,2).EQ.IDMAX(NOT))
128         9GO TO 125
129           IF(ILINE(NO).EQ.0) GO TO 30
130           IF(ILINE(NOT).EQ.0) GO TO 30
131    125 CONTINUE
132           IF(NO.NE.NOT) GO TO 14
133     13 CONTINUE
134           GO TO 30
135     14 CONTINUE
136           IF(NO.LE.NOT) GO TO 400
137           NOMIN=NOT
138           NOC=NO
139           GO TO 140
140    400 NOMIN=NO
141           NOC=NOT
142    140 IF(IDMAX(NO)-IDMAX(NOT))
143         A141,142,143
144    141 NOB=NOT
145           GO TO 144
146    142 IAX=IXMAX(NO)
147           IBX=IXMAX(NOT)
148           IXMAX(NOMIN)=IOKII(IAX,IBX,I)
149           ICX=IXMIN(NO)
150           IDX=IXMIN(NOT)

151           IXMIN(NOMIN)=IHISAI(ICX,IDX,I)
152           IAY=IYMAX(NO)
153           IBY=IYMAX(NOT)
154           IYMAX(NOMIN)=IOKII(IAY,IBY,NOY)
155           ICY=IYMIN(NO)
156           IDY=IYMIN(NOT)
157           IYMIN(NOMIN)=IHISAI(ICY,IDY,NOY)
158           IAZ=IDMAX(NO)
159           IBZ=IDMAX(NOT)
160           ICZ=ID(I,2)
161           IDMAX(NOMIN)=IOKII(IAZ,IBZ,ICZ)
162           ILINE(NOMIN)=1
163           IF(ID(I,2).GT.IDMAX(NO)) GO TO 147
164           ICOUNT(NOMIN)=ICOUNT(NO)
165         B+ICOUNT(NOT)
166           GO TO 148
167    147 ICOUNT(NOMIN)=0
168           IDMAX(NOMIN)=ID(I,2)
169           IXMAX(NOMIN)=I
170           IXMIN(NOMIN)=I
171           IYMAX(NOMIN)=NOY
172           IYMIN(NIMIN)=NOY
173           ILINE(NOMIN)=1
174    148 IF(IDMAX(NOMIN).EQ.ID(I,2))
175         1ICOUNT(NOMIN)=ICOUNT(NOMIN)+1
176           ISIGN(NOMIN)=0
177           GO TO 146
178    143 NOB=NO
179    144 IF(IDMAX(NOB).LT.ID(I,2))
180         1GO TO 147
181           ICOUNT(NOMIN)=ICOUNT(NOB)
182           IF(IDMAX(NOB).EQ.ID(I,2))
183         EICOUNT(NOMIN)=ICOUNT(NOMIN)+1
184           IDMAX(NOMIN)=IDMAX(NOB)
185           IXMAX(NOMIN)=IXMAX(NOB)
186           IXMIN(NOMIN)=IXMIN(NOB)
187           IYMAX(NOMIN)=IYMAX(NOB)
188           IYMIN(NOMIN)=IYMIN(NOB)
189           ILINE(NOMIN)=0
190           ISIGN(NOMIN)=1
191           NOB=NOD-1
192           IDMAXT(NOMIN)=ID(I,2)
193           ID(I,2)=NOMIN
194           DO 15 K=NOC,NOB
195           IXMAX(K)=IXMAX(K+1)
196           IXMIN(K)=IXMIN(K+1)
197           IYMAX(K)=IYMAX(K+1)
198           IYMIN(K)=IYMIN(K+1)
199           IDMAX(K)=IDMAX(K+1)
200           IDMAXT(K)=IDMAXT(K+1)

201           ILINE(K)=ILINE(K+1)
202           ICOUNT(K)=ICOUNT(K+1)
203           ISIGN(K)=ISIGN(K+1)
204     15 ICOUNT(K+1)=0
205           DO 18 K=1,IX
206           IF(ID(K,1)-NOC)18,16,17
207     16 ID(K,1)=NOMIN
208           GO TO 18
209     17 ID(K,1)=ID(K,1)-1
210     18 CONTINUE
211           DO 21 K=1,IX
212           IF(ID(K,2)-NOC)21,19,20
213     19 ID(K,2)=NOMIN
214           GO TO 21
215     20 ID(K,2)=ID(K,2)-1
216     21 CONTINUE
217           GO TO 45
218     22 IF(ISIGN(NO).NE.1) GO TO 30
219           NOB=NO
220           NOT=9999
221           DO 221 K=1,3
222           IT=I+K-2
223           IF(IT.LT.1.OR.IT.GT.IX) GO TO 221
224           NO=ID(IT,1)
225           IF(IDP(IT)-ID(I,2))222,220,112
226    222 IF(IDMAX(NO).NE.1) GO TO 221
227    220 IF(NOT.LT.NO) GO TO 221
228           NOT=NO
229    221 CONTINUE
230           NO=NOT
231           IF(NO.EQ.NOB) GO TO 30
232           IF(NO-9999) 224,228,224
233    224 IF(ILINE(NOB).EQ.0) GO TO 229
234           NO=NOB
235           NOT=NO
236           GO TO 14
237    229 NOT=0
238           DO 223 K=1,3
239           IT=I+K-2
240           IF(IT.LT.1.OR.IT.GT.IX)
241         1GO TO 223
242           IF(ID(IT,1).EQ.NOB.AND.
243         1IDP(IT).EQ.ID(I,2))
244         2NOT=1
245    223 CONTINUE
246           IF(ID(I,2).EQ.IDMAX(NOB).AND.
247         1 ID(I,2).LE.IDMAX(NO).AND.
248         2 NOT.EQ.1)
249         2 GO TO 227
250           GO TO 30

251    227 NOT=NOB
252           GO TO 14
253    228 ID(I-1,2)=NOD+1
254           GO TO 5
255     30 MAXT=IDMAXT(NO)
256           MAX=IDMAX(NO)
257           ILINE(NO)=1
258           IDMAXT(NO)=ID(I,2)
259           IF(MAXT.EQ.IDMAXT(NO)) GO TO 31
260           IF(MAX.EQ.ID(I,2)) GO TO 33
261           IF(MAX.GT.ID(I,2)) GO TO 112
262     34 IDMAX(NO)=ID(I,2)
263           IXMAX(NO)=I
264           IXMIN(NO)=I
265           IYMAX(NO)=NOY
266           IYMIN(NO)=NOY
267           ID(I,2)=NO
268           ISIGN(NO)=0
269           ICOUNT(NO)=1
270           GO TO 45
271     31 IF(MAX.LT.ID(I,2)) GO TO 34
272           IF(MAX.EQ.ID(I,2))
273         1ICOUNT(NO)=ICOUNT(NO)+1
274           ISIGN(NO)=0
275           ID(I,2)=NO
276    302 IF(IDMAX(NO).GT.IDMAXT(NO))
277         FGO TO 32
278           IF(IXMAX(NO).LT.I) IXMAX(NO)=I
279           IF(IXMIN(NO).GT.I) IXMIN(NO)=I
280           IF(IYMAX(NO).LT.NOY) IYMAX(NO)=NOY
281           IF(IYMIN(NO).GT.NOY) IYMIN(NO)=NOY
282           GO TO 45
283     32 ISIGN(NO)=1
284           ILINE(NO)=0
285           GO TO 45
286     33 IDMAX(NO)=ID(I,2)
287           ISIGN(NO)=0
288           ID(I,2)=NO
289           ICOUNT(NO)=ICOUNT(NO)+1
290           GO TO 302
291     45 CONTINUE
292           DO 47 I=1,IX
293           IDP(I)=IDPT(I)
294     47 ID(I,1)=ID(I,2)
295           DO 48 I=1,NOD
296     48 ISIGN(I)=0
297           GO TO 2
298     50 DO 80 J=2,IY2
299           DO 86 K=1,130
300     86 LP(K)=0
```

```
FOR131   EY100 SOURCE PROGRAM   COUNT2   PROGRAM

301            DO 81 I=1,NOD
302            IF(.NOT.(J.GE.IYMIN(I).AND.
303           &J.LE.IYMAX(I))) GO TO 81
304            IXMN=IXMIN(I)
305            IXMX=IXMAX(I)
306            DO 82 K=IXMN,IXMX
307            LP(K)=LD(K,J)
308        82 CONTINUE
309        81 CONTINUE
310            DO 83 K=1,IX
311            IF(LP(K).LE.0) GO TO 84
312            IF(LP(K).GE.10) GO TO 85
313            WRITE(20,106) LP(K)
314       106 FORMAT(I1)
315            READ(20,107) LP(K)
316       107 FORMAT(A1)
317            GO TO 83
318        84 LP(K)=BLANK
319            GO TO 83
320        85 LP(K)=ISTRSK
321        83 CONTINUE
322            WRITE(6,108) (LP(K),K=1,IX)
323       108 FORMAT(' ',130A1)
324        80 CONTINUE
325            WRITE(6,103)
326       103 FORMAT(1H1//1H ,'    NO.',5X,
327           #'X-MIN',5X,'X-MAX',5X,'Y-MIN',5X,
328           #'Y-MAX'#7X,'MAX',9X,'X',9X,'Y',
329           #5X,'COUNT')
330            REWIND 14
331            DO 52 I=1,NOD
332            IXT=IXMAX(I)-IXMIN(I)+1
333            IYT=IYMAX(I)-IYMIN(I)+1
334            WRITE(14,105) I,IXMIN(I),IXMAX(I),
335           #IYMIN(I),IYMAX(I),IDMAX(I),
336           #IXT,IYT,ICOUNT(I)
337        52 WRITE(6,104) I,IXMIN(I),IXMAX(I),
338           IIYMIN(I),IYMAX(I),IDMAX(I),
339           JIXT,IYT,ICOUNT(I)
340            REWIND 14
341       105 FORMAT(I5,8I10)
342       104 FORMAT(1H ,I5,8I10)
343            GO TO 2030
344     5000 CONTINUE
345            STOP
346            END
```

```
FOR131   EY100 SOURCE PROGRAM

347            SUBROUTINE DISTNC(IX1,IY1,IP,MX,MY,JD)
348            IMPLICIT INTEGER*2(I-N)
349            DIMENSION IDN(130)
350            COMMON ID(130,130)
351            DATA IBLANK,ISTRSK,IDN/' ','*',
352           #130#0/
353            IJK=0
354            IX=IX1+2
355            IY=IY1+2
356            IY2=IY1+1
357            IX2=IX1+1
358     2000 REWIND13
359            DO 1 JJ=1,IY1
360            J=IY1-JJ+1
361            DO 400 I=1,IX1
362       400 ID(I+1,J+1)=0
363         1 READ(13,101) (ID(I+1,J+1),I=1,IX1)
364            REWIND 13
365            IF(IJK)700,701,700
366       701 MAX=ID(2,2)
367            MIN=MAX
368            DO 1009 I=2,IX2
369            DO 1009 J=2,IY2
370            IF(ID(I,J)-MAX)1111,1111,1100
371     1100 MAX=ID(I,J)
372            GO TO 1009
373     1111 IF(MIN-ID(I,J))1009,1009,1102
374     1102 MIN=ID(I,J)
375            IJK=IJK+1
376     1009 CONTINUE
377       700 DO 24 I=1,IX
378            ID(I,1)=0
379            ID(1,I)=0
380            ID(I,IY1+2)=0
381        24 ID(IX1+2,I)=0
382            WRITE(6,200) IP,MX,MY
383       200 FORMAT('1 FILM NO. ',I2,'   XO=',
384           #I5,'   YO=',I5)
385            WRITE(6,120)IX1,IY1,MAX,MIN
386            READ(1,110)ICU1,ICU2,LIM
387       110 FORMAT(2I3,I2)
388            ICU=ICU1*100/ICU2
389     5001 DO 5000 I=2,IX2
390            DO 5000 J=2,IY2
391     5000 ID(I,J)=((MAX-ID(I,J))*100-MAX*ICU)/100
392            IF(JD.EQ.2) GO TO 1028
393            MEX=((MAX-MIN)*100-MAX*ICU)/100
394            DO 140 I=1,IX
395            DO 140 J=1,IY
396       140 ID(I,J)=ID(I,J)*LIM/MEX+1
```

```
FOR131   EY100 SOURCE PROGRAM   DISTNC   SUBROUTINE

397            GO TO 71
398     1028 DO 3 J=1,IY
399            DO 2 I=1,IX
400            IF(ID(I,J)) 60,60,17
401        17 ID(I,J)=1
402            GO TO 2
403        60 ID(I,J)=0
404         2 CONTINUE
405         3 CONTINUE
406            DO 13 J=1,IY
407            DO 4 I=1,IX
408            IF(ID(I,J).EQ.0) GO TO 4
409            IF(I.EQ.1) GO TO 18
410            IF(J.EQ.1) GO TO 19
411            GO TO 41
412        18 IF(J.EQ.1) GO TO 29
413            ID(I,J)=ID(I,J-1)+1
414            GO TO 4
415        19 ID(I,J)=ID(I-1,J)+1
416            GO TO 4
417        29 ID(I,J)=1
418            GO TO 4
419        41 IF(ID(I,J-1)-ID(I-1,J)) 77,78,78
420        77 ID(I,J)=ID(I,J-1)+1
421            GO TO 4
422        78 ID(I,J)=ID(I-1,J)+1
423         4 CONTINUE
424        13 CONTINUE
425            JT=IY+1
426            MOK=ID(1,1)
427            MTE=ID(1,1)
428            DO 6 J=1,IY
429            IT=IX+1
430            JT=JT-1
431            DO 5 I=1,IX
432            IT=IT-1
433            IF(IT.EQ.IX) GO TO 43
434            IF(JT.EQ.IY) GO TO 42
435            IF(ID(IT+1,JT)-ID(IT,JT+1))
436           L88,89,89
437        88 IDT=ID(IT+1,JT)+1
438            GO TO 45
439        89 IDT=ID(IT,JT+1)+1
440            GO TO 45
441        42 IDT=ID(IT+1,JT)+1
442            GO TO 45
443        43 CONTINUE
444            IF(JT.EQ.IY) GO TO 44
445            IDT=ID(IT,JT+1)+1
446            GO TO 45
```

```
FOR131   EY100 SOURCE PROGRAM   DISTNC   SUBROUTINE

447        44 IDT=1
448        45 IF(ID(IT,JT).LE.IDT) GO TO 1000
449            ID(IT,JT)=IDT
450     1000 CONTINUE
451            IF(ID(IT,JT).LT.MOK) GO TO 1001
452            MOK=ID(IT,JT)
453     1001 CONTINUE
454            IF(ID(IT,JT).GT.MTE) GO TO 5
455            MTE=ID(IT,JT)
456         5 CONTINUE
457         6 CONTINUE
458        71 CONTINUE
459            IF(JD.EQ.2) GO TO 631
460            MOK=LIM
461            MTE=0
462       631 REWIND 14
463            DO 67 J=1,IY
464            DO 66 I=1,IX
465            IDN(I)=ID(I,J)-1
466            ID(I,J)=IDN(I)
467            IF(IDN(I)) 660,66,66
468       660 IDN(I)=0
469        66 CONTINUE
470        67 WRITE(14,101)(IDN(I),I=1,IX)
471            REWIND 14
472       101 FORMAT(60I5)
473            WRITE(6,121) ICU1,ICU2,ICU,MOK,MTE
474            IF(MOK.LT.10) GO TO 1010
475            DO 32 J=1,IY
476            DO 32 I=1,IX
477            DO 600 K=1,10
478            IF(ID(I,J).GE.(K*10)) GO TO 600
479            ID(I,J)=ID(I,J)-(K-1)*10
480            GO TO 32
481       600 CONTINUE
482        32 CONTINUE
483     1010 DO 35 J=1,IY
484            DO 34 I=1,IX
485            IF(ID(I,J).LT.0) GO TO 33
486            IF(ID(I,J).GE.10) GO TO 8
487            WRITE(20,102) ID(I,J)
488       102 FORMAT(I1)
489            READ(20,103) ID(I,J)
490       103 FORMAT(A1)
491            GO TO 34
492        33 ID(I,J)=IBLANK
493            GO TO 34
494         8 ID(I,J)=ISTRSK
495        34 CONTINUE
496        35 CONTINUE
```

```
FOR131   EY100 SOURCE PROGRAM  DISTNC   SUBROUTINE

497       38 INW=IX/130+1
498          IF(MOD(IX,130).EQ.0) INW=INW-1
499          DO 40 IW=1,INW
500          WRITE(6,104) IW
501          WRITE(6,199)
502          IT=130*(IW-1)+1
503          JT=130*IW
504          IF(IW.EQ.INW) JT=IX
505          DO 39 J=1,IY
506          DO 99 K=1,9
507          IF(J.LT.(K*100)) GO TO 98
508          IF(J.EQ.(K*100)) GO TO 97
509       99 CONTINUE
510       97 JJ=0
511          WRITE(6,165) JJ
512          WRITE(6,166) JJ,(ID(I,J),I=IT,JT)
513          GO TO 39
514       98 JJ=J-((K-1)*100)
515          WRITE(6,105) JJ,(ID(I,J),I=IT,JT)
516       39 CONTINUE
517      105 FORMAT(' ',I2,130A1)
518          WRITE(6,199)
519       40 CONTINUE
520      165 FORMAT(1H ,I1)
521      166 FORMAT(1H+,1X,I1,130A1)
522      120 FORMAT(' X=',I3,'    Y=',I3,
523         *'   BLACK=',I3,'    WHITE=',I3)
524      121 FORMAT('+',41X,'CUTLEV1=',I4,
525         *'   CUTLEV2=',I4,
526         *'   CUTLEV1/CUTLEV2=',I4,
527         *'   MAX=',I3,'   MIN=',I3)
528      104 FORMAT(' ',100X,'NO.',I2)
529      199 FORMAT('     ....''....''....''',
530         *'....''....''....''....''',
531         *'....''....''....''....''',
532         *'....''....''....''....''',
533         *'....''....''....''....''',
534         *'....''....''....''....''',
535         *'....''....''....''')
536          RETURN
537          END
```

```
FOR131   EY100 SOURCE PROGRAM

538          INTEGER FUNCTION IOKII*2(I,J,K)
539          IMPLICIT INTEGER*2(I-N)
540          IF(I.GE.J) GO TO 1
541          IF(J.GE.K) GO TO 2
542          IOKII=K
543          RETURN
544        1 CONTINUE
545          IF(I.GE.K) GO TO 3
546          IOKII=K
547          RETURN
548        2 IOKII=J
549          RETURN
550        3 IOKII=I
551          RETURN
552          END
```

```
FOR131   EY100 SOURCE PROGRAM

553          INTEGER FUNCTION IHISAI*2(I,J,K)
554          IMPLICIT INTEGER*2(I-N)
555          IF(I.LE.J) GO TO 1
556          IF(J.LE.K) GO TO 2
557          IHISAI=K
558          RETURN
559        1 CONTINUE
560          IF(I.LE.K) GO TO 3
561          IHISAI=K
562          RETURN
563        2 IHISAI=J
564          RETURN
565        3 IHISAI=I
566          RETURN
567          END
```

V.

Three-dimensional Observations in Electron Microscopy

Scanning Electron Microscopy in Histology and Cytology

Tsuneo FUJITA and Junichi TOKUNAGA

A burst of studies has been published on biomedical application of the scanning electron microscope (SEM) since this instrument became available in the middle of the 1960's. As it is almost impossible to give an overall review of the rapidly increasing literature in this field, the purpose of this paper will be to introduce new methods in specimen preparation for scanning electron microscopy and to review what has been revealed in the fields of tissue and cell research by using these methods. Further discussion will be presented on future fields and problems in cytological and biomedical applications of SEM.

On the current trend of studies on biomedical application of the SEM, references are given for the reviews by Hayes and Pease (1968), Boyde and Wood (1969), Tokunaga et al. (1969), Carr (1970), Fujita et al. (1971), Hollenberg and Erickson (1973) and Echlin (1971, 1973). A very comprehensive list of the literature in this field was prepared by Boyde et al. (1973).

Specimen Preparation for Scanning Electron Microscopy

General remarks

Since every specimen is placed in the vacuum chamber of the SEM, it must be dried beforehand. Generally the tissue and cells are fixed with an aldehyde and/or a heavy metal salt fixative in order to preserve their living state and to reduce shrinkage during drying. The shrinkage and deformation of specimens during drying procedures have been the most serious problem in SEM studies of biological materials and various methods have been devised which will be reviewed below.

Dried specimens generally are coated with metals in order that they obtain electrical conductivities and that ample secondary emissions can be taken from the metal coat. The metal coat, however, produces a "snow-covered landscape" of the specimen surface in which minute subcellular reliefs are poorly reproduced. Thus, specimens are becoming available with complete or partial omittance of this coating as will be mentioned later.

Although the object of most SEM observation is the surface of a material, they are not always restricted to the free surfaces of tissues and cells. Artificially produced surfaces such as by cutting, cracking, tearing and crushing are equally valid. Cracking of frozen or resin embedded materials produces valuable surfaces and, moreover, may allow visualization of the internal structures of the specimens. Ion-etching has been applied to expose certain elements within cells.

Replicas of different materials also are suited for SEM observation. They may afford visualization of the living surface state free of the effects of fixation and drying. More important are the casts of cavities and vessels, which are useful especially for study of the microcirculation.

A few routine procedures in specimen preparation now will be briefly reviewed, before some new and less widely known methodologies are treated in detail.

Routine Techniques for Ordinary Tissue Blocks

Fixation and washing

When the tissue surface, either natural or artificially produced, is covered by mucous or a protein-rich fluid, this should be eliminated as completely as possible before fixation. Some enzymes such as papain (plus EDTA, FERGUSON & HEAP 1970) and N acetyl L cysteine (BOYDE & WOOD 1969) have been applied for the elimination of mucous substances.* Mechanical treatment, such as a low frequency ultrasonic vibration (KUWABARA 1970) may be effective. Generally, however, careful and intensive washing with a jet of saline gives good results. This washing is usually performed before fixation but in many materials rewashing after the aldehyde fixation is effective.

Vascular perfusion with warmed saline followed by perfusion fixation is also a valuable method as not only the vascular lumen but also the surfaces of surrounding tissues are washed by the oozing saline (MIYOSHI et al. 1970; ANDREWS 1974).

Glutaraldehyde solutions of principally 2.5% in 0.1 M phosphate buffer, pH 7.3, are most widely used for fixation of tissues. We believe, however, that the simple formaldehyde fixative (5%) may have an advantage over glutaraldehyde in some specimens, e.g. intestinal surface, because of its higher penetration rate and poorer fixing ability on some of the mucous coating of tissue surfaces.

The aldehyde fixed tissue after being trimmed into an appropriate size and shape, generally is postfixed in OsO_4, or, as BOYDE and WOOD (1969) recommend, in PARDUCZ's OsO_4-$HgCl_2$ mixture. This procedure of postfixation hardens the tissue and reduces its shrinkage and deformation to some extent during drying on the one hand, and on the other hand gives the tissue a certain degree of conductivity. Thus a thinner coating may suffice for the postfixed specimen than for the simply aldehyde-fixed one.

Drying

Simple air drying, i.e., dehydration of materials directly from their water containing state, was tried in early studies but did not give good results because of the high surface tension of water which excessively suppressed surface fine structures. However, this method may still be useful for certain fields of study if the object is stiff enough to combat the pressure of surface tension and where promptness in specimen preparation is important.

Drying from a volatile medium has been adopted since 1968 by BARBER and BOYDE and by FUJITA et al. in order to reduce the surface tension of the liquid medium which damages the object when it dries. Volatile media such as ether, propylene oxide, acetone, ethanol and amyl acetate are available but acetone has been favored in many laboratories. Tissue blocks, spread preparations and culture specimens are dehydrated after fixation through ascending concentrations of acetone and a few changes of 100% acetone and simply dried in the air. Free cellular specimens are dehydrated and suspended in a small amount of 100% acetone and dropped by a pipette on small glass slides which later are mounted on

---

* More extensive elimination of tissue elements than just the removal of mucous substances may be attempted by the use of some enzymes. ZEEVI and LEWIS (1970) reported that they could remove the outer ganglionic sheath of the *Aplysia* ganglion with a mixture of pronase and elastase in order to expose the ganglion cells to be observed. Such procedure may perhaps better be included in the category of chemical etching.

the specimen stubs. The suspension may be dropped directly on the specimen stubs.

A number of valuable studies on the stereo fine structure of tissues and cells have been and are still being made by the air drying method through a volatile medium. Considerable artifacts have been noticed after this method which were mainly caused by severe shrinkage of the specimen and by suppression of delicate subcellular processes on the surface of the specimen. For instance cilia and the hair-like microvilli by and large do not stand up but fall flat on the cell surface. To avoid these artifacts freeze-drying and critical point drying methods were adopted which will now be described.

Freeze-drying was applied by BOYDE and BARBER (1969) to the SEM studies of soft and fragile objects. Using a protozoon *Spirostomum*, they demonstrated that this method could preserve the cilia in their life-like position.

Generally, specimens are previously fixed and rapidly frozen by immersion in liquid nitrogen or better in a quencher such as Freon 12 cooled by liquid nitrogen and finally dried by vacuum sublimation. Demerits of the freeze-drying method are its time consuming procedure and frequent damage to specimens with ice crystals. Measures to reduce these damages were discussed in detail by BOYDE and WOOD (1969).

Critical-point drying method is now most widely used in specimen preparation for SEM (BOYDE & WOOD 1969; HORRIDGE & TAMM 1969). The method itself was first applied by ANDERSON as early as 1951 for preserving blood and sperm cells in their natural form, which he observed under the transmission electron microscope without sectioning as was usual at that time.

Critical-point drying is a method used to avoid the surface tension at the liquid/gas boundary which inevitably passes through the specimen when it is dried and damages its structure. CO_2 is most convenient to use as the liquid medium because its critical point is 31.4 °C, 72.9 atm. The fixed specimens are dehydrated in ethanol and dipped in amyl acetate, which then is replaced by several changes in liquid CO_2 in a high-pressure chamber. The enclosed chamber then is warmed to about 35 °C.

Freon (COHEN et al. 1968) and N_2O (KOLLER & BERNHARD 1964) have been used in critical-point drying. Freon has the advantage of a lower critical pressure and temperature (Freon 13: 38.2 atm, 28.9 °C; Freon 116: 32.6 atm, 24.3 °C) but the disadvantage of being expensive. The results are identical with those obtained with CO_2. N_2O is mixed with water, though not very easily. The fact that the critical pressure of N_2O is as high as 69 atm (critical temperature: 36.5 °C) and that the gas is poisonous are disadvantages.

Although liquid CO_2 is usually used in the critical point drying, a modification using dry ice (solid CO_2) was proposed by TANAKA and KAWAKAMI (1973). After the specimens previously dipped in amyl acetate are placed into the chamber of the critical-point drying apparatus, the chamber is filled with a cylinder of dry ice and tightly closed. The chamber is then kept at 10–20 °C for 10 min and the dry ice converts to liquid CO_2 which gently soaks the specimens. Following procedures are the same as in the ordinary method.

The fact that the cytoplasmic structures can be beautifully preserved after critical-point drying (CO_2 or Freon 13) was demonstrated by MELLER et al. (1973) who, after embedding (Epon-Araldite) and sectioning the specimens prepared for SEM, observed them under the transmission electron microscope.

Coating

Although some relatively hard tissues can be observed at a low accelerating voltage (1–5 KV) in the SEM, biological specimens generally are evaporated with a metal coating (Ag, Al, Au, C, Cu, Pd, Pt or their alloys), which insures the conductivity of the specimen surface and gives a rich secondary emission. The specimens are rotated and tilted in two mutually perpendicular directions while being evaporated in order to attain a uniform and continuous coverage (BARBER & BOYDE 1968).

The elevated resolution in the field emission type SEM demands metal coatings of fine texture and even distribution. Gold which has been widely used because of the ease of handling is now recognized as being unsuitable for high resolution observation since it forms coarse aggregations on specimen surface. Gold-palladium (60: 40) is recognized as forming a much smoother coating layer than gold (Fig. 1). The use of platinum-palladium (80: 20) also appears recommendable (NAGATANI & SAITO 1974).

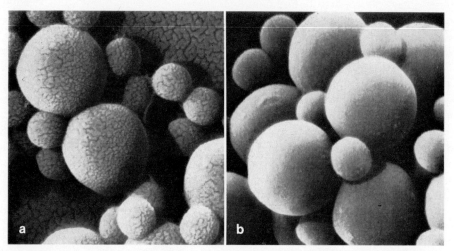

Fig. 1 Latex particles coated with 100 A of gold (a) and with 100 A of gold-palladium (6:4) (b). (HFS-2, ×5,600) Courtesy of Dr. T. NAGATANI and Dr. M. SAITO (Hitachi).

Special and New Methodologies

Treatment of free cells

Free floating cells in the body fluid are treated also by aldehyde fixatives. Our studies, using red blood cells which are especially sensitive to osmotic pressure, indicated that 1% glutaraldehyde (0.1 M phosphate buffer) preserves the shape of the cells closest to their natural form in the circulation. A 0.5% glutaraldehyde solution in the above buffer which was recommended by SALSBURY and CLARKE (1967) has an osmotic pressure closer to that of the plasma than the 1% solution but causes rather swollen forms of red blood cells for unknown reasons (HATTORI et al. 1969).

It is important to obtain "naked" cells free of proteinaceous substances covering them for preparations of blood cells, peritoneal free cells and other floating cells. This is especially necessary with the spermatozoa embedded in viscous sperm plasma. The simplest method is to drop a small amount (a few drops) of blood or semen into a large amount (20 ml or more) of aldehyde fixative while shaking. The blood and semen are thus fixed and simultaneously extensively diffused. Substances of the blood and sperm plasma are eliminated by discarding the supernatant after gentle centrifugation. They would otherwise cover the cells after drying. The method was first adopted by us for blood cells (TOKUNAGA et al. 1969; HATTORI et al. 1969) and then for spermatozoa (FUJITA et al. 1970). A similar method was later adopted by LUSE (1970). Subcellular fractions made by routine methods of biochemistry can be treated with the same technique as free cells (KURAHASHI et al. 1969).

Specimens are carefully rinsed and processed for drying after fixation. Simple air drying on a piece of glass slide is applicable only in restricted specimens and other such purposes as the clinical examination of red blood cells (Hattori 1972). Acetone drying of free cells is still useful, if one disregards the insufficient preservation of the surface fine structures of the cells.

Freeze-drying or critical-point drying is inevitable if one intends a visualization of cells as they appear in nature. In the former method, a droplet containing the cells is frozen and dried. In critical-point drying, cell suspensions are dehydrated by repeated centrifugations which are tedious. Finally, the cells suspended in amyl acetate are dropped on pieces of glass slides and placed in the critical-point drying chamber while they are still wet. Though this method gives beautiful images of cells, many cells are lost by floating away with the stream of liquid CO_2.

Centrifugation of a cell suspension either before (after washing) or after fixation onto a piece of glass slide placed in the bottom of a centrifuge tube results in the attachment of well dispersed cells on the glass slide. The cell carrying slide thus made can then be dehydrated and critical-point dried quickly and safely. Because of the suppressing force during centrifugation the cells become attached to the glass with an area rather than a point as produced by the former method; thus they give much less of a charging effect in the SEM.

Treatment of small fragile bodies

Very small organisms (e.g. certain parasites and protozoa) and isolated tissue elements (e.g. ovum, glomerulus and giant neuron) are difficult to treat in the drying procedures and especially to mount on the metal stub. Some workers used small nylon mesh bags to hold tiny specimens during the processes of fixation, dehydration and drying (Horridge & Tamm 1969). Marchant (1972) used millipore filters which did not dissolve in dehydrating solvents. A variety of algal cells and protozoa were collected on the filters by gentle suction after which they were further treated and brought to the SEM in their attached state on the filter.

We recently discovered a good method using the usual copper grid for transmission electron microscopy as the specimen holder. We will introduce by way of an example this technique for the preparation of a specimen of a single renal glomerulus for SEM.

A kidney which has been previously fixed in glutaraldehyde is minced into small cubes (1 mm in size) and a few glomeruli are gently squeezed out into a buffer solution under a stereoscopic magnifier by using two sharp forceps. A single glomerulus or, if disired, a larger number of glomeruli thus isolated are sucked into a glass capillary and blown out on the backside of a copper grid (180 mesh for human glomeruli and 300–400 mesh for rat) which has previously been shadowed with carbon and laid in a Petri dish. From the margin of the dish 50% ethanol is gently introduced with a pipette and this is successively replaced by a series of ascending concentrations of ethanol. During this procedure the glomeruli become attached to the carbon covered grid so that they will not become detached in later procedures. The absolute alcohol is then replaced by amyl acetate and the grid is placed into the critical-point drying chamber. The dried grid is covered with metals by the usual method and mounted on a metal stub to be observed in the SEM (Fig. 2).

Cracking methods

Simple tearing or cracking of fixed tissues either before or after drying is useful for exposing the surfaces of cells as well as some intercellular structures which have been covered by other elements. Very beautiful images of cellular construction of the retina (frog), for instance, have been produced by cracking the tissue after critical-point drying (Steinberg

Fig. 2 Single renal glomerulus of the rat on a copper grid. The vascular pole is seen on the upper left border. Critical point dried and coated with carbon and gold. (JSM-2, ×1,200)

1973). It is difficult to obtain an intracellular image by this method since the cleavage occurs usually along the intercellular phase.

Using a Smith-Farquhar tissue sectioner, MAKITA and SANDBORN (1971) demonstrated granules in the glandular cells of the hen oviduct. Freeze-fracturing has been applied by some researchers (HAGGIS 1970) to a study of the retina which was frozen, cleaved and then freeze-dried.

Better results were obtained with a "frozen resin cracking method" introduced by TANAKA (1972). The fixed tissues were dehydrated with ethanol and propylene oxide and then placed in gelatine capsules filled with Cemedine 1500 (epoxy resin, Cemedine Co., Tokyo)* without adding any catalysts. The resin in the capsules was hardened for 1–2 hrs at −30 °C and finally cracked with a chisel and hammer. The broken surface glittered and was smooth. The resin was then removed from the tissue with propylene oxide and the tissue transferred to isoamyl acetate and dried by the critical-point method. On the cracked surfaces produced by this method TANAKA and his associates succeeded in

---

* Araldite GY260, GY255 or GY250 is also useful.

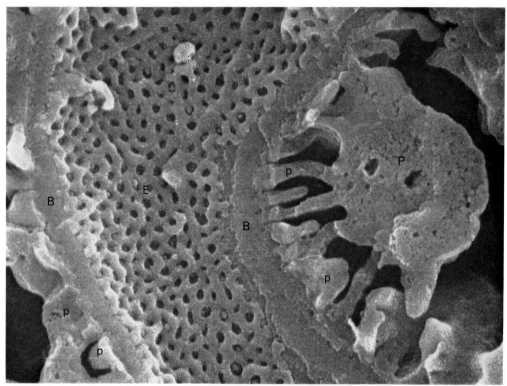

Fig. 3 Part of a rat renal glomerulus fractured in frozen amyl acetate.
A podocyte process (P) with its end feet(p), a basement membrane of homegeneous appearance (B) and an endothelial sheet (E) with pores are seen forming a blood-urine barrier. Critical-point dried and coated with carbon and gold-palladium. (HFS-2, ×23,000) (From Tokunaga J. et al. (1974) Arch. histol. jap., *37*: 165–182)

demonstrating various intracellular organelles including mitochondria, secretory granules and the endoplasmic reticulum (Tanaka 1972). This cracking method also gave good results in studies of the inner ear (Tanaka et al. 1973), thyroid gland (Kobayashi 1973) and cartilage (Tanaka 1973).

A "frozen liquid cracking method" was also proposed by Tanaka's group (Hamano et al. 1973). Tissues in ethyl alcohol were frozen and cracked to subsequently be processed by amyl acetate-CO_2 critical point drying. The authors reported that they could demonstrate intracellular structures such as the mitochondrial cristae and the endoplasmic reticulum with this method. Recently it was recognized by us (Tokunaga et al. 1974) that specimens immersed in isoamyl acetate, as well as in ethanol, could be frozen by being dipped in liquid nitrogen or in Freon 22 cooled by liquid nitrogen, and then cracked (Fig. 3). Tissue blocks immersed in DMSO (dimethyl sulfoxide) were also found to be frozen and cracked in the same quencher and to produce fracture planes occurring along the cell surfaces and intracellular membrane systems. These fractured specimens were processed by CO_2 critical-point drying.

Frozen specimens (cryo-scanning electron microscopy)

Recently attempts were made to observe frozen specimens directly in the SEM. Frozen biological materials have good conductivity and give sufficient secondary emissions for

Fig. 4 Conidial head of *Aspergillus nidulans* as revealed by cryo-SEM (no fixation and no coating). (JSM-2 and SUM3-CRU, ×900) Note the long rows of well-kept conidia.

SEM. Materials having wet surfaces are not suited for this purpose, although a certain amount of ice covering the specimen may be removed by sublimation within the microscope column.

ECHLIN et al. (1970) developed a versatile temperature-controlled stage for SEM. They quench-froze biological materials in either liquid nitrogen or Freon 22 cooled with liquid nitrogen and then transferred them to a cold stage held at −185 °C. Transfer devices (ECHLIN & MORETON 1973) prevented frost formation. ECHLIN and his associates (1970) using primarily plant leaves, demonstrated that the specimens thus frozen could be examined in the SEM without a metal coating on the specimen surface. ECHLIN and MORETON (1973) further reported a method of shadow evaporating carbon and metal on the surface of frozen specimens. They stressed with great emphasis that care should be taken not to allow the specimen to rise above −130 °C which was the recrystallization point of pure ice.

NEI and his associates (1971, 1972) also observed frozen (liquid nitrogen) animal and plant tissues (e.g. hamster tongue and iris petals) on a cold stage within the SEM. They cracked frozen specimens in the pre-evacuation chamber and observed the resultant surface at relatively low magnification.

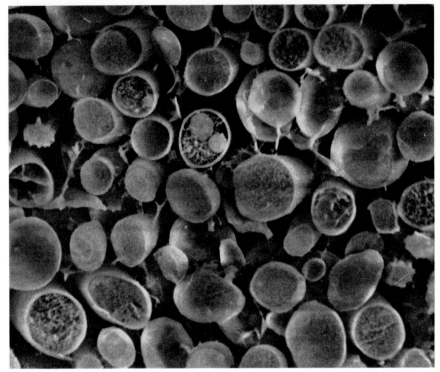

Fig. 5 *Candida albicans* colony freeze-fractured in a cryo-SEM (no fixation and no coating). (JSM-2 and SUM3-CRU, ×3,500)

Using the originally devised cold stage and transfer apparatus (JEOL, SUM3-CRU), we observed frozen (Freon 22) *Aspergillus* and some other fungi. The process of conidium formation in *Aspergillus* was demonstrated with excellent preservation of structure and without the separation of the beaded conidia which could not be avoided in other specimen preparation methods (Fig. 4). By fracturing the fungi we could reveal some of their intracellular organelles (Fig. 5). To stress the value of this procedure, the *Aspergillus* once frozen and electron bombarded in the SEM remained *viable* as demonstrated by reculturing (TOKUNAGA & FUJITA 1973).

Skin is a suitable subject for study with cryo-SEM. Human skin from various portions of the body taken from cadavers fixed by formalin perfusion through the femoral artery was excised and frozen to be observed in the SEM without coating. In this procedure we were able to minimize the shrinkage of tissue which was unavoidable in the other methods used for processing skin. The epidermal cells looked rather "wet" instead of the dried and lifted-up appearance after other techniques of specimen preparation (Fig. 6; TOKUNAGA & FUJITA 1973).

Avoidance of metal coating

The metal coating which insures an ample secondary emission and conductivity of the specimen, disturbs the visualization of the fine surface structures. This problem is serious now that the field-emission SEM has been able to produce observations of much higher resolution than was previously possible. The use of an antistatic spray may be effective (SYLVESTER-BRADLEY 1971) but its utility is limited.

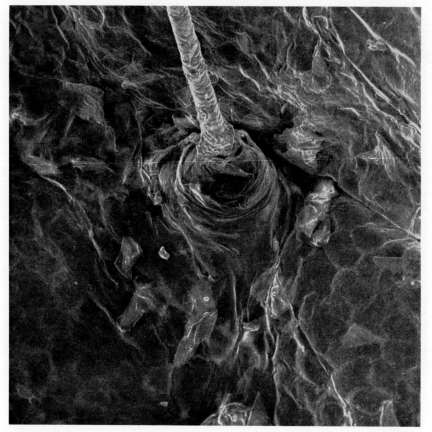

Fig. 6 Human skin with a hair shown by cryo-SEM. Note the hexagonal epidermal cells and the smooth area (left bottom corner) representing the Haarscheibe of Pinkus. Formalin fixed piece of skin (scapular region) was frozen and observed without any coating. (JSM-2 and SUM3-CRU, ×360)

GOLDMAN and LEIF (1973) proposed the designation, *wet chemical methods*, for the treatment of specimens in chemical solutions for the sake of conductivity and secondary emissions. The attempt of PANESSA and GENNARO (1972) in dipping a specimen in solutions of metallic salts such as lead acetate and glycerin belongs in this category of methods. GOLDMAN and LEIF (1973) themselves reported that free cells laid on a conductive polyethylene plate obtained sufficient conductivity and gave good surface images in the SEM.

MURAKAMI (1973) proposed a tannic acid-osmium method, which was recently further revised (MURAKAMI 1974). The revised method, tentatively referred to as the sucrose-tannic acid-osmium method, is as follows: Tissues fixed in glutaraldehyde are suspended for 16 hrs in an aqueous solution composed of 2% glycine, 2% sodium glutamate and 2% sucrose (pH 6.2), rinsed for 30 min then immersed for 24 hrs in 2% tannic acid in distilled water (pH 4.0). The specimens are then washed in distilled water for 15 hrs and treated with 2% OsO_4 for 18 hrs. After being washed in distilled water, the specimens are processed by the drying procedures. MURAKAMI believes that ample osmium is combined with the specimen proteins across the interbridges formed by the amino acids, sugar and tannic acid. In any event sufficient conductivity of the specimens is secured by this wet chemical method entirely without coating, and one can obtain clear images of high magnifications with an accelerating voltage as high as 20 KV (Fig. 7).

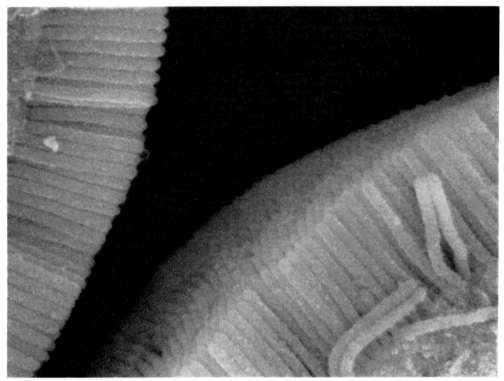

Fig. 7 Microvilli of rat duodenal epithelium. Glutaraldehyde fixed tissue was treated with the sucrose-tannic acid-osmium method and viewed without any coating. (HFS-2, ×25,000)

With the technical progress along this line, one may eventually reveal unknown surface structures as minute as those approximating the resolution of the field-emission type SEM on the one hand, and on the other hand, one may be able to microdissect specimens in the SEM exposing their deeper portions without adding a metal coat to them.

Quite recently KELLEY and his associates (1973) utilized the osmium-binding properties of thiocarbohydrazide to bind additional osmium on the surfaces of animal tissues. Glutaraldehyde-fixed material is post-fixed in 1% OsO_4 (3–4 hrs) and thoroughly rinsed. It is then dipped into a saturated aqueous solution (about 1%, w/w) of thiocarbohydrazide for 10 min or more, and after being rinsed, returned to OsO_4 to be kept there for 30–60 min. The treatments in thiocarbohydrazide and OsO_4 may be repeated.

The relatively low power (up to ×900) scanning electron micrographs shown by KELLEY et al. (1973) give evidence of the utility of this simple method. Conductivity of the specimens is sufficient and their surface views are beautiful. However, since this method produces a surface coat of osmium black of considerable thickness (40–200 Å), this method apparently should be excluded from the attempts to eliminate metal coatings. However, the possibility that the layer of osmium black does not represent an artificially added coat may not be excluded but that it may actually correspond to a biological layer such as the sugar coating of cells.

Ion-etching

Freeze-etching, chemical-etching (cf. PANESSA and GENNARO, 1973) and ion-etching are the main methods which are now available in SEM for the purpose of eliminating some

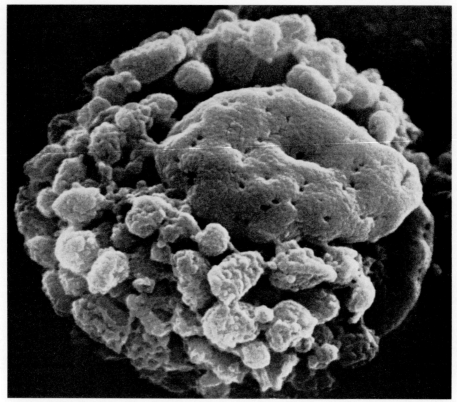

Fig. 8 Leukocyte (probable eosinophil) ion-etched for 3 hrs using a low vacuum glow discharge (see text). A lobulated nucleus with pores are exposed together with cytoplasmic granules. Coating with gold-palladium. (HFS-2, ×20,000) (From T. Fujita, T. Nagatani and A. Hattori (1974) Arch. histol. jap., *36*: 195)

elements in the material surface and "digging out" certain interior structures resistant to etching. Among these methods ion-etching is being most actively studied and represents the most promising method.

Ion-etching is achieved by the ejection of atoms from the material surface bombarded by ions. The rate of atom ejection depends: (1) upon the mass and velocity of the ions, (2) upon the mass of the atoms to be ejected, and (3) upon the energy combining the atoms to the material surface.

The significance of the ion etching for biological materials had been disputed. Lewis et al. (1968, 1970) etched red blood cells with ions of argon, hydrogen, oxygen or aqueous vapor in a radio-frequency ion source (13.5 mHz) with a vacuum of 10^{-3}–10^{-6} Torr, and demonstrated radially arranged ridges which they declared to be an intrinsic, hitherto unknown structure of the erythrocyte. On the other hand, Hodges and his associates (1972) were able to find only identical images of cony projections and pores in different kinds of cultured cells by the same method, and concluded that ion-etching was of no use for biological studies.

In a series of model experiments using biological and nonbiological materials, Fulker and his associates (1973) demonstrated that after ion bombardment in a uniform direction, low angle cones were produced on the material surface which could be avoided by rotating and tilting the specimen during the ion bombardment. In the epithelium of the urinary bladder those authors demonstrated certain internal structures by this method such as the apparent tonofibrils.

An attempt was made by our group to avoid the unidirectional cone formation by using ions of different energies and directions. In a simple glow discharge apparatus, an alternating current discharge (50 Hz) was made between two aluminum plates under very low vacuum (0.05–0.5 Torr). The electrode voltage was about 700 V and the current about 2 mA. Ions mainly of nitrogen and oxygen thus produced were applied for 1–2 hrs to the specimens placed on one of the electrode plates. By this method we were able to eliminate the unidirectional effect of etching without tilting the specimens and thus could "dig out" some intrinsic structures of the cells.

In human red cells etched by our method we found porous and reticular structure which conspicuously differed from the etched images of erythrocytes shown by LEWIS et al. (1968, 1970). In human white cells we exposed some cytoplasmic granules and the nucleus which showed small pits presumably corresponding to the chromatin lacking portion under the nuclear pores (Fig. 8). In human spermatozoa the acrosome, postnuclear cap and mitochondria were exposed and the usefulness of ion-etching for morphological analysis of sterile spermatozoa was suggested.

Replica

Single or double stage replicas taken from material surfaces to be examined are available for SEM, provided they are electrically conductive or given conductivity with a metal coating.

CHAPMAN (1967) reported that polystyrene replicas of leaf surfaces could be examined by SEM. BERNSTEIN and JONES (1969) demonstrated beauitful scanning micrographs of double stage replicas of the skin. Negative replicas (molds) were made of silicone rubber and the positives (casts) with polyethlene.

Independently we found a much more simple method using celluloid impressions in a study of some biological objects, in particular, the skin surface (FUJITA & TOKUNAGA 1969). A celluloid plate was attached with a drop of amyl acetate and immediately pushed against a point on the skin to be examined. After the celluloid was hardened, which took a few minutes, the plate was removed and a beautiful impression of the skin was visible. However, one must add to be precise that this specimen and that of BERNSTEIN and JONES (1969) were not genuine replicas but were covered to a large extent by desquamated epidermal cells and one could only observe partially genuine replicas under the SEM, but for the most part one saw backside of the most superficial epidermal cells. The scanning images of the celluloid impressions were valuable especially because one could examine given portions of the skin in living subjects quickly and easily and also because one could avoid the unpleasant shrinkage of tissue which occurred with excised skin.

Replica methods were adopted by several other researchers with different impression materials and for different objectives (CLARKE 1971; WATTERS & BUCK 1971; PAMEIJER et al. 1972; BARNES 1972).

Casts

While some attempts had been made to prepare casts of hollowed biological structures such as dentinal canaliculi with polyester resin (MIKAMI et al. 1969) and bronchial ramifications with latex (NOWELL et al. 1970; TYLER et al. 1970) for SEM, MURAKAMI extended a systematic study on the preparation of vascular casts using methyl methacrylate possessing a viscosity considerably less than that in the materials previously used for the macro and microscopic studies of blood vessels.

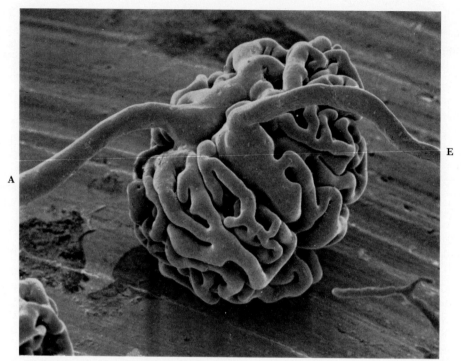

Fig. 9 Methacrylate vascular cast of a rat renal glomerulus. A: vas afferens, E: vas efferens. Courtesy of Dr. T. Murakami. Coating with gold. (JSM-2, ×450)

A method simplified from the one originally reported by Murakami (1971) is as follows (Murakami 1973): Methyl methacrylate ester monomer (100 ml) to which is added 1–1.5 ml 2, 4-dichlorobenzoyl peroxide (catalyst) is warmed to 60–65 °C. As polymerization is initiated it will spontaneously reach 85–95 °C in a few minutes. This half-polymerized resin is then rapidly cooled to a temperature lower than 30 °C. One and a half to 2 g benzoyl peroxide is then added and after it is gently stirred for 5–10 min it should have a suitable viscosity (slightly less viscous than glycerin) at which time 1.5 ml dimethyl aniline (accelerator) is added to the mixture. After the animal or desired organ is arterially perfused with Ringer solutions of gradually ascending temperatures (37–50 °C), the mixture is injected with a moderate amount of pressure. The animal or organ is immersed in a hot water bath (60–70 °C) and placed in an oven of the same temperature for 24 hrs in order to polymerize the resin. The animal tissue is then macerated with 20% NaOH and the resin cast is washed thoroughly and coated with metals. Under the SEM, one may see beautiful arborizations of the blood vessels up to the capillary network (Fig. 9). By alternate scanning and microdissection of the cast one may examine the casts successively from the superficial to the deeper portions (Murakami 1972; Fujita & Murakami 1973).

Recently Murakami et al. (1973) reported that treatment of methacrylate casts with osmium gas gave excellent conductivity and beautiful images of the casts with no charging effects and without any metal coating. Microdissection of the vascular casts for successive scanning could be greatly facilitated by this method. By using the Murakami's vascular casting method, valuable findings of the microcirculation of different organs have been obtained which will be introduced below. Methacrylate casting of blood vessels apparently similar to the method introduced by Murakami was independently adopted by

SCHÄFER and his associates (1973) for their studies on the arteriovenous anastomoses in the carotid body.

Prospects in Various Fields of Study

The SEM has revealed and is revealing three-dimensional structures of almost every part of the body and our understanding of histology and cytology has become much firmer than previously attainable. The objects of study cover the tissues and cells of man and laboratory animals and extend to different orders of lower animals. In this section only a few fields of study will be selected in which our knowledge is markedly advancing especially with SEM or where great progress in our study seems promising by the application of this method.

Embryology and developmental anatomy

Embryology is a field which requires three-dimensional visualization especially of structures and SEM is expected to be applied very effectively. Previous difficulties in drying the water-rich embryos have been removed through the use of critical point drying. According to WATERMAN (1972) acetone drying generally causes as much as 55% shrinkage of the original dimensions of the raw embryos, whereas critical point drying produces only 20%.

Attempts have been made to examine the surface fine structures of embryos in their earliest stages of development, such as the amphibian and the mammalian blastocyste and neurula (TARIN 1971; BERGSTRÖM 1972).

An important contribution with the SEM is being made in visualizing the development of organs and their relationships in the later stages of embryonic development. ARMSTRONG (1971) proposed a method of embedding glutaraldehyde fixed embryos in paraffine, cutting them with a microtome down to a desired level under exact orientation, and, after removal of the paraffine with warmed toluene, drying by the critical-point method. One can visulize a wide-open field for this kind of technique, if one only glances at the beautiful "inside view" of a chick embryo showing the structure and topography of different organs under development (ARMSTRONG & PARENTI 1973). It is expected that large portions of the illustrative materials in textbooks of embryology will be replaced by scanning electron micrographs in the near future.

SEM analysis of the surface fine structure of cells which are concerned in the movement, fusion and fission of tissues will also be a promising field for study. The work of WATERMAN, ROSS and MELLER (1973) on the mouse palate shelves before and during palatal fusion made it clear that the epithelial cells which were to be fused underwent specific cytological changes shortly before fusion. WATERMAN and MELLER (1973) beautifully and unequivocally demonstrated the process of nasal pit formation in the hamster.

Hematology

Erythrocyte morphology and its pathological changes have been studied very extensively with SEM such that there already are numerous papers available in this field (SALSBURY & CLARKE 1967; TOKUNAGA et al. 1969; BULL & KUHN 1970; KAYDEN & BESSIS 1970; OSS & MOHN 1970; BESSIS 1972; HATTORI 1972; HATTORI et al. 1972; CARTEAUD et al. 1973).

Leukocytes have been observed less systematically. HATTORI et al. (1972) succeeded in identifying the main cell types under the SEM by directly comparing them with their stained light microscopic views. Although HATTORI et al. (1973, unpublished) also have demonstrated the SEM images of leukocytes moving on a glass surface with the characteristic pseudopod, the surface morphology of the rounded-up leukocytes first needs to be established.

Platelets provide the most suitable objects for study with SEM in hematology. Numerous papers are being published on the processes of platelet adhesion, aggregation and fibrin formation both *in vivo* (SHOOP et al. 1970) and *in vitro* (CLARKE et al. 1969; HATTORI et al. 1969; SCARBOROUGH et al. 1969; LARRIMER et al. 1970; BARNHART et al. 1972; HATTORI, 1972; HATTORI et al. 1974). Forms of the platelets and their processes will be investigated as interesting models of the structural cellular response to various stimuli.

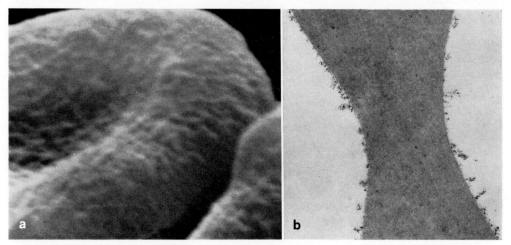

Fig. 10 Type A erythrocyte labeled with clusters of ferritin-anti-A antibody. a. SEM view after gold coating. (SSM-2, ×18,000) b. Section of the same specimen. (×18,000) The pictures were produced by collaboration with Y. MATSUKURA and T. INAGE.

Immunology

Since it is known that many kinds of antigens are located on the cell surfaces, they thus are potential objects for SEM investigation. LOBUGLIO et al. (1972) used latex particles of 0.23 μm diameter as immunologic markers. They coated the latex particles with certain antibodies and, after appropriate incubation with erythrocytes and lymphocytes, demonstrated that cells with the reactive antigens were densely covered with the particles.

Stimulated by this work we were able to confirm the usefulness of this method using latex particles coated with anti-bloodtype antibodies. If several markers of different sizes and/or shapes would become available, we could simulatanously demonstrate the sites of different antigens on cells with the SEM. We tried to observe erythrocytes that had been reacted with ferritin-bound antibodies (TOKUNAGA & FUJITA 1973). Individual ferritin particles were too small for the resolution of the SEM. Fortunately they tended to be aggregated in clusters which could be seen as wart-like elevations on the cell surface (Fig. 10). We are now re-examining the usefulness of this methodology by using a field emission SEM with an elevated resolution.

Another direction of immunological studies with SEM is the morphological identification of cells having some relation to immunological reactions and the observation of their reactions to other cells. It had been proposed by LIN et al. (1973a), who observed lymphocytes

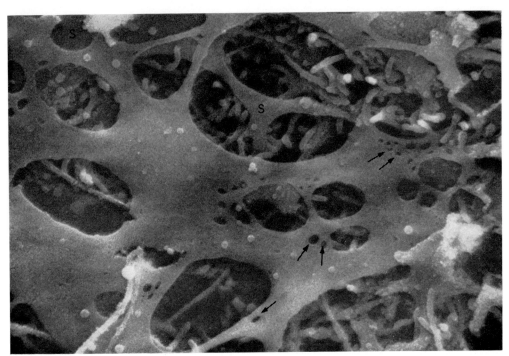

Fig. 11 Surface view of the rat liver sinusoid. The sheath of endothelial cytoplasm is perforated showing tiny fensters (arrows) besides very large ones. Beneath this sheath another cytoplasmic sheath (S) may be seen. Finger-like processes partly projecting through the endothelial perforations are the microvilli of hepatic cell. This picture was produced by collaboration with M. Muto and K. Adachi. Sucrose-tannic acid-osmium method and critical drying. Very thinly coated with gold-palladium. (HFS-2, ×15,000)

from rat thymus and spleen, that the T and B cells were different from each other in their surface structures. In their later SEM studies on human lymphocytes, Lin et al. (1973b) identified T cells by their ability to form rosettes with sheep red cells, and B cells to form rosettes with complement-coated human red cells. On the basis of this cell identification, it was found under the SEM that the B cells were generally larger in size and covered more densely with longer microvilli than the T cells. It was further shown that the microvilli were an important mode of primary contact between lymphocytes and the outside world.

A number of papers are being published on the identification, distribution and immunopathological significance of the B and T cells using SEM. Some authors remark, however, that the morphological criteria by Lin and his associates may be applied only to a majority of lymphocytes and a considerable part of the cells show atypical or even reverse feature of cell processes (Polliack et al. 1974).

Study of the reticulo-endothelial system (RES)

In connection with immunological reactions of the body, the RES is being regarded with greater importance. Nevertheless, the fine structure of the tissues and cells of this system has only been poorly understood since the transmission electron microscope studies of thin sections were unsuited for elucidation of such complicated and delicate structures in their three-dimensional extensions.

If one could only succeed in the complete elimination of blood (in the spleen and bone marrow) or lymph (in lymph nodes) by vascular perfusion, one might obtain scanning images of the RES tissues which are so clearcut and understandable with regard to the identification of certain cellular elements and their relationship to each other that many of the long disputed problems might be simply resolved.

The structure of the sinus wall of the spleen has been unequivocally elucidated especially with regards to the constantly present openings between the rod cells. It is indicated that the early light microscopists understood the sinus structure more precisely than the modern transmission electron microscopists (MIYOSHI et al. 1970; MIYOSHI & FUJITA 1971; FUJITA 1974).

What seems important for RES studies is that the SEM has revealed the independent nature of the macrophages as distinguished from the reticular cells. In the perfused spleen and lymph nodes the former are represented by round cells covered with tentacle-like processes, whereas the latter by spider-shaped smooth surfaced cells forming the reticular skelton of the organ together with the reticulum fibers, and there are recognized no transitional forms between both cell types (MIYOSHI et al. 1970; MIYOSHI & FUJITA 1971; FUJITA & MIYOSHI 1972; FUJITA 1974).

One of the most exciting tasks for SEM is to explore the visualization of various kinds of cells such as macrophages, T and B type lymphocytes *(vide supra)*, plasma cells, moving and transforming according to their immunological and other reactions within a jungle of reticular cells.

The fine structure of the bone marrow largely remains unknown, and SEM is also expected to settle many problems here, such as the true structure of the wall of the sinus and the process of penetration of differentiated blood cells.

Circulation research

The endothelial surface of the blood vessels represents a suitable object for SEM study and numerous papers have been published on its normal and pathological structures. Only quite recently, however, the application of critical point drying has enabled the demonstration of long and delicate processes of the endothelial cells in certain parts of the mammalian blood vessels (SMITH et al. 1971; TOKUNAGA et al. 1973; EDANAGA 1974). The occurrence of these hair-like projections, whose significance is completely unknown, should now be examined in various vessels of different animals and man.

SEM can easily demonstrate the endothelial pores in some organs (FUJITA et al. 1970). The overviews of the much disputed occurrence and distribution of endothelial fensters and pores in the hepatic sinusoids (Fig. 11) are now being obtained (ITOSHIMA et al. 1974).

Anatomical studies of the microcirculation are now being revolutionized by the application of the vascular casting technique to SEM originated by MURAKAMI *(vide supra)*. The lobulation and reticular connections of glomerular capillaries as well as the absence of an inter-arteriolar anastomosis were evidenced (rat, MURAKAMI 1972). The vascular distribution of the pancreas, especially the occurrence of a single vas afferens of the islet and numerous vasa efferentia ("insulo-acinar portal vessels") was demonstrated (monkey, FUJITA & MURAKAMI 1973). Findings supporting the "closed theory" of the spleen, i.e. a direct connection of the arteriolar termination with the splenic sinus, were obtained in the rat (MURAKAMI et al. 1974) though the studies in the human spleen support the "open theory" (FUJITA 1974). Structure and connections of the bone marrow sinuses are now being elucidated (IRINO et al. 1974).

SEM studies of vascular casts seem to be one of the most promising fields of microscopic anatomy and, in the very near future, will definitely influence the progress of the study of physiology and pathophysiology of the circulatory system.

Nephrology

Our knowledge of the structure of the glomerulus has become much more accurate than was previously known since the SEM came to be applied in this field. Buss and Krönert (1969) were the first to publish SEM observations of the glomerular surface (rat). Although they demonstrated the fern-leaf pattern of the podocyte foot processes, they proposed an incorrect conclusion that the foot processes of one and the same cell might be interdigitated with each other, presumably because of a thick metal coating and inadequate drying.

In a series of studies in different animals we proposed a rule that the foot processes of different cells were always alternately juxtaposed on the capillary wall, and that the processes from one and the same cell never became interdigitated (Fujita et al. 1970). This rule was fully supported by a recent paper by Andrews and Porter (1974). We further evidenced in a study on the differentiation of podocytes (rat, Miyoshi et al. 1971) that, from the very beginning of the process formation, different cells interdigitate with each other and the concept of Buss (1970) and some previous authors that the foot processes were first produced within a single cell as the basal incisures developed was no longer tenable.

By these and other studies the arborization, extension and mutual relation of the renal podocytes that had been visualized on the basis of two-dimensional transmission electron micrographs have been largely revised. Furthermore, the morphological changes, especially the process of disappearance of the podocyte interdigitations in animal experimental and human pathological conditions, are being elucidated through the use of SEM (Arakawa 1970, 1971). Now that we are able to bring, with ease and sureness, a very small number of glomeruli obtained through biopsy into the SEM *(vide supra)*, the diagnostic utility of this microscope in clinical nephrology will be increased.

The SEM study of the urinary tubules, on the other hand, has been enriched by a beautiful paper recently published by Andrews and Porter (1974).

Other fields of study

One may count other numerous fields, such as tissue culture, otology, reproductive science and parasitology, to which SEM has markedly contributed, but it seems sufficient to give concrete instances of what has been revealed and what is now expected in the biomedical fields with the application of SEM.

Although there are various methods using other informational patterns from an electron-beamed specimen, e.g. cathode-luminescence, X-ray and transmission electrons (high resolution scanning transmission electron microscopy by Crewe 1971a, b), the present paper is restricted to the methods and studies using the secondary emission of the non-transmission type, which have been the main interests in SEM studies of the biomedical fields.

REFERENCES

[1] Andrews P. M. (1974) A scanning electron microscopic study of the extrapulmonary respiratory tract. Amer. J. Anat., *139:* 399.

[2] Andrews P. M. and Porter K. R. (1974) A scanning electron microscopic study of the nephron. Amer. J. Anat., *140:* 81.

[3] Arakawa M. (1970) A scanning electron microscopy of the glomerulus of normal and nephrotic rats. Lab. Invest., *23:* 489.

[4] Arakawa M. (1971) A scanning electron microscope study of the human glomerulus. Amer. J. Pathol., *64:* 457.

[5] Armstrong P. B. (1971) A scanning electron microscope technique for study of the internal microanatomy of embryos. Microscope, *19:* 281.

[6] ARMSTRONG P. B. and PARENTI D. (1973) Scanning electron microscopy of the chick embryo. Dev. Biol., *33:* 457.

[7] BARBER V. C. and BOYDE A. (1968) Scanning electron microscopic studies of cilia. Z. Zellforsch., *84:* 269.

[8] BARNES I. E. (1972) Replica models for the scanning electron microscope. A new impression technique. Br. dent. J., *133:* 337.

[9] BARNHART M. I., WALSH R. T. and ROBINSON J. A. (1972) A three-dimensional view of platelet responses to chemical stimuli. Ann. N. Y. Acad. Sci., *201:* 360.

[10] BERGSTRÖM S. (1972) Shedding of the zona pellucida in normal pregnancy and in various hormonal states in the mouse. Z. Anat. Entw.-Gesch., *136:* 143.

[11] BERNSTEIN E. O. and JONES C. B. (1969) Skin replication procedure for the scanning electron microscope. Science, *166:* 252.

[12] BESSIS M. (1972) Red cell shapes. An illustrated classification and its rationale. Nouv. Rev. franc. hématol., *12:* 721.

[13] BOYDE A. and BARBER V. C. (1969) Freeze-drying methods for the scanning electron-microscopical study of the protozoon *Spirostomum ambiguum* and the statocyst of the cephalopod mollusc *Loligo vulgaris*. J. Cell Sci., *4:* 223.

[14] BOYDE A., JONES S. J. and BAILEY E. (1973) Bibliography on biomedical applications of scanning electron microscopy. *In:* Scanning Electron Microscopy/1973, pp. 697–734, JOHARI O. & CORVIN I. (eds.) IIT Research Institute, Chicago.

[15] BOYDE A. and WOOD C. (1969) Preparation of animal tissues for surface-scanning electron microscopy. J. Microscopy, *90:* 221.

[16] BULL B. S. and KUHN I. N. (1970) The production of schistocytes by fibrin strands (A scanning electron microscope study). Blood, *35:* 104.

[17] BUSS H. (1970) Die morphologische Differenzierung des visceralen Blattes der Bowmanschen Kapsel. Raster- und durchstrahlungselektronenmikroskopische Untersuchungen am Nierenglomerulum der Ratte. Z. Zellforsch., *111:* 346.

[18] BUSS H. and KRÖNERT W. (1969) Zur Struktur des Nierenglomerulum der Ratte. Rasterelektronenmikroskopische Untersuchungen. Virchows Arch. Abt. B Zellpathol., *4:* 79.

[19] CARR K. E. (1971) Applications of scanning electron microscopy in biology. Int. Rev. Cytol., *30:* 183.

[20] CARTEAUD J. P., MUIR M. D. and CATAIGNE G. (1973) A morphological investigation in the SEM of haemagglutination with influenza and vaccinia viruses. *In:* Scanning Electron Microscopy/1973, pp. 497–504, JOHARI O. and CORVIN I. (eds.) IIT Research Institute, Chicago.

[21] CHAPMAN B. (1967) Polystyrene replicas for scanning reflexion electron microscopy. Nature, *216:* 1347.

[22] CLARKE I. C. (1971) A method for the replication of articular cartilage surfaces suitable for the scanning electron microscope. J. Microscopy, *93:* 67.

[23] CLARKE J. A., HAWKEY C. and SALSBURY A. J. (1969) Surface ultrastructure of platelets and thrombocytes. Nature, *223:* 401.

[24] COHEN A. L., MARLOW D. P. and GARNER G. E. (1968) A rapid critical point method using fluorocarbons ("Freons") as intermediate and transitional fluids. J. Microscopie, *7:* 331.

[25] CREWE A. V. (1971a) High resolution scanning microscopy of biological specimens. Phil. Trans. Roy. Soc. Lond., *261:* 61.

[26] CREWE A. V. (1971b) A high-resolution scanning electron microscope. It has succeeded in resolving individual atoms for the first time. Sci. Amer., *224* (4): 26.

[27] ECHLIN P. (1971) The applications of scanning electron microscopy to biological research. Phil. Trans. Roy. Soc. Lond., (B) *261:* 51.

[28] ECHLIN P. (1973) The scanning electron microscope and its applications to research. Microsc. Acta, *73:* 97.

[29] ECHLIN P. and MORETON R. (1973) The preparation, coating and examination of frozen biological materials in the SEM. *In:* Scanning Electron Microscopy/1973, pp. 325–332, JOHARI O. and CORVIN I. (eds.) IIT Research Institute, Chicago.

[30] ECHLIN P., PADEN R., DRONZEK B. and WAYTE R. (1970) Scanning electron microscopy of labile biological material maintained under controlled conditions. *In:* Scanning Electron

Microscopy/1970, pp. 49–56, Johari O. and Corvin I. (eds.) IIT Research Institute, Chicago.

[31] Edanaga M. (1974) A scanning electron microscope study on the endothelium of the vessels. I. Fine structure of the endothelial surface of aorta and some other arteries in normal rabbits. Arch. histol. jap., *37:* 1.

[32] Ferguson D. R. and Heap P. F. (1970) The morphology of the toad urinary bladder: A stereoscopic and transmission electron microscopical study. Z. Zellforsch., *109:* 297.

[33] Fujita T. (1974) A scanning electron microscope study of the human spleen. Arch. histol. jap., *37:* 187.

[34] Fujita T., Inoue H. and Kodama T. (1968) Scanning electron microscopy of the normal and rheumatoid synovial membranes. Arch. histol. jap., *29:* 511.

[35] Fujita T., Miyoshi M. and Murakami T. (1972) Scanning electron microscope observation of the dog mesenteric lymph node. Z. Zellforsch., *133:* 147.

[36] Fujita T., Miyoshi M. and Tokunaga J. (1970) Scanning and transmission electron microscopy of human ejaculate spermatozoa with special reference to their abnormal forms. Z. Zellforsch., *105:* 483.

[37] Fujita T. and Murakami T. (1973) Microcirculation of monkey pancreas with special reference to the insulo-acinar portal system. A scanning electron microscope study of vascular casts. Arch. histol. jap., *35:* 255.

[38] Fujita T., Tokunaga J. and Inoue H. (1969) Scanning electron microscopy of the skin using celluloid impressions. Arch. histol. jap., *30:* 321.

[39] Fujita T., Tokunaga J. and Inoue H. (1971) Atlas of Scanning Electron Microscopy in Medicine. Igaku Shoin, Tokyo and Elsevier, Amsterdam-London-New York.

[40] Fujita T., Tokunaga J. and Miyoshi M. (1970) Scanning electron microscopy of the podocytes of renal glomerulus. Arch. histol. jap., *32:* 99.

[41] Fulker M. J., Holland L. and Hurley R. E. (1973) Ion etching of organic materials. *In:* Scanning Electron Microscopy/1973, pp. 379–386, Johari O. and Corvin I. (eds.) IIT Research Institute, Chicago.

[42] Goldman, M. A. and Leif, R. C. (1973) A wet chemical method for rendering scanning electron microscopy samples conductive and observations on the surface morphology of human erythrocytes and Ehrlich ascites cells. Proc. Nat. Acad. Sci., *70:* 3599.

[43] Haggis G. H. (1970) Cryofracture of biological material. *In:* Scanning Electron Microscopy/1970, pp. 97–104, Johari O. and Corvin I. (eds.) IIT Research Institute, Chicago.

[44] Hamano M., Otaka T., Nagatani T. and Tanaka K. (1973) A frozen liquid cracking method for high resolution scanning electron microscopy (Abstr.). J. Electron Microsc., *22:* 298.

[45] Hattori A. (1972) Scanning electron microscopy of human peripheral blood cells. Acta haematol. jap., *35:* 457.

[46] Hattori A., Ito S. and Matsuoka M. (1972) Scanning electron microscopy of reticulocytes. Arch. histol. jap., *35:* 37.

[47] Hattori A., Ito S., Tsukada T., Koike K. and Matsuoka M. (1974) Platelet adhesion and aggregation in the glass bead column method (Hellem-II). A scanning electron microscope study. Arch. histol. jap., *36:* 323.

[48] Hattori A., Sugawara A., Matsuoka M., Tokunaga J. and Fujita T. (1971) Preparation of blood cells for scanning electron microscopy. J. Electron Microsc., *20:* 222.

[49] Hattori A., Tokunaga J., Fujita T. and Matsuoka M. (1969) Scanning electron microscopic observations on human blood platelets and their alterations induced by thrombin. Arch. histol. jap., *31:* 37.

[50] Hayes T. L. and Pease R. F. W. (1968) The scanning electron microscope: Principles and applications in biology and medicine. Adv. biol. med. Phys., *12:* 85.

[51] Hodges G. M., Muir M. D., Sella C. and Carteaud A. J. P. (1972) The effect of radiofrequency sputter ion etching and ion-beam etching on biological material. A scanning electron microscope study. J. Microscopy, *95:* 445.

[52] Hollenberg M. J. and Erickson A. M. (1973) The scanning electron microscope: potential usefulness to biologists. A review. J. Histochem. Cytochem., *21:* 109.

[53] Horridge G. A. and Tamm S. L. (1969) Critical point drying for scanning electron microscopic study of ciliary motion. Science, *163:* 817.

[54] IRINO S., ONO T., HIRAKI K. and MURAKAMI T. (1974)   A new approach to the scanning electron microscopy of the vascular structure of bone marrow by the use of injection replica method. (In Japanese)   Igaku no Ayumi, *88:* 293.

[55] ITOSHIMA T., KOBAYASHI T., SHIMADA Y. and MURAKAMI T. (1974)   Fenestrated endothelium of the liver sinusoids of the guinea pig as revealed by scanning electron microscopy. Arch. histol. jap., *37:* 15.

[56] KAYDEN H. J. and BESSIS M. (1970)   Morphology of normal erythrocyte and acanthocyte using Nomarski optics and the scanning electron microscope.   Blood, *35:* 427.

[57] KELLEY R. O., DEKKER R. A. F. and BLUEMINK J. G. (1973)   Ligand-mediated osmium binding: its application in coating biological specimens for scanning electron microscopy. J. Ultrastr. Res., *45:* 254.

[58] KOBAYASHI S. (1973)   Rasterelektronenmikroskopische Untersuchungen der Shilddrüse (Scanning electron microscope observations of the thyroid).   Arch. histol. jap., *36:* 107.

[59] KOLLER T. and BERNHARD W. (1964)   Séchage de tissus au protoxyde d'azote (N$_2$O) et coupe ultrafine sans matière d'inclusion.   J. Microscopie, *3:* 589.

[60] KURAHASI K., TOKUNAGA J., FUJITA T. and MIYAHARA M. (1969)   Scanning electron microscopy of isolated mitochondria I.   Arch. histol. jap., *30:* 217.

[61] KUWABARA T. (1970)   Surface structure of the eye tissue.   *In:* Scanning Electron Microscopy/1970, pp. 185–192, JOHARI O. and CORVIN I. (eds.) IIT Research Institute, Chicago.

[62] LARRIMER N. R., BALCERZAK S. P., METZ E. N. and LEE R. E. (1970)   Surface structure of normal human platelets.   Amer. J. med. Sci., *259:* 242.

[63] LEWIS S. M., OSBORN J. S. and STUART P. R. (1968)   Demonstration of an internal structure within red blood cell by ion etching and scanning electron microscopy.   Nature, *220:* 614.

[64] LEWIS S. M. and STUART P. R. (1970)   Ultrastructure of the red blood cell.   Proc. Roy. Soc. Med., *63:* 465.

[65] LIN P. S., TSAI S. and WALLACH D. F. H. (1973a)   Major differences in the surface morphology of thymocytes and peripheral lymphocytes are revealed by scanning electron microscopy.   *In:* 2nd Int. Symp. on Metabolism and Membrane Permeability of Erythrocytes, Thrombocytes and Leukocytes. GERLACH E. et al. (eds.) Georg Thieme, Stuttgart.

[66] LIN P. S., COOPER A. G. and WORTIS H. H. (1973b)   Scanning electron microscopy of human T-cell and B-cell rosettes.   New Engl. J. Med., *289:* 548.

[67] LOBUGLIO A. F., RINEHART J. J. and BALCERZAK S. P. (1972)   A new immunological marker for scanning electron microscopy.   *In:* Scanning Electron Microscopy/1972, pp. 313–320, JOHARI O. and CORVIN I. (eds.) IIT Research Institute, Chicago.

[68] LUSE S. A. (1970)   Preparation of biologic specimens for scanning electron microscopy. Proc. 3rd Ann. Stereoscan Colloq. Kent Cambridge Scientific, Morton Grove, pp. 149–157.

[69] MAKITA T. and SANDBORN E. B. (1971)   Identification of intracellular components by scanning electron microscopy.   Exp. Cell Res., *67:* 211.

[70] MARCHANT H. J. (1972)   Processing small delicate biological specimens for scanning electron microscopy.   J. Microscopy, *97:* 369.

[71] MELLER S. M., COPPE M. R., ITO S. and WATERMAN R. E. (1973)   Transmission electron microscopy of critical point dried tissue after observation in the scanning electron microscope. Anat. Rec., *176:* 245.

[72] MIKAMI K., SATO H., SAITO T., KOBAYASHI Y. and IMAI M. (1969)   Scanning electron microscopic observations of polyester resin polymerized in dentinal tubules.   Nihon Univ. dent. J., *43:* 736.

[73] MIYOSHI M. and FUJITA T. (1971)   Stereo-fine structure of the splenic red pulp. A combined scanning and transmission electron microscope study on dog and rat spleen.   Arch. histol. jap., *33:* 225.

[74] MIYOSHI M., FUJITA T. and TOKUNAGA J. (1970)   The red pulp of the rabbit spleen studied under the scanning electron microscope.   Arch. histol. jap., *32:* 289.

[75] MIYOSHI M., FUJITA T. and TOKUNAGA J. (1971)   The differentiation of renal podocytes. A combined scanning and transmission electron microscope study in rats.   Arch. histol. jap., *33:* 161.

[76] MURAKAMI T. (1971)   Application of the scanning electron microscope to the study of the fine distribution of the blood vessels.   Arch. histol. jap., *32:* 445.

[77] MURAKAMI T. (1972) Vascular arrangement of the rat renal glomerulus. A scanning electron microscope study of corrosion casts. Arch. histol. jap., *34:* 86.

[78] MURAKAMI T. (1973) A metal impregnation method of biological specimens for scanning electron microscopy. Arch. histol. jap., *35:* 323.

[79] MURAKAMI T. (1974) A revised tannin-osmium method for non-coated scanning electron microscope specimens. Arch. histol. jap., *36:* 189.

[80] MURAKAMI T., MIYOSHI M. and FUJITA T. (1973) Closed circulation in the rat spleen as evidenced by scanning electron microscopy of vascular casts. Experientia, *29:* 1374.

[81] MURAKAMI T., UNEHIRA M., KAWAKAMI H. and KUBOTSU A. (1973) Osmium impregnation of methyl methacrylate vascular casts for scanning electron microscopy. Arch. histol. jap., *36:* 119.

[82] NAGATANI T. and SAITO M. (1974) Structure analysis of evaporated films by means of TEM and SEM. *In:* Scanning Electron Microscopy/1974, pp. 51–58, JOHARI O. and CORVIN I. (eds.) IIT Research Institute, Chicago.

[83] NEI T., YOTSUMOTO H., HASEGAWA Y. and NAGASAWA Y. (1971) Direct observation of frozen specimens with a scanning electron microscope. J. Electron Microsc., *20:* 202. (In Japanese)

[84] NEI T., YOTSUMOTO H., HASEGAWA Y. and NAGASAWA Y. (1972) Electron microscopic observation of biological specimens in their native state by employing cryogenic techniques. Proc. 5th Eur. Congr. Electron Microsc., pp. 252–253.

[85] NOWELL J. A., PANGBORN J. and TYLER W. S. (1970) Scanning electron microscopy of the avian lung. *In:* Scanning Electron Microscopy/1970, pp. 249–256, JOHARI O. and CORVIN I. (eds.) IIT Research Institute, Chicago.

[86] VAN OSS C. J. and MOHN J. F. (1970) Scanning electron microscopy of red cell agglutination. Vox Sang., *19:* 432.

[87] PAMEIJER C. H. and STALLARD R. E. (1972) Application of replica techniques for use with scanning electron microscopes in dental research (Abstr.). J. dent. Res., *51:* 672.

[88] PANESSA B. J. and GENNARO J. F., JR. (1972) Preparation of fragile botanical tissues and examination of intracellular contents by SEM. *In:* Scanning Electron Microscopy/1972, pp. 327–335, JOHARI O. and CORVIN I. (eds.) IIT Research Institute, Chicago.

[89] PANESSA B. J. and GENNARO J. F., JR. (1973) Use of potassium iodide/lead acetate for examining uncoated specimens. *In:* Scanning Electron Microscopy/1973, pp. 395–402, JOHARI O. and CORVIN I. (eds.) IIT Research Institute, Chicago.

[90] POLLIACK A., FU S. M., DOUGLAS S. D., BENTWICH Z., LAMPEN N. and DE HARVEN E. (1974) Scanning electron microscopy of human lymphocyte-sheep erythrocyte rosettes. J. Exp. Med., *140:* 146.

[91] SALSBURY A. J. and CLARKE J. A. (1967) New method for detecting changes in the surface appearance of human red blood cells. J. clin. Pathol., *20:* 603.

[92] SCARBOROUGH D. E., MASON R. G., DALLDORF F. G. and BRINKHOUS K. M. (1969) Morphologic manifestations of blood-solid interfacial reactions. Lab. Invest., *20:* 164.

[93] SCHÄFER D., SEIDL E., ACKER H., KELLER H. -P. and LÜBBERS D. W. (1973) Arteriovenous anastomoses in the cat carotid body. Z. Zellforsch., *142:* 515.

[94] SHOOP R., BALCERZAK S. P., LARRIMER N. R. and LEE R. E. (1970) Surface morphology of the early hemostatic reaction. Amer. J. med. Sci., *260:* 122.

[95] SMITH U., RYAN J. W., MICHE D. D. and SMITH D. S. (1971) Endothelial projections as revealed by scanning electron microscopy. Science, *173:* 925.

[96] STEINBERG R. H. (1973) Scanning electron microscopy of the bullfrog's retina and pigment epithelium. Z. Zellforsch., *143:* 451.

[97] SYLVESTER-BRADLEY P. C. (1971) The reaction of systematics to the revolution in micropalaeontology. *In:* Scanning Electron Microscopy, Systematic and Evolutionary Applications. HEYWOOD V. H. (ed.) pp. 95–111, Academic Press, London & New York.

[98] TAKATA K. (1974) Scanning electron microscopic studies on rat epiphyseal growth plates by the resin cracking method. Arch. histol. jap., *36:* 281.

[99] TANAKA K. (1972) Freezed resin cracking method for scanning electron microscopy of biological materials. Naturwiss., *59:* 77.

[100] TANAKA K. and KAWAKAMI S. (1973) Some technical improvement in preparing biological specimen for scanning electron microscopy (Abstr.). J. Electron Microsc., *22:* 299.

[101] TANAKA T., KOSAKA N., TAKIGUCHI T., AOKI T. and TAKAHARA S. (1973) Observation on the cochlea with SEM. *In:* Scanning Electron Microscopy/1973, pp. 427–434, JOHARI O. and CORVIN I. (eds.) IIT Research Institute, Chicago.

[102] TARIN D. (1971) Scanning electron microscopical studies of the embryonic surface during gastrulation and neurulation in *Xenopus laevis*. J. Anat., *109:* 535.

[103] TOKUNAGA J. and FUJITA T. (1973) Application of scanning electron microscopy to biomedical fields. Its present and future aspects. J. Electron Microsc., *22:* 105.

[104] TOKUNAGA J., FUJITA T. and HATTORI A. (1969) Scanning electron microscopy of normal and pathological human erythrocytes. Arch. histol. jap., *31:* 21.

[105] TOKUNAGA J., FUJITA T. and INOUE H. (1969) Application of scanning electron microscopy to medical studies: Its state and prospect. Igaku no Ayumi, *68:* 485. (In Japanese)

[106] TOKUNAGA J., OSAKA M. and FUJITA T. (1973) Endothelial surface of rabbit aorta as observed by scanning electron microscopy. Arch. histol. jap., *36:* 129.

[107] TOKUNAGA J., TOKUNAGA M. and Application Department, Electron Optics Division, JEOL Ltd. (1973) Cryo-scanning microscopy of conidiospores formation in *Aspergillus niger*. JEOL News, *11e:* 1.

[108] TOKUNAGA J., TOKUNAGA M., YOTSUMOTO H. and HATABA Y. (1973) Cryo-scanning electron microscopy of conidiospores formation in *Aspergillus niger*. Jap. J. Mycol., *14:* 39. (In Japanese)

[109] TYLER W. S., NOWELL J. A. and PANGBORN J. (1970) Techniques for scanning electron microscopy of pulmonary tissues and replicas. Proc. 7me Congr. Int. Microsc. Electron., Grenoble, pp. 477–478.

[110] WATERMAN R. E. (1972) Use of the scanning electron microscope for observation of vertebrate embryos. Dev. Biol., *27:* 276.

[111] WATERMAN R. E. and MELLER S. M. (1973) Nasal pit formation in the hamster: A transmission and scanning electron microscopic study. Dev. Biol., *34:* 255.

[112] WATERMAN R. E., ROSS L. M. and MELLER S. M. (1973) Alterations in the epithelial surface of A/Jax mouse palatal shelves prior to and during palatal fusion: A scanning electron microscopic study. Anat. Rec., *176:* 361.

[113] WATTERS W. B. and BUCK R. C. (1971) An improved simple method of specimen preparation for replicas or scanning electron microscopy. J. Microscopy (Oxf.), *94:* 185.

[114] ZEEVI Y. Y. and LEWIS E. R. (1970) A new technique for exposing neuronal surfaces for viewing with the scanning electron microscope. Proc. 7me Congr. Int. Microsc. Electron., Grenoble, pp. 481–482.

Three-dimensional Observations of the Cellular Fine Structure by Means of High Voltage Electron Microscopy

Kiyoshi HAMA

The high voltage electron microscope has been used primarily in the field of material science and mineralogy. Especially in mineralogy, the advantage of the higher penetrating power of electrons at the higher accelerating voltage which enables the observation of a material in bulk, is enormous and the high voltage device becomes indispensable.

Recently, the remarkable merits of the high voltage operation in the field of biology have been recognized by several investigators (Dupouy et al. 1960; Dupouy 1968; Hama & Porter 1969; Hama & Nagata 1970 a, b; Nagata & Hama 1970; Cosslett 1971 a, b; Porter et al. 1973; Hama 1973, 1974; Ramburg & Marraud 1973).

The following advantages can be expected using the high voltage electron microscope in the observation of biological materials:

1) Thicker specimens can be observed at the higher accelerating voltage because of the higher penetrating power of the electrons.

2) The practical resolution obtainable from a specimen of a given thickness is expected to improve mainly because of a decrease in chromatic aberration.

3) Damage to the specimen caused by the electron beam is expected to be smaller at higher accelerating voltages. The first merit mentioned above is especially useful for the observation of biological specimens. In the present paper, the theoretical and experimental bases for this point will be discussed briefly, and some examples of three dimensional observations of the cellular fine structure taking advantage of the higher penetrating power of electrons at a high voltage operation will be shown.

Specimen Penetration

The maximum thickness of a specimen observable in an electron microscope may be defined in terms of both the fraction of the incident beam collected in the objective aperture (thus determining the overall image brightness), and the energy lost by the beam in transversing the specimen which limits the resolution obtained in the image owing to chromatic aberration. If one considers only image brightness, the maximum thickness observable may be very large even at a relatively low voltage using a high current density and a larger objective aperture. However in this case, beam damage on the specimen and chromatic aberration cause serious problems. Consequently there is a practical limit on the maximum specimen thickness observable at a given accelerating voltage. Theoretically, it is known that the penetrating power of electrons is proportional to β^2 where β equals the velocity of electrons at a given voltage divided by the velocity of light, when α/λ is kept constant (α=semi angle of objective aperture, λ=wave length of electrons) (Fig. 1) (Hashimoto 1964; Fujita et al. 1967; Dupouy, 1968; Nagata & Hama 1970).

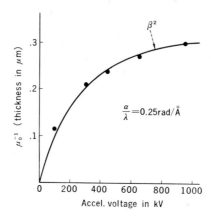

Fig. 1 This figure shows the relation between accelerating voltages and penetrating power of electrons using glass fibers as the test specimen. μ_0 is the absorption coefficient.

This means that an increase in the penetrating power of electrons with a rise in the accelerating voltage is remakable up to a certain voltage and the rate of increase of the penetrating power becomes smaller finally reaching a plateau. The accelerating voltage where the rate of rise of the penetrating power becomes smaller should be determined experimentally. The actual increase of the maximum usable thickness with a rise in accelerating voltage has been found to be more remarkable than was postulated (UEDA & NONOYAMA 1967; DUPOUY 1968). According to our own experiments, 1 μm, 1.5 μm and 2 μm thick Epon embedded biological sections could be observed at 500, 800 and 2000 KV respectively with a resolution high enough to clearly resolve the "unit membrane structure" (HAMA & PORTER 1967; HAMA & NAGATA 1970a). We tried observing extremely thick biological specimens using a very high accelerating voltage (2500 KV) and it was found that in observations on specimens thicker than 3 μm the image was bright enough but the thermal effects on the specimen, drift of the specimen and sublimation of the stain and embedding substance, became important. A good image quality could not be expected under these conditions. Thus, it seems that about 3 μm would be the limit of the observable specimen thickness even at a very high accelerating voltage when resonable image quality is required.

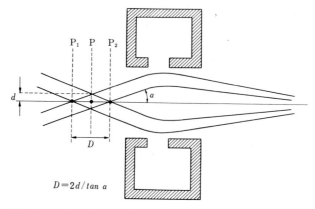

Fig. 2 This figure shows the relation between the aperture angle and the depth of the focus of the optical system.

Depth of Field

The depth of field is inversely proportional to the aperture angle of the optical system and is given by the formula

$$D = \frac{2d}{\tan a}$$

where d is the resolution required and a is the semiangle of the objective aperture (Fig. 2). With the electron microscope the aperture of the objective lens is very small being $3\text{--}4 \times 10^{-3}$ radian. As a consequence, over 2 μm in the depth of field can be expected at a resolution of better than 20 A which is capable of resolving the unit menbrane structure. The high penetrating power of electrons at the higher voltage operation together with the large depth of field of the electron microscope enables the observation of a thick specimen by the high voltage electron microscope. However this causes a serious problem. When the specimen is thick, images from the entire depth of the specimen are superimposed and a detailed analysis of the cellular fine structure cannot be expected.

Observation of a Thick Specimen

As mentioned before the use of the high voltage electron microscope has enabled the observation of specimens thicker than 1 μm. The merit of a thick specimen observation in biology will be shown in the following two figures. Fig. 3 shows an ordinary electron micrograph of a salivary gland cell of a *Drosophila* larva. This cell is known to have giant chromosomes which exhibit beautiful band patterns in the light microscope. However, in this micrograph, the images of the giant chromosomes appears as dissociated electron dense masses and do not show the characteristic band patterns because the specimen is too thin for an integration of the overlapped band pattern to occur. The thickness of the specimen is about 80 nm. Fig. 4 shows the giant chromosomes of a similar specimen observed at 800 KV using a 1 μm thick section. The sections was cut from the same block as the one shown in Fig. 3. The band pattern of a giant chromosome is clearly seen at a much better resolution than that obtained with ultraviolet or light microscopy. As shown here, the advantage of a thick specimen observation is apparent in the field of biology, however, at the same time, there is the problem of overlapping of images through the depth of the specimen. Although the band pattern is clearly seen in this micrograph, details of the three dimensional distribution of materials constituting the band pattern cannot be clarified. Methods of unscrambling this confusion of images caused by the superimposition at high voltage electron steroscopy are required.

High Voltage Stereoscopy

The procedure for electron stereoscopy is as follows: First, observe a favorable field then tilt the specimen stage in one azimuth by 8–10 degrees and take one photograph. Then take another picture of the same field after tilting the specimen stage by 8–10 degrees from the original position in the opposite azimuth. In other words, two photographs are taken by tilting the specimen stage 16–20 degrees in opposite azimuth between the two exposures. The pair of photographs thus prepared are examined under a stereo viewer.

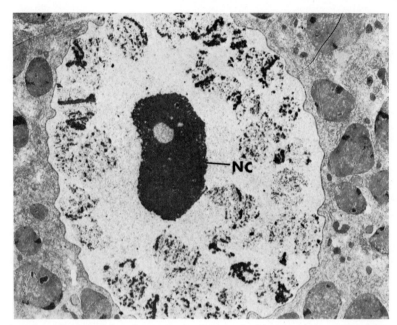

Fig. 3 A nucleus of a drosophila salivary gland cell is seen in this thin section. The photograph was taken at 75 KV. The familiar band patterns of the giant chromosomes cannot be seen in this thin section picture. (×5,000)

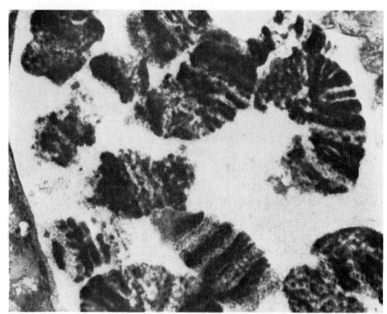

Fig. 4 Giant chromosomes are observed at 800 KV using a thick section (1 μm thick) cut from the same block as the previous picture. Band patterns of the giant chromosomes are clearly seen in this high voltage electron micrograph. (×10,000)

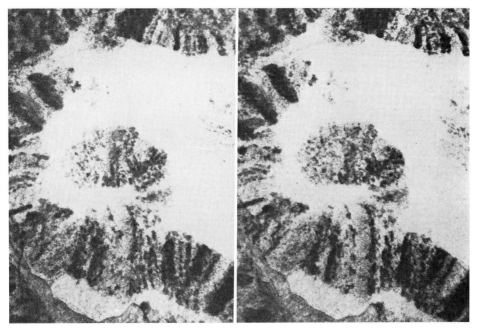

Fig. 5 A stereo pair of a portion of a giant chromosome from the salivary gland cell of the drosophila. The band patterns are seen to be composed of a three-dimensional accumulation of irregular dense granules. ($\times 13,000$)

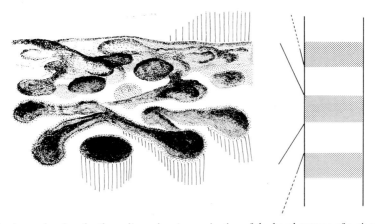

Fig. 6 A scheme showing the three dimensional organization of the band pattern of a giant chromosome.
The band pattern seen in light and high voltage electron microscopy are overlap images of the irregular dense masses.

Some examples of the application of high voltage stereoscopy will be shown. As mentioned before, a beautiful band pattern is seen in a thick section high voltage electron micrograph. However the three dimensional organization of the giant chromosome is obscured by the superimposition of images in the depth dimension of the specimen. Fig. 5 is a stereo pair of a portion of a giant chromosome from a salivary gland of a *Drosophila* larva. The specimen is similar to the one shown in Figs. 3 and 4, fixed with glutaraldehyde and osmium and stained *en bloc* with uranyl acetate before dehydration. One μm thick sections were

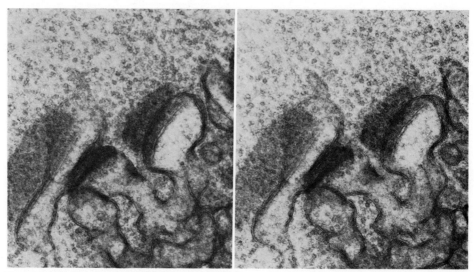

Fig. 7 A stereo pair showing the synapse between a cone and second order neurons of the tortoise retina. (×15,000)

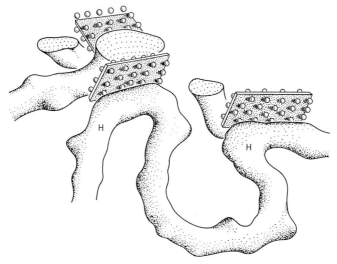

Fig. 8 A schematic drawing based on the finding of the previous stereo pair. The processes of the horizontal cell (H) run tortuous courses and make synaptic contact with the receptor cell more than once.
The plasma membrane of the receptor cell is omitted because of simplicity.

cut and stained with uranyl acetate and lead citrate and examined with a high voltage electron microscope operated at 800 KV. A stereo pair of photographs of a favorable field was prepared and observed under a stereo viewer. The results were schematized in Fig. 6. It was discovered that the dense band seen in the high voltage electron micrograph or light micrograph was an overlap of images of ill-defined electron dense masses accumulated in the plane forming a coarse band pattern. The fine filaments of about 7 nm in diameter were found interconnecting the electron dense masses. These filaments ran roughly parallel to the long axis of the giant chromosome.

Fig. 7 shows a stereo pair of a cone and second order neuron synapses seen in the tortoise

Fig. 9 A stereo pair showing the surface of a mouse trachea taken at 800 KV. The specimen thickness is about 1 μm. Three-dimensional organization of basal bodies, kinocilia, microvilli and a secretory granule is seen in this stereo pair. (×40,000)

retina. The specimen was prepared in the same way as described in the previous paragraph. The stereo image was schematized in Fig. 8 in which the plasma membrane of the receptor cell was omitted for simplicity. As it is shown here, the process of the horizontal cell runs a tortuous course and makes contact with the same receptor cell more than once. Along its course there are seen many short processes on the surface. The synaptic ribbons found in the presynaptic cell at the synaptic sites are seen to be quadrate forms covered with synaptic vesicles. The receptor second order neuron synapse of the retina is a very complicated structure and it has been very difficult to comprehend its three-dimensional organization, although many have tried three-dimensional reconstructions by means of serial thin sections. As shown here, with the aid of high voltage stereoscopy using a thick section, it has become possible to clearly visualize the three dimensional organization of the entire synaptic structure.

Fig. 9 is a stereo pair showing the surface of a ciliated epithelium from a mouse trachea. The three-dimensional organization of cilia, microvilli, secretory granules and basal bodies is clearly seen in the depth of the specimen.

Three-dimensional Reconstruction Using Thick Serial Sections

Reconstruction using serial sections has been employed in understanding the three-dimensional organization of a biological structure. With the aid of high voltage electron microscopy it becomes possible to reconstruct the cell structures using smaller numbers of thick serial sections. The procedures for reconstruction were as follows: Ten to fifteen serial sections of one μm thick were cut and mounted on collodion coated slit meshes. The sections were stained with uranyl acetate and lead citrate and finally examined in a

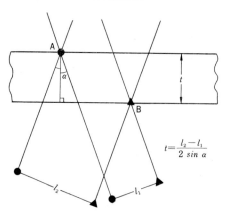

Fig. 10 This figure shows a method of estimating the section thickness from a stereo pair. Both l_1 and l_2 are distances between the two points, A and B, projected on each of the stereo pair. a is the tilting angle from the original position.

high voltage electron microscope. A photograph was taken of a favorable field then stereo pair pictures were prepared of the same field tilting the specimen stage ± 8 degrees in the opposite azimuths. The same field was selected in the next section and three photographs were taken as mentioned before. The procedures were repeated on each section and the series of photographs in the normal position and the series of stereo pair pictures were prepared. The series of normal position pictures were used for the reconstruction while the series of stereo pairs were used as references. For the three-dimensional reconstruction accurate estimations of the section thicknesses were required. The thickness of each section was estimated by means of a microtome scale, interference color, interferometer and other methods. The thickness judged with the microtome scale was not entirely accurate. The accuracy with the interference color or interferometer was rather high for thin specimens but not for thick specimens over 1 μm. In our experiments, the thicknesses of the specimens were estimated in the following way (Fig. 10). Two photographs were taken by tilting the stage in the opposite azimuths of a given angle $\pm 8°$. Two image points A and B on an opposite surfaces of the specimen were selected. The projected distances between the two points l_1 and l_2 were measured on each of the two photographs. The thickness of the section (t) was estimated from the formula:

$$t = \frac{l_1 - l_2}{2 \sin a}$$

The accuracy of this method primarily depended upon the accuracy of the microscopic magnification.

Fig. 11 is a high voltage stereo micrograph showing one of the serial cross sections of sensory hairs of the lateral line canal organ of a sea eel. The thickness of the section was 1 μm. The picture was taken at 800 KV. The sensory hair group consisted of a kinocillium and forty to fifty microvilli called stereocilia. In this picture the entire ciliary structures were traced from the top surface of the section to the bottom surface. Fig. 12 is a photograph of a plastic reconstruction of the sensory hairs using thick serial sections. A kinocilium was cut because it was too long. The sterocilia situated next to the kinocilium were the tallest. They became shorter as they receded further away. Those located in the central area were shorter than those at the periphery. The tallest stereocilia were about 6 μm long, thus, if one cut thick serial sections over 1 μm the entire length of the stereocilia could be included within 5–6 serial sections. The total surface area and volume of the stereocilia

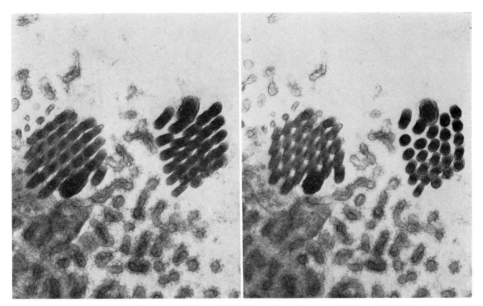

Fig. 11 A stereo pair showing one of the serial thick sections of sensory hairs of the lateral line canal organ. These stereo pairs are used as references for the three-dimensional reconstruction. ($\times 25{,}000$)

Fig. 12 A plastic reconstruction of sensory hairs of the lateral line canal organ.

and kinocilium were estimated using the serial sections. They were 51–72 μm^2 and 2.7–3.2 μm^3 for stereocilia and 17–18 μm^2 2–3 μm^3 for the kinocilium respectively (HAMA & KAMINO 1973).

Fig. 13 A photograph showing the outer view of the 3000 KV electron microscope. The upper part of the picture shows a pressure tank containing the accelerator unit. The lower part of the picture shows a part of the column and the operation desk.

Other Applications of High Voltage Stereoscopy to the Study of Biological Materials

The three dimensional network of the T-system in skeletal muscle was observed by means of high voltage stereoscopy. The T-system was marked by peroxidase and the electron dense network of the system was clearly visualized in the three-dimensional scale (YAMADA & ISHIKAWA in this book). A similar technique was applied in the three-dimensional analyses of the Golgi apparatus from liver and spinal ganglion cells (RAMBOURG & MARRAUD 1973). The forming face of the Golgi apparatus was impregnated with osmium and the GERL system was marked by the acid phosphatase technique. Thick sections were prepared and stereo pairs were prepared with a 3000 KV electron microscope. The Golgi apparatus was classfied by their three-dimensional organization. High voltage stereoscopy was also effectively applied in the whole cell observation of the cultured material. The intracytoplasmic distribution of mitochondria, endoplasmic reticulum and filamentous materials were clearly observed (PORTER 1973). There might be other possible ways for the application of high voltage stereoscopy in the field of biology and fruitful results could be expected in the future. Finally, a photograph of the 3000 KV electron microscope employed at the Osaka University is shown (Fig. 13). The height of the microscope is about 12 meters and its weight is about 70 tons. The resolution achieved at 2500 KV is 2 A.

REFERENCES

[1] COSSLETT V. E. (1971a) High voltage electron microscopy and its application in biology. Phil Trans. B., *261:* 35–44.
[2] COSSLETT V. E. (1971b) The scope and limitations of high voltage electron microscopy. J. Ultrast. Res., *37:* 255–257.
[3] DUPOUY G. (1968) Electron microscopy at very high voltages. *In:* Advances in Optical and Electron Microscopy, BARER R. and COSSLETT V. E. (eds.) Vol 2, pp. 168–250, Academic Press, New York.
[4] DUPOUY G., PERRIER F. and DURRIRU L. (1960) Microscopie électronique. L'observation de la matiére vivante au moyen d'um microscope électronique fonctionnant suos trés haute tension. C. R. Acad. Sc., *251:* 2836–2841.
[5] FUJITA H., FURUBAYASHI E., KAJIWARA S., KAWASAKI Y. and TAOKA T. (1967) Metallurgical investigation with 500KV electron microscope, Japan. J. appl. Physics, *6:* 214–230.
[6] HAMA K. (1973) High voltage electron microscopy. *In:* Advanced Techniques in Biological Electron Microscopy, KOEHLER J. K. (ed.) pp. 275–297. Springer Verlag, Berlin-Heidelberg-New York.
[7] HAMA K. and KAMINO T. (1974) Three-dimensional observation of the sensory hairs of the leteral line canal organ of the sea eel by means of high voltage stereoscopy using thick serial sections. Proc. Royal Micro. Soc.: Third International Conference on High Voltage Electron Microscopy, Oxford. pp. 423–425, Academic Press, New York.
[8] HAMA K. and NAGATA F. (1970a) High resolution observation of biological sections with a high voltage electron microscope. J. Electron Microscopy, *19:* 170–175.
[9] HAMA K. and NAGATA F. (1970b) A stereoscopic observation of tracheal epithelium of mouse by means of the high voltage electron microscope. J. Gell Biol., *45:* 654–659.
[10] HAMA K. and PORTER K. R. (1969) An application of high voltage electron microscopy to the study of biological materials J. de Microscope, *8:* 149–158.
[11] HASHIMATO H. (1964) Energy dependence of extinction distance and transmissive power for electron waves in crystals. J. appl. physics, *35:* 277–290.
[12] NAGATA F. and HAMA K. (1970) Observation of biological specimens with a high voltage electron microscope. 27th Annual Proceedings EMSA.
[13] PALAY S. L. and PALAY V. C. (1972) High voltage electron microscopy of the central nervous system in Golgi preparations. J. Microscopy, *97:* 41–47.
[14] PORTER K. (1973) Biological Application.
[15] RAMBOURG A. and MARRAUD A. (1973) Three-dimensional aspects of the forming face of the Golgi apparatus. Proc. Royal Micro. Soc.: Third International Conference on High Voltage Electron Microscopy. Oxford.
[16] UEDA R. and NONOYAMA M. (1967) The observation of thick specimens by high voltage electron microscopy. Experiment with molybdenite films at 50–500 KV. Jap. J. appl. physics, *6:* 557–566.

High Voltage Electron Microscopy Combined with Molecular Tracers: Observations on the Cardiac Muscle T-system and Renal Epitheliocyte

Eichi YAMADA *and* Harunori ISHIKAWA

One of the advantages of high voltage electron microscopy arises from the great penetration of its electron beam, which permits the examination of thick sections of resin-embedded biological materials (Hama & Porter 1969). However, when the stained thick sections are examined under a high voltage electron microscope, all the structural details within the entire thickness are brought into focus as superimposed images. Hence, certain structures are not easily observable in such a picture. In other words, due to the confusion of the superimposition, useful informations for particular studies cannot be readily collected from the superimposed images, even from a stereo view combined with the specimen tilting technique and stereoscopy.

To overcome this difficulty, molecular tracers, ruthenium red and peroxidase, were introduced into high voltage electron microscopy in an attempt to selectively visualize certain structures. This technique can provide a clear-cut stereo image of certain structures which are stained or filled with the tracers.

Materials and Methods

Both ruthenium red and peroxidase were used for this study, since they are useful molecular tracers for demonstrating the extracellular space and the invaginations of the plasma membrane.

Ruthenium red. Various tissues of young mice were fixed by perfusion through the left ventricle of the heart with 3% glutaraldehyde in 0.1 M sodium cacodylate buffer (pH 7.4), containing 500 ppm ruthenium red (Luft 1966). Tissues were then dissected out for continued fixation in the same fixative. Tissue blocks were postfixed for 2 hr in 1% OsO_4 again containing ruthenium red.

Peroxidase. Adult mice were intravenously given 10 mg of horseradish peroxidase (Sigma) dissolved in physiological saline, and second injections of the same amount of peroxidase were made 20 min later. Five min after the second injections, the animals were sacrificed. Cardiac muscles were dissected out and fixed in 2% paraformaldehyde and 2.5% glutaraldehyde in 0.1 M sodium cacodylate buffer (pH 7.2), for 2 hr. Tissues were teased into small bundles of muscle fibers. Specimens were incubated for cytochemistry according to the method described by Graham and Karnovsky (1966) after rinsing in the same buffer. Incubation was continued for 50 min which was long enough to get a substantial quantity of the reaction product. Specimens were then postfixed in 1% OsO_4 for 1 hr.

All the specimens were dehydrated in graded concentrations of ethanol and finally

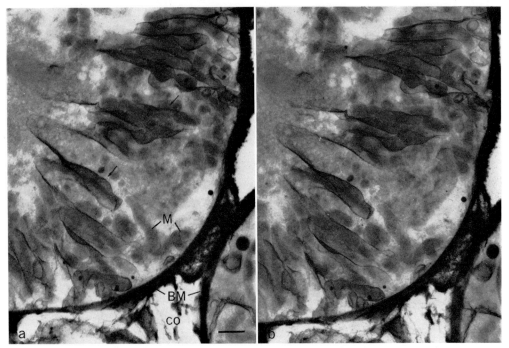

Fig. 1 A stereo pair of high voltage electron micrographs of a proximal tubule treated with ruthenium red. The basal infoldings (arrows) of the epitheliocytes are seen as a continuous sheet. The basal membrane (BM) is also strongly stained with ruthenium red. The lumen and brush border of the tubule are at the upper left. M: Mitochondria. CO: Fibrous connective tissue. 1 μm thick section. (For all figures, the scales represent 1 μm)

embedded in Epon. Thick sections ranging from 0.5 to 3 μm were cut with glass knives on a Porter-Blum MT-2 ultramicrotome. Serial thick sections were also made at thickness settings of 1 μm or 2 μm. The sections were mounted on grids coated with Neoprene and examined in a Hitachi HU-1000 or a JEOL JEM-1000 high voltage electron microscope with an accelerating voltage of 100 KV without staining. Pictures were routinely taken as stereo pairs using a tilting stage device at +10° and −10°. Printed micrographs of the stereo pairs were viewed and examined under a Sokkisha Stereoscope MS-27.

Observations

Various tissues stained with both extracellular tracers were examined in thin sections as well as thick sections. It turned out that ruthenium red and peroxidase were best suited to examine the renal epithelia and the cardiac T-system respectively. Therefore, these two typical examples will be described in this report. The uniformity and consistency of selective staining for the expected structures were checked on thin sections under the conventional voltage (75 KV) electron microscope.

High voltage electron microscopy of thick sections also demonstrated that the tracer substances were clearly visible and sufficiently contrasty as compared to other organelles in the background. Although the sections were not stained with either lead or uranyl acetate, the background organelles such as the plasma membranes, mitochondria and myofibrils were discernible just enough to spatially locate the tracer-stained structures. Inter-

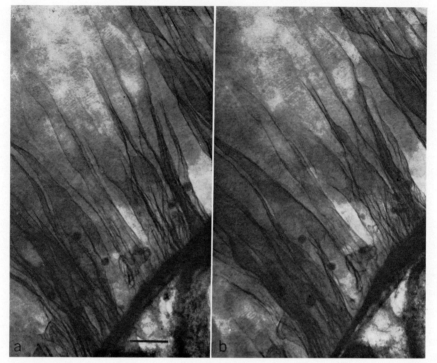

Fig. 2 A stereo pair demonstrating the basal infoldings of the proximal tubule treated with ruthenium red. 3 μm thick section.

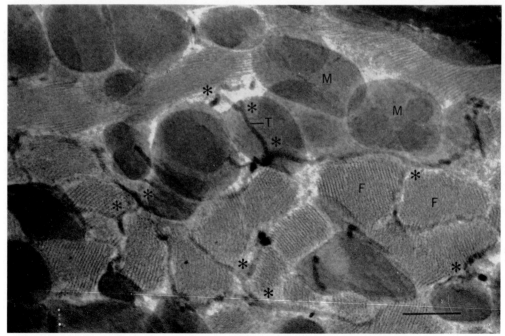

Fig. 3 A high voltage electron microgragh of a transversely sectioned cardiac muscle cell cytochemically treated for exogeneous peroxidase. The T-system tubules (T) are filled with the reaction product, run around the myofibrils (F) and show many branchings (stars). M: Mitochondria. 1.5 μm thick section.

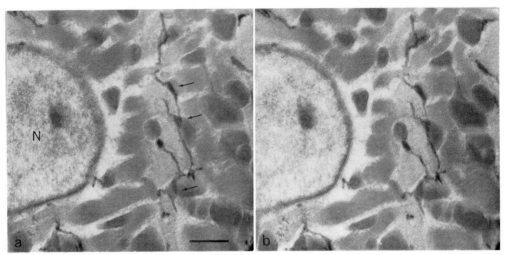

Fig. 4 A stereo pair demonstrating the spatial distribution of the T-system. The T-system forms locally dilated cisternae along its course (arrows), and shows branchings. N: Nucleus. 0.5 μm thick section.

estingly, the fused stereo images under the stereoscope exhibited a better contrast and sharper contour of the tracer-stained structures than did the individual micrographs. One could also obtain fused stereo images viewed with the upper and the lower parts reversed in portion without the stereoscope. Relatively sharp pictures were taken even with the 3 μm thick sections with an accelerating voltage of 1000 KV.

Ruthenium red for the renal epithelia

The basal region of the epitheliocytes of the proximal and distal tubules is characterized by lengthy, complex infoldings of the plasma membrane. In thin sections, the basal infoldings are seen as apposed plasma membranes with an extracellular space between them. Ruthenium red stains and fills the extracellular space of the basal infoldings. The tracer is never seen in the intracellular matrix nor in cytoplasmic organelles. Some lysosomal granules with intrinsic densities (probably due to osmium) are seen in the cytoplasm.

The extracellular space of the basal infoldings is seen as a continuous sheet of ruthenium red in the thick sections examined in the high voltage electron microscope (Figs. 1 and 2). Such sheets show varying densities, which depend on the degree of the cross-sectional orientation to the axis of the electron beam. Under the stereoscope, the sheets are viewed as a three-dimensional extention within the sections. Mitochondria with lower contrasts are observed as wrapped by sheets of the tracer. The sheets of the tracer arise from the base of the cell and extend deep into the cell. The intercellular space can also be traced up to the zonula occludens near the lumen.

Peroxidase for the cardiac T-system

In mouse cardiac muscles, the T-system tubules are filled with the peroxidase reaction product, since the space within the T-tubule is continuous with the extracellular space (Fig. 3). Under the stereoscope, a three-dimensional image is easily obtained from a stereo pair of a thick section, showing the spatial distribution of the T-system (Figs. 4–6). The T-system tubules in the cardiac muscle do not possess such uniform sizes along their entire length nor regular positioning at the level of the Z-disc as has been generally believed (Figs. 4 and 6). In the thick sections, the course of the T-system tubules can easily be

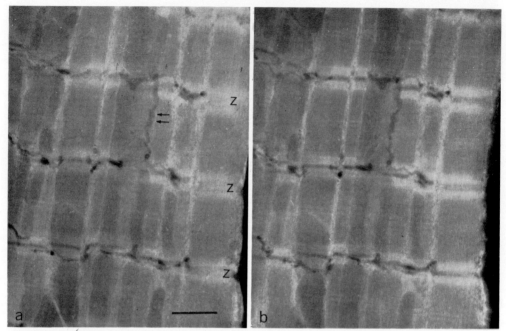

Fig. 5 A stereo pair of a longitudinally sectioned cardiac muscle cell. The T-system occasionally takes a longitudinal course (arrows), connecting between the adjacent transversely running tubules at the level of the Z discs (Z). 1 μm thick section.

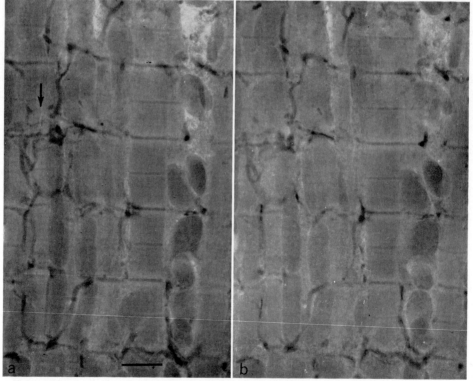

Fig. 6 A stereo pair of a longitudinally sectioned cardiac muscle cell. The T-system takes a longitudinal and oblique course and show many branchings to form a network (arrow). 1 μm thick section.

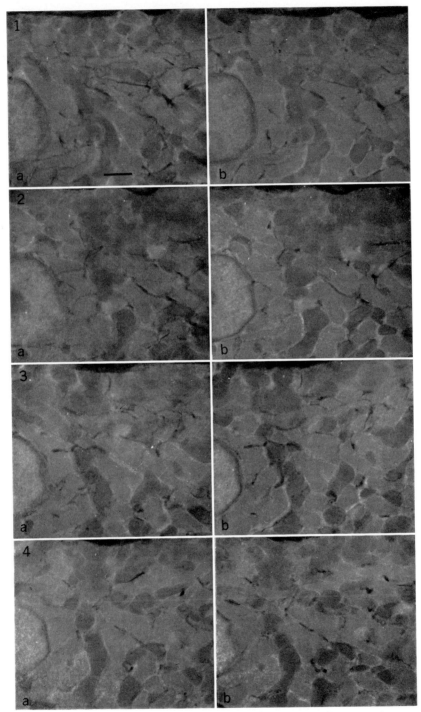

Fig. 7 Serial stereo pairs of a transversely sectioned cardiac muscle cell. The stereo pairs are arranged in such a way that Figs. 7·1a and b come to the top of the serial section and Figs. 7·4a and b to the bottom under the stereoscope. 1 μm thick sections.

traced as continuous tubules around the myofibrils. The T-system often forms, besides the typical large channels, locally dilated cisternae as well as narrow beaded tubules (Fig. 4). The branchings of the T-system are seen occasionally forming loose networks (Figs. 3 and 6). The T-system tubules are occasionally cul-de-sacs. With the aid of a stereoscope, serial thick sections were examined to reconstruct a three-dimensional distribution of the T-system. Since the top or bottom of a thick section is easily recognizable under the stereoscope, it is possible to follow the T-system from one section to the next using serial thick sections (Fig. 7). Five to ten serial thick sections are used for this analysis at a thickness of 1 μm or 2 μm. It turns out that the transverse sections of muscle cells are more convenient for a three-dimensional reconstruction (Fig. 7). In this way, the analysis can cover a comprehensive thickness and field.

A reaction product is also seen in the lumen of the capillaries providing a luminal surface view and many pinocytotic vesicles or caveolae in the endothelial cells.

Discussion

From the observations described here, it is obvious that an examination of thick biological sections with the high voltage electron microscope can selectively provide a clear-cut image of certain structures when combined with molecular tracers. Under the stereoscope the image so stained can be viewed in three-dimension. Instead of an elaborate task of serial thin sectioning and reconstruction, one can easily obtain a stereo image of the selectively stainable structures and organelles such as the T-system tubules of striated muscle fibers. Although not many structures can be selectively stained with the molecular tracers, this technique may be useful in studying many other structures in the broad field of cytology.

The selective staining technique for the high voltage electron microscopy has recently been well documented by FARVARD and CARRASO (1973). Silver or osmium impregnation and cytochemical staining for particular organelles involving the selective deposition of heavy metals are also applicable to thick sections combined with high voltage electron microscopy. As such, more structures and organelles will be examined stereologically with thick sections. These approaches are one of the ways of exploring the values of high voltage electron microscopy.

Similar cytochemical staining for relatively thick sections has been used to obtain a three-dimensional view of the Golgi apparatus of the rat neurons (NOVIKOFF et al. 1971). Structures so stained can be visualized in 0.5–1.0 μm thick sections under a conventional 100 KV electron microscope (RAMBOURG 1969; NOVIKOFF et al. 1971). However, application of the high voltage electron microscope for thick sections can provide a greater transmission of the electron beam, so that one can examine sections thicker than 3 μm with much less beam damage and with higher resolution (HAMA & PORTER 1969; HAMA 1973; FARVARD & CARRASO 1973).

Without any staining one can selectively visualize structures of intrinsic density with osmium fixation, such as certain pigments and granules in thick sections (HAMA & PORTER 1969). In using the selective staining technique including molecular tracer methods, the spatial organization of certain structures can be analyzed. However, this is only useful under conditions where the particular structures to be examined are all uniformly and selectively stained throughout the section. Therefore, one should check the staining property by thin sectioning and conventional electron microscopy.

It should be realized that in some cases, but not rarely, conventionally stained thick sections are limited in use for structures to be examined due to the over-all high density and superimposition of unwanted structures. Therefore, the interpretation of such images is often difficult in stained thick sections even with a stereo viewer. Caution is also needed in view of the fact that thick sections are not easy to stain. The penetration of the stains should be checked by the reembedding method (FARVARD & CARRASO 1973).

In the present paper, the T-system of the mouse cardiac muscles was examined with thick sections by high voltage electron microscopy combined with peroxidase as a tracer. Horseradish peroxidase has been proven to be most helpful in the visualization of T-tubules. With the aid of a stereoscope, the spatial distribution of the T-tubules were examined and reconstructed. Analysis is also possible on the skeletal muscle T-system, since the T-tubules are well stained cytochemically with peroxidase (EISENBERG & EISENBERG 1969). Ruthenium red was found not to be suited for the T-system due to uncertainities of uniform staining of the T-tubules and the positive staining of the lipid droplets in cardiac muscles (see also LUFT 1972).

The T-system in mammalian cardiac muscles are much larger than the skeletal muscles (SIMPSON & OERTELIS 1962; NELSON & BENSON 1963). Since the T-tubules are tubular invaginations of the sarcolemma and usually accompanied by basement membranes, its size and the presence of accompanied basement membranes have often been the markers with which the T-tubules could be distinguished. However, it is now obvious that the size and course of the T-tubules are not as uniform and regular as has been generally believed. The T-tubules are usually large but vary in size along their entire length. Some are just similar in size to that of the skeletal muscles. Therefore, it may be possible that some of the T-tubules, if they are small in size and not accompanied by the basement membranes, are overlooked or misinterpreted as the part of the sarcotubular (sarcoplasmic reticulum) system (PAGE 1967; SIMPSON & RAYNS 1968). Recently, it has been shown that the T-system has branchings and takes a connecting longitudinal course between adjacent transversely oriented tubules (FORSSMANN & GIRARDIER 1966; FAWCETT & McNUTT 1969; SPERELAKIS & RUBIO 1971). The present paper comfirms, from a three-dimensional analysis with thick sections, that the T-tubules form an extensive interconnected latticework.

Since the top or bottom of a thick section is easily recognizable under a stereo viewer as mentioned earlier, it is possible to follow the particular structure from one section to the next, by using high voltage electron microscopy of serial thick sections. Thus, one can reconstruct the certain structures such as the T-system of striated muscles covering a comprehensive thickness and field. From this analysis the exact information about their quantitative data may be obtained. A determination of the thickness of the sections is required for an exact reconstruction and quantitation of the structures. Some efforts are being made to estimate the section thickness from a stereo pair under a stereoscope (HAMA 1973; PEACHEY, personal communication). Quantitative studies on the T-system are underway in this laboratory and will be reported in the near future.

REFERENCES

[1] EISENBERG B. and EISENBERG R. S. (1968) Selective disruption of the sarcotubular system in frog sartorius muscle. A quantitative study with exogenous peroxidase as a marker. J. Cell Biol., *39:* 451.

[2] FARVARD P. and CARASSO N. (1973) The preparation and observation of thick biological sections in the high voltage electron microscope. J. Microscopie, *97:* 59.

[3] FAWCETT D. W. and McNUTT N. S. (1969) The ultrastructure of the cat myocardium. I. Ventricular papillary muscle. J. Cell Biol., *42:* 1.

[4] Forssmann W. G. and Girardier L. (1966) Untersuchungen zur Ultrastruktur des Rattenherzmuskels mit besonderer Berücksichtigung des sarcoplasmischen Retikulums. Z. Zellforsch., *72:* 249.

[5] Graham R. C. and Karnovsky M. J. (1966) The early stages of absorption of injected horseradish peroxidase in the proximal tubules of mouse kidney: ultrastructural cytochemistry by a new technique. J. Histochem. Cytochem., *14:* 291.

[6] Hama K. (1973) High voltage electron microscopy. *In:* Advanced Techniques in Biological Electron Microscopy. Koehler J. K. (ed.) p. 275, Springer, Berlin-Heidelberg-New York.

[7] Hama K. and Porter K. R. (1969) An application of high voltage electron microscopy to the study of biological materials. J. Microscopie, *8:* 149.

[8] Luft J. H. (1966) Ruthenium red staining of the striated muscle cell membrane and the myotendinal junction. Sixth Intern. Cong. EM., Kyoto, *2:* 65.

[9] Luft J. H. (1971) Ruthenium red and violet. II. Fine structural localization in animal tissues. Anat. Rec., *171:* 369.

[10] Nelson D. A. and Benson E. S. (1963) On the structural continuities of the transverse tubular system of rabbit and human myocardial cells. J. Cell Biol., *16:* 297.

[11] Novikoff P. M., Novikoff A. B., Quintana N. and Hauw J.-J. (1971) Golgi apparatus, GERL and lysosomes of neurons in rat dorsal ganglia, studied by thick section and thin section cytochemistry. J. Cell Biol., *50:* 859.

[12] Rambourg A. (1969) L'appareil de Golgi examen en microscopie electronique de coupes epaisses (0.5–1 μ) colorees par le melnage chlorhydrique phosphotungustique. C. R. Acad. Sci., *D269:* 2125.

[13] Rubio R. and Sperelakis N. (1972) Penetration of horseradish peroxidase into the terminal cisternae of frog skeletal muscle fibers and blockade of caffeine contracture by Ca^{++} depletion. Z. Zellforsch., *124:* 57.

[14] Simpson F. O. and Oertelis S. J. (1962) The fine structure of sheep myocardial cells; sarcolemmal invaginations and the transverse tubular system. J. Cell Biol., *12:* 91.

[15] Simpson F. O. and Rayns D. G. (1968) The relationship between the transverse tubular system and other tubules at the Z disc levels of myocardial cells in the ferret. Amer. J. Anat., *122:* 193.

[16] Sommer J. R. and Johnson E. A. (1968) Cardiac muscle. A comparative study of Purkinje fibers and ventricular fibers. J. Cell Biol., *36:* 497.

[17] Yamada E. and Ishikawa H. (1972) High voltage electron microscopy for thick sections of biological materials combined with molecular tracers. 30th Ann. Meeting EMSA, 480.

Index

A

α-dendritic cell 290, 291
ABERCROMBIE's method 292
absorbing effect of specimen toward β-ray 171
absorption 14
— efficiency 168
A-cell 116
acidic polysaccharide 13
acinar cell 111
actinomycin D 109
actomyosin 41, 43, 45
A/D converter 306
adrenocorticotropic hormone (ACTH) 189
— — secreting cell 277, 278
adrenal cortex 91
A face 244
alkaline phosphatase 15
alloxan 109
amphioxus 182
anterior pituitary 277
antifreeze agent 226
areal percentage 287
arrowhead 43, 45
arterial occlusion 33
artificial membrane 6
ascidian 182, 186
— endostyle 183
aspergillus 326, 327
asymmetrical synapse 30
ATPase 9, 15
— reaction 290
automatic image analyzer 260
autonomic nervous system 31
autophagic vacuole 115, 116
axial ratio 102

B

β-ray absorption 156
— —, elemental 156
— — in water 156
— — rate by carbon film 158
ballistic cryofixation 225
basement membrane 197
basic function 3
basket cell 29
B-cell 118
— under SEM 335
B face 244
bimolecular phospholipid micell 4
binary representation 308
biosynthesis of testosterone 148
birefringence 96
black widow spider venom (BWSV) 37
blood platelet under SEM 334
— vessel and SEM 336

C

calcitonin 189
calculation of cells per mm^2 of epidermis 292
carbon film 158
cardiac T-system 357
cast for SEM 331, 336
castration 281
catalase 133
catecholamine 32
cell kinetics 291
— surface 18
cellular membrane 3
— size of thymic small lymphocyte 261
chicken gizzard 40
cholinergic synapse 32
chromatic aberration 343
circulation research by SEM 336
city block distance 308
coated vesicle 37
coating of SEM specimen 321
collapse 229
collodion stabilized copper grid 155
colloid droplet 183
— reabsorption 185
complementary double replica 233
complete random sampling 255
complex vesicle 37
computerization 305
concanavalin A 19
conoscopic figure 96
contact point theory 171
coolant 225
cooling rate 219, 225
— velocity 219
cored vesicle 26
corticoid 279
corticotroph 277
counting granule number 277
cracking of SEM specimen 323
critical-point drying 321
— — of free cell 323
cryoinjury 224
cryoprotectant 226
cryo-sectioning 202
cryo-SEM 235, 236, 325
cryotechnique 213
crystalline lattice 96
crystallization of ice 215
crystalloid of Charcot-Böttcher 85
— of Lubarsch 82
— of Spangaro 85
C-type 31
cyclostome 179
cylindrical pattern 70
cystic fibrosis 109, 118
cytochrome oxidase 10

cytometric analysis 253
cytonet 26, 37
cytopathogenesis of vitiligo 293

D

damage to specimen 343
decrease of melanocyte 296
degeneration of skeletal muscle 49
dehydroascorbic acid 109
dendritic cell count 291
dense body 41, 45–47
density level 307
depth of field 345
determination of size of spherical object 253
diacrine secretion 127
difference between thymic cortical and medullary small lymphocyte 263
diffusible substance 172
digital computer 306
diiodotyrosine 175
disaccharidase 14, 15
dislocation 91
distance function 305
dog thyroid 197
dopa reaction 290
dorsal root peptide 33, 34
dot pattern 98
double synapse 34
drying of SEM specimen 320
dual innervation 31
Dubreuilh's precancerous melanosis 294

E

effect of section thickness 259
effective latent image at various β energy 160
electron microscopic quantitation 291
— microscopy 201
— stereoscopy 345
— transfer chain 7
— — particle 9
elementary particle 7
Elon ascorbic acid (EAA) 172
— — — developer 147
embryology and SEM 333
emiocytosis 125
endocrine tissue 190
endoplasmic reticulum 4
endostylar lumen 186
endostyle 181, 186
endothelial cell by SEM 336
energy dispersion 202, 204
— spectrum of tritium β-ray 156
— transducing system 7
environmental cell 213
enzyme-labeled antibody 189
epidermis 290
EPSPs 26, 29
equation of BETHE-BLOCK 156
eruptocrine secretion 125
erythrocyte 330, 333
evaluation of statistical significance 255
exocytosis 37
exposure theory of silver haloid grain 158

extracellular freezing 217
extrinsic protein 9

F

ferritin-bound antibody under SEM 334
ferritin-conjugated antibody 19
field emission type SEM 322
final resolution, EM-AUT 172
fine-grain development 172
fixation of SEM specimen 320
fluid mosaic model 6, 19, 20
flying spot densitometer 306
follicle stimulating hormone (FSH) 189
— — — cell 281
follicular lumen 177, 178, 183, 186
free cell for SEM observation 322
freeze-dried section 172
freeze-drying 228, 321
— of free cell 323
freeze-etching 232
— technique 6, 13, 16, 244
freeze-fracturing 16, 232
freeze-replication 232
freeze-sectioning 231
freeze-substitution 230
freezing pattern 216
— process 216
— technique 213
frequency distribution 257
— of sectional circle 256
frozen liquid cracking method 325
— resin cracking method 324
— specimen for SEM 325
F-type 29, 31, 33
functional maturation of lymphocyte 264

G

GABA 32–34
gap junction 247
gastrin 189
geometrical relation in EM-AUT 161
giant chromosome 345
glucose-6-phosphatase 133
glutaraldehyde 190
glycine 33
glycoprotein 19
Golgi apparatus 4, 133, 352
gonadotroph 281
— type 1 281
green membrane 10
growth hormone 189

H

^{3}H-cholesterol in trophoblast 148
hagfish 179, 181, 185
half distance (HD) 147
hematology and SEM 333
hepatectomy 136
hepatic sinusoid by SEM 335, 336
hexagonal crystalline core 123
— — lattice 104
— pattern 70

hexagonal system   101
high voltage electron microscope   124, 354
higher accelerating voltage   343
hippocampus   30
homogeneous pattern   73
horizontal cell   349
hormone   189
horse radish peroxidase (HRP)   37
human antral mucosa   197
— chorionic gonadotropin (HCG)   189
— placenta   197
— somatomammotropin hormone (HCS)   189
hydrolysis of thyroglobulin   184
hypophyseal portal system   273
hypothalamic neurosecretory system   272
hypothalamo-hypophyseal system   273, 277
hypothalamo-infundibular system   277

I

ice   214
— crystal   215
image quality   344
immunologic marker under SEM   334
immunology and SEM   334
increase of α-dendritic cell   296
inorganic iodide   176, 179
insulin   189
intact secretion   128
interface   306
intermediate filament   40–43, 45–47
— lobe   192
intestinal epithelial cell   13
intracellular freezing   217
iodination of thyroglobulin   177, 179, 186
ion-etching   329
IPSPs   26, 29
islet cell nucleus   116

J

junctional complex   248

K

kinocilium   350

L

lamprey   179
Langerhans cell   290
larval lamprey   181–183, 186
latent image   170
— — in silver haloid crystal   148, 158
lent filament   40
leucine aminopeptidase   15
leukocyte   330, 331, 333
line pattern   99
linear algorithm   308
lipoprotein complex   10
lipovitellin   105
liver cell membrane   16
local maximum   308
luteinzing hormone (LH)   189
lymphocyte under SEM   335

lysosomal crystalline inclusion   115
lysosome   4, 184
lysosome-like dense body   186

M

macromolecular repeating unit   5
macrophage under SEM   336
mast cell   69
maturation process of mast cell granule   78
maximum usable thickness   344
measurement of granule size   272
median eminence   275
melanocyte   290
— stimulating hormone (MSH)   189
melanosome   291
membrane fluidity   18
— recycle   37
metal coating of SEM specimen   321
methacrylate cast for SEM   332
microanalysis   201
microbody   4
micropinocytotic mechanism   184
microvillus   13, 329
— membrane   13
miniature endplate potential   24
mitochondrial membrane   7
mitochondrion   4, 7, 115
mitosis   133
Mizuhira-Kurotaki's solution   155
model for hexagonal and parallel crystalline structure   78
— for scroll-cylindrical lamellar structure   78
molecular model   4, 12, 18
— organization   3, 17
— tracer   354
monoiodotyrosine   175
morphometry   272, 305
multivesicular body   186
Murakami's method (tannin osmium)   328
myotube   64

N

nephrology and SEM   337
neurosecretory granule   275
nuclear envelope   137
— membrane   4
— size of thymic small lymphocyte   261
nucleus-cell ratio   260
nucleus field   292
number of point or profile necessary for point-counting volumetry   256

O

oblate spheroid   106
oligomycin-insensitive ATPase   9
optimal solution   267
orthoscope   96
oxidative phosphorylation   7

P

packing space   105

pancreatic acinar cell   305
paper chromatography   176
parallel crystalline pattern   70
— fiber   29
parasympathetic nerve   32
— nervous system   31
paraventricular nucleus   275
path length of silver bromide crystal   170
pattern analysis   297
penetrating power of electron   343
peptidase   14
peripheral lymphoid organ   264
peroxidase   179, 354
peroxisome   134
p-formaldehyde   190
p-formaldehyde-picric acid   190
phagocytic procedure   183
phospholipid   6
— micell   6
phosvitin   105
photodensitometric analysis   296
phylogenetic aspect   179, 185
piglet thyroid   197
planimeter   281
plasma membrane   4, 13
platelet under SEM   334
podocyte under SEM   337
point-counting method   254
— — of Weibel   272
— planimetry   281
— volumetry   255
polynomial approximation   267
population density   279, 283
posterior pituitary   273, 275
— — gland   197
premelanosome   291
presumptive myoblast   49, 66
presynaptic inhibition   33, 34
primary afferent   33
— — depolarization   33
profile area on sections   257
progressive depigmentation   296
prolactin   189
protein-integrated lipid bilayer model   17, 18, 20
protochordate   186
psoriatic epidermis   296
P-type   31
Purkinje cell   29

Q

quantal nature   24
quantitative analysis   201, 205, 207, 296
— morphology   305

R

random sampling   255
reabsorption of colloid   183
receptor   19
reconstitution   7
recrystallization   215, 219
red blood cell   330, 333
reference level   309

regeneration of skeletal muscle   49
Reinke's crystal   87
relaxing medium   41, 43
relay neuron   33
renal epithelia   357
— epitheliocyte   354
— glomerulus preparation for SEM   323–325, 332, 337
repeating particle   15
— unit model   5
replica method for SEM   331
reserpine   32
resolution in EM-AUT   147
— of freeze-dried section   172
reticular structure   73
reticulo-endothelial system and SEM   335
rhombic dodecahedron   124
— hexahedron   124
ribbon   40, 41, 45
ribonucleoprotein particle   26
ribosome   305
Rosenfeld and Pfaltz   305
ruthenium red   13, 354

S

sampling error in point-counting volumetry   255
satellite cell   49, 65
scroll-like pattern   70
secretory granule   266
section thickness   350
selective stain   202
self-absorption   171
— to β-ray   147
sensory hair   350
serial synapse   34
Sertoli cell   85
S-F hypothesis   30, 34
signet-ring cell   286
size distribution of spherical object   256
sliding mechanism   46, 47
small intestine   244
spatial (digital) filtering   305
— resolution   201
specialization of membrane system   3
spinal cord   30
spleen under SEM   336
spray method   225
square distance   309
stereocilia   350
stereological method   253
stereopaired analysis of height   299
stereoscopy   354
STH-cell   266
stress   278
stretch receptor   30
structure of water in biological system   214
S-type   29, 31, 33
substance P   33
sucrose-tannic acid-osmium method   328
supraoptic nucleus   275
symmetrical synapse   30
sympathetic nervous system   31
synapse   348
synaptic cleft   26

synaptic complex 26
— ribbon 349
— transmission 37
— vesicle 24, 26, 37, 349
syncytiotrophoblast 197
systematic random sampling 255

T

tannin osmium method for SEM 328
T-cell under SEM 335
tetraiodothyronine (=thyroxine) (T_4) 175
thermal effect 344
thiamine pyrophosphatase 133
thickness of carbon film 158
— of silver bromide 156
thiocarbohydrazide for osmification of SEM specimen 329
three-dimensional reconstruction 349
— structure 213
— — of crystalline B-cell granule 124
thymic small lymphocyte 253
thymus 253
thymus-dependent area 264
thymus-derived lymphocyte 253
thyroglobulin 175
thyrotropic hormone (TSH) 176, 189
thyroxine 184
tight junction 244
tilting goniometer stage 69
tonoactomyosin 43
triiodothyronine (T_3) 175, 184
trilaminal structure 5
tritium absorption rate in specimen 168
transmission electron microscope 206, 210
transmitter substance 24, 26
T-system 352, 354

U

ultracryotomy 231

uniaxial crystal 101
unit cell 105
— membrane hypothesis 5
urinary tubule under SEM 337

V

vascular casting for SEM 331, 336
vesicle hypothesis 37
vasopressin 189
vitamin A 109
— D_2 109
vitiligo vulgaris 290
vitrification 215, 219
volume of entire cell body 287
— percentage 287
volumetry 255
— of cell organelle 281

W

washing of SEM specimen 320
water 214
WEIBEL and ELIAS 305
wet chemical method for SEM specimen 328
white blood cell 330, 331, 333

X

X-ray 201–209
— diffraction 99

Y

yolk platelet 95

Z

zonula occludens 244